High Energy Density Laboratory Astrophysics 2008

Tomasz Plewa
Editor

High Energy Density Laboratory Astrophysics 2008

Previously published in *Astrophysics and Space Science*
Volume 322, Issues 1–4, 2009

Springer

Editor
Tomasz Plewa
Florida State University
Department of Scientific Computing
Tallahassee, FL 32306-4120, USA
tplewa@fsu.edu

ISBN 978-90-481-9998-3
Springer Dordrecht Heidelberg London New York

Library of Congress Control Number: 2010934316

Printed on acid-free paper

springer.com

Contents

Astrophys Space Sci (2009) 322: 1
DOI 10.1007/s10509-009-0041-7

Introduction to High Energy Density Laboratory Astrophysics 2008

Tomek Plewa

Received: 21 May 2009 / Accepted: 21 May 2009 / Published online: 9 June 2009

The 7th International Conference on High Energy Density Laboratory Astrophysics, HEDLA-08, was held in St. Louis, Missouri, on April 11–15, 2008. The meeting has been jointly organized by HEDLA and the American Physical Society, and shared its venue with the 1st International Conference on High Energy Density Physics. The conference was organized by Bruce Remington (LLNL, Chair), R. Paul Drake (Univ. of Michigan) Paul Bellan (Caltech), George Kyrala (LANL), Tomasz Plewa (Florida State Univ.), Sergey Lebedev (Imperial College), Hideaki Takabe (Osaka Univ.), Roberto Mancini (Univ. of Nevada-Reno), Ron Davidson (Princeton Univ.), Steve Rose (Imperial College), Kenichi Nomoto (Univ. of Tokyo), Jonathan Fortney (NASA-Ames), Edison Liang (Rice Univ.), Ryosuke Kodama (Osaka Univ.), Manuel Hegelich (LANL), Peter Norreys (RAL), Todd Ditmire (Univ. Texas-Austin), Barbara Jacak (SUNY-Stony Brook), and Tetsuo Hatsuda (Univ. of Tokyo).

The HEDLA-08 conference was a continuation of a series of biennial conferences first held in Pleasanton, California, in 1996, and reflected on progress made during recent years in designing, conducting, and analyzing the results of laboratory experiments, theoretical work, and computer simulations relevant to high-energy density (HED) astrophysics. The astrophysics setting for HED experiments is extremely diverse and includes jets, the mechanism of their generation and interaction with the ambient medium; stellar evolution with focus on turbulence, hydrodynamic instabilities, and mixing of nuclear species; radiative supersonic flows associated with stellar explosions and jets; radiative processes in plasma; equation of state and material properties; and last but not least magnetized and relativistic plasmas.

The HEDLA community continues enjoying support of and access to a wide range of the most technologically advanced experimental facilities including the Z Machine at the Sandia National Laboratory, the Omega laser at the University of Rochester, the LULI2000 laser at Ecole Polytechnique in France, and heavy ion beams accelerators. There is a strong expectation that new exciting results will soon become available thanks to OMEGA EP (Extended Performance) upgrade, and vigorous work has already started on designing experiments for the National Ignition Facility in the US and the Laser Megajoule in France. Also smaller but more accessible experimental facilities (Astra and Vulcan lasers at the Rutherford Appleton Laboratory, UK, Nike laser at Naval Research Laboratory, LFEX laser at the Institute of Laser Engineering, Osaka University, Japan, or Shen-Guang family of lasers at the Research Center of Laser Fusion, China Academy of Engineering and Physics) are proving extremely useful in helping us understand physics of high energy density systems.

The conference has been organized into 13 oral sessions with about 60 review and original talks. There were two poster sessions providing ample time for just under 200 participants to discuss more than 100 presentations. This volume brings a smaller but representative subset of the HEDLA-08 meeting with 35 peer-reviewed papers.

I would like to thank the authors for their interest and contributions and the reviewers for investing their personal time and helping improving the original presentations. Special thanks go to the Editorial Office at Springer for their help and advise in managing my editorial work. Thank you!

Guest Editor
Tallahassee, May 2009

T. Plewa (✉)
Department of Scientific Computing, Florida State University, Tallahassee, FL, USA
e-mail: tplewa@fsu.edu

Astrophys Space Sci (2009) 322: 3–9
DOI 10.1007/s10509-009-9984-y

ORIGINAL ARTICLE

Launching mechanisms of astrophysical jets

Attilio Ferrari · Petros Tzeferacos · Claudio Zanni

Received: 5 June 2008 / Accepted: 12 January 2009 / Published online: 28 January 2009

Abstract The combination of accretion disks and supersonic jets is used to model many active astrophysical objects, viz., young stars, relativistic stars, and active galactic nuclei. However, existing theories on the physical processes by which these structures transfer angular momentum and energy from disks to jets through viscous or magnetic torques are still relatively approximate. Global stationary solutions do not permit understanding the formation and stability of these structures; and global numerical simulations that include both the disk and jet physics are often limited to relatively short time scales and astrophysically out-of-range values of viscosity and resistivity parameters that are instead crucial to defining the coupling of the inflow/outflow dynamics. Along these lines we discuss self-consistent time-dependent simulations of the launching of supersonic jets by magnetized accretion disks, using high resolution numerical techniques. We shall concentrate on the effects of the disk physical parameters, and discuss under which conditions steady state solutions of the type proposed in the self-similar models of Blandford and Payne can be reached and maintained in a self-consistent nonlinear stationary state.

Keywords Jets · Accretion disks · Magnetohydrodynamics · Numerical simulations

A. Ferrari (✉)
Department of Astronomy and Astrophysics, University of Chicago, Chicago, USA
e-mail: ferrari@oddjob.uchicago.edu

A. Ferrari · P. Tzeferacos
Dipartimento di Fisica Generale, Università di Torino, Turin, Italy

C. Zanni
Laboratoire d'Astrophysique de Grenoble, Grenoble, France

1 Introduction

Supersonic outflows and collimated jets have been detected in many astrophysical objects: young stellar objects (YSO), microquasars, X-ray binaries, gamma-ray bursts (GRB), extended radio galaxies and active galactic nuclei (AGN). In all these scenarios the production of outflows is correlated with the presence of inflows in the form of disks accreting onto a central compact object. The outflow velocities appear to be of the order of the escape velocity from the central attractors; therefore in the most active environments, where the central attractors are most likely black holes, the outflow velocities are relativistic. Although supersonic velocities can be reached by purely hydrodynamic mechanisms or by radiation pressure, outflows with Lorentz factors substantially large can only be explained by the intervention of electromagnetic effects. Therefore, the most acknowledged models for interpreting the origin of astrophysical supersonic and relativistic jets are based on the interaction of large scale magnetic fields with an accreting quasi-Keplerian disk.

In seminal works Lovelace (1976) and Blandford (1976) showed that starting from a force-free poloidal field anchored in a Keplerian disk it is possible to extract energy and angular momentum creating a Poynting flux jet that accelerate a plasma current along the disk funnel up to relativistic speeds. This model has been adapted to the MHD regime, including the matter's inertia, by Blandford and Payne (1982), who showed how the Poynting jet can transfer its energy and momentum to the kinetics of a matter outflow through a magneto-centrifugal mechanism capable of reaching super-fast-magnetosonic speeds. In these works the disk was considered as a fixed boundary and no self-consistent treatment of the whole inflow/outflow dynamics was attempted.

While the Poynting flux solutions may apply to the extended radio sources and AGN jets whose densities appear much lower than in the ambient medium, the MHD case with relatively dense jets carrying inertia is in fact of more general application for all types of jets. In fact most of the works in the last 20 years make reference to the Blandford and Payne model aiming to include the dynamics of the accretion disk to show for which parameters the MHD launching of the jet may occur. Ferreira and Pelletier (1995) and Casse and Ferreira (2000) have derived analytic "cold" steady state outflow solutions linked to accreting slim disks including viscous and resistive effects in the disks and allowing for anisotropic magnetic diffusivity between poloidal and toroidal fields; their conclusion is that super-fast-magnetosonic outflows can be obtained with magnetic fields around equipartition, for a limited range of Prandtl numbers and larger toroidal diffusivity.

Ogilvie and Livio (2001) investigated the effect of the vertical structure of a cold disk in equilibrium with an isothermal corona considering also different topologies of the poloidal field; they showed that substantial outflow rates are allowed only for magnetic fields around equipartition values and poloidal components exiting at angles around 35–45° from the disk.

Supercomputer simulations have been used to investigate time-dependent solutions, often still limited to the study of the origin of the outflow and treating the disk as a boundary condition (Ustyugova et al. 1999; Ouyed and Pudritz 1997a, 1997b; Krasnopolsky et al. 1999), or treating the entire disk-jet system but for very short timescales (Uchida and Shibata 1985; Kato et al. 2002). Casse and Keppens (2002) have followed the evolution of an accretion-ejection system for longer timescales but replacing the energy equation by a simple polytropic equation of state; more recently they updated their work to include non-adiabatic effects (Casse and Keppens 2004).

At present the stability of configurations of jet-driving magnetized disks is still under discussion. An interesting approach on this issue has been proposed by Königl (2004) working out a stability criterion that show the existence of a class of stable solutions of the self-similar Blandford and Payne model.

In this review we address the same inflow/outflow dynamics stability problem based on compressible MHD numerical simulations and analyze the effects of the physical parameters with respect to the possibility of reaching steady state solutions over long time scales of integration. Simulations have been performed making use of two different numerical codes, the FLASH (Fryxell et al. 2000) and the PLUTO (Mignone et al. 2007) codes. We shall concentrate our discussion on the non-relativistic dynamics of a jet-driving disk. The interaction with the central object and the outflows which could possibly emerge from this part of the system will be neglected.

2 The MHD simulation setup

The dynamics of MHD accretion/ejection is solved numerically in terms of the MHD equations; we shall neglect the physical viscosity but include resistivity: in fact for Prandtl numbers close to unity and magnetic fields not too far from equipartition the effect of viscosity is negligible (Meliani et al. 2006). For reference we report here the equations used, with the standard meaning of the physical quantities.

$$\frac{\partial \rho}{\partial t} + \nabla\cdot(\rho \mathbf{u}) = 0$$

$$\frac{\partial (\rho \mathbf{u})}{\partial t} + \nabla\cdot\left[\rho \mathbf{u}\mathbf{u} + \left(P + \frac{\mathbf{B}\cdot\mathbf{B}}{2}\right)\mathbf{I} - \mathbf{B}\mathbf{B}\right] + \rho \nabla \Phi_g = 0,$$

$$\frac{\partial e}{\partial t} + \nabla\cdot\left[\left(e + P + \frac{\mathbf{B}\cdot\mathbf{B}}{2}\right)\mathbf{u} - (\mathbf{u}\cdot\mathbf{B})\mathbf{B} + \bar{\bar{\eta}}\mathbf{J}\times\mathbf{B}\right] = -\Lambda_{cool},$$

$$\frac{\partial \mathbf{B}}{\partial t} = \nabla \times \mathbf{E}, \quad \mathbf{E} = -\mathbf{u}\times\mathbf{B} + \bar{\bar{\eta}}\mathbf{J}, \quad \mathbf{J} = \nabla\times\mathbf{B}.$$

Here $\Phi_g = -GM/\sqrt{r^2+z^2}$ is the gravitational potential, Λ_{cool} is a cooling function and $e = P/(\gamma - 1) + \rho \mathbf{u}\cdot\mathbf{u}/2 + \mathbf{B}\cdot\mathbf{B}/2 + \rho\Phi_g$ is the total energy density. The system was closed by the equation of state for ideal gas $P = nKT$ with polytropic index $\gamma = 5/3$. The magnetic resistivity is a diagonal tensor η_{ij} whose non zero components are $\eta_{\phi\phi} = \eta_m$ and $\eta_{rr} = \eta_{zz} = \eta'_m$; the anisotropy between η_m and η'_m will be henceforth denoted by the free parameter χ_m defined as $\eta'_m = \chi_m \eta_m$. The cooling function Λ_{cool} will be taken equal to the Ohmic heating term $\Lambda_{Ohm} = \bar{\bar{\eta}}\mathbf{J}\cdot\mathbf{J}$: i.e. the Ohmic heating is radiated away instead of being dissipated locally inside the disk.

The simulations started from a thin disk rotating at slightly sub-Keplerian speed and threaded by a purely poloidal magnetic field defined by its flux function ensuring $\nabla\cdot\mathbf{B} = 0$:

$$\Psi = \frac{4}{3} B_{z0} r_0^2 \left(\frac{r}{r_0}\right)^{3/4} \frac{m^{5/4}}{(m^2 + z^2/r^2)^{5/8}},$$

where the parameter m determines the initial bending of the field at the disk surface and the coefficients B_{z0}, r_0 can be derived by normalization at the disk's midplane; in particular B_{z0} is used to define the magnetization parameter $\mu = B^2/2P$, i.e. the strength of the magnetic field at the midplane. While the radial behavior of the flux function has been chosen in order to have a constant magnetization μ along the midplane of the disk, its bending and curvature have been imposed by testing that with such a radial gradient the magnetic field naturally tends to open and bend to balance its radial pressure gradient with its curvature and tension. Pressure, density and rotation velocity of the disk have been determined by solving for the equilibrium of the disk balancing the gradient of thermal pressure, Lorentz force, gravity and centrifugal force.

In particular the thermal height scale becomes $H = \epsilon r$, where $\epsilon = c_s / V_K$ is the ratio of the sound speed $c_s = \sqrt{P/\rho}$ and the Keplerian velocity $V_K = \sqrt{GM/r}$, both evaluated at the disk's midplane.

We assumed that the toroidal component of the resistivity tensor follows the paradigm of Shakura and Sunyaev (1973) for viscous diffusion, i.e. can be written in the form:

$$\eta_m = \alpha_m V_A H \exp\left(-2\frac{z^2}{H^2}\right),$$

where α_m is a free parameter, while $V_A = B_z/\sqrt{\rho}$ is the Alfvén speed. The vertical and radial components of the resistivity tensor η'_m are then defined through the anisotropy parameter $\chi_m = \eta'_m/\eta_m$. This is the inverse of the analogous parameter defined in Ferreira (1997) and is taken as a free parameter slightly larger than unity. This anisotropy is justified by the fact that the resistivity acting on the toroidal component of the field is dominated by the turbulence generated in the disk by such instabilities as magneto-rotational and/or shearing instabilities, while the diffusivity in the poloidal plane is less affected by these processes.

Above the disk we have imposed a radially stratified hot atmosphere, initially in hydrostatic equilibrium. The region where the atmospheric and disk pressures balance defines the disk initial surface. The density, linked to the pressure by a polytropic relationship, is assumed a factor δ smaller than the corresponding density of the disk.

The resulting model has in total six arbitrary non dimensional parameters: ϵ, μ, m, α_m, χ_m and δ. However, as shown by Ferreira (1997), in order to obtain a trans-Alfvénic solution, the thermal scale height is constrained by $0.1 > \epsilon > 5 \times 10^{-4}$; for convenience we assumed a value corresponding to the maximum. The initial bending of the field lines, set through a value $m = 0.4$, satisfies the Blandford and Payne criterion for magneto-centrifugally driven outflows. The constant δ has been set $\approx 10^{-4}$. The remaining parameters have been the subject of our analysis to verify the possibility of launching jets by magneto-centrifugal acceleration.

Starting from these initial conditions the simulations were then performed allowing the system to build up a toroidal magnetic field by the action of the differential rotation. To solve the resistive MHD equations we employed both a modified version of the MHD module provided with the public code FLASH with Adaptive Mesh Refinement (AMR) capabilities and the MHD module of the PLUTO code. They belong to the class of high-resolution shock-capturing schemes which are best suited to follow supersonic flows. The simulations have been carried out in 2.5 dimensions, i.e. cylindrical geometry is used in the coordinates r and z assuming axisymmetry around the rotation axis of the disk-jet system.

Besides restricting our study to axisymmetric structures we also assumed that the system is symmetric with respect to the midplane of the disk. The computational domain covered a rectangular region with a radial extent $[0, 40r_0]$ and $[0, 120r_0]$ along the z direction: axisymmetry was imposed on the rotation axis $r = 0$ while planar symmetry was imposed at the disk midplane $z = 0$. Suitable outflow boundary conditions were assumed on the outer sides of the domain, and an inner boundary was placed inside the computational box to avoid the singularity of the potential well and of the initial setup at the origin of the axis: we defined a rectangular box with a size $r \times z = [0, r_0] \times [0, 0.5r_0]$ which was excluded from the computation and on which inflow boundary conditions appropriate to maintain reasonable physical configurations. A detailed discussion of the method used, also for ensuring the solenoidality of the magnetic field and the conservation of the angular momentum and poloidal current flux can be found in Zanni et al. (2007).

3 MHD simulations and physical resistivity

In a first set of simulations we analyzed the effect of resistivity on configurations that produce collimated super-fast-magnetosonic jets extracting angular momentum from accretion disks. For these simulations we set a magnetization parameter $\mu = 0.3$. In Fig. 1 we give the time evolution of density and poloidal magnetic field for two simulations at large $\alpha_m = 1$ and low $\alpha_m = 0.1$ resistive diffusivity; in the high diffusivity case we also consider a poloidal diffusivity 3 times smaller than toroidal. In both cases a robust outflow emerges from the underlying accretion disk: jets appear to be hollow, as the axial underdensity corresponds to the "sink" region $r < r_0$. The hollow structure is obviously a consequence of the assumptions of the model, since the physics of the central object and its interaction with the surrounding accretion disk are not taken into account. The outflows are not completely collimated at the end of the runs, as matter outflows from the computational box from the outer cylinder at $r = 40r_0$. Nevertheless all the solutions cross the Alfvénic and fast-magnetosonic critical surfaces well inside the domain.

The structure of the magnetic field defines two classes of solutions: the cases with large α_m develop an "ordered" magnetic configuration, while the poloidal field lines become strongly distorted for small α_m; for $\alpha_m < 0.1$ the outflow becomes irregular and intermittent and reaches a lower terminal speed inside the integration domain.

Despite of these morphological differences, the mechanism which drives the outflows from the disk is always

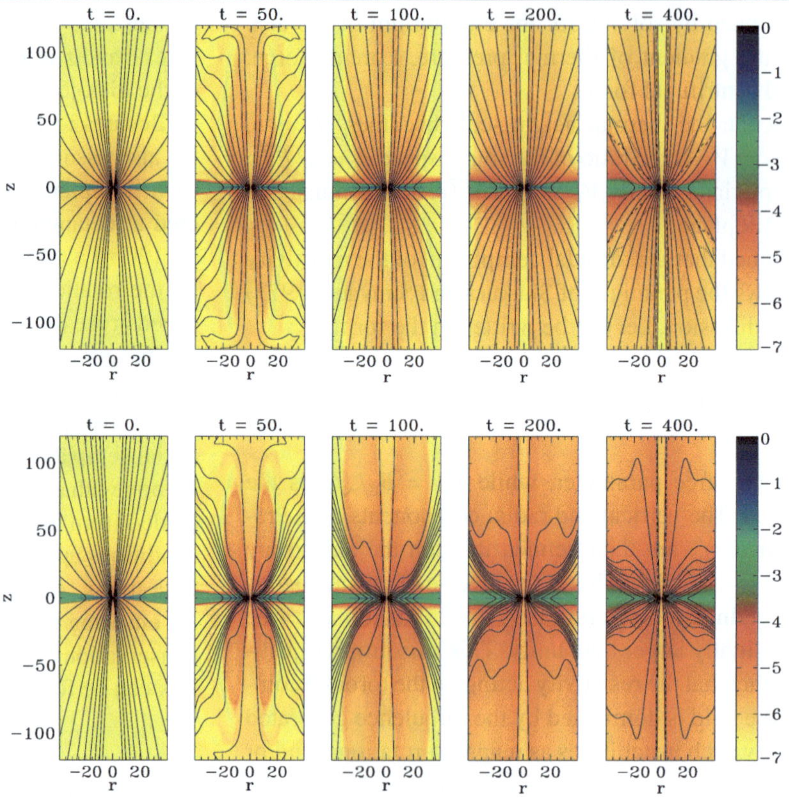

Fig. 1 Time evolution of density maps in logarithmic scale of two simulations with different resistivities: *upper panel* $\alpha_m = 1$, $\chi_m = 3$; *lower panel* $\alpha_m = 0.1$, $\chi_m = 1$. Time is given in units of $t_0 = r_0/V_{K0}$. In these units the Keplerian period at the inner radius of the disk $r = r_0$ is equal to 2π. Superimposed are sample poloidal magnetic field lines: the distance between the field lines is proportional to the intensity of the field. More details in Zanni et al. (2007)

magneto-centrifugal. We add some details of our simulations to clarify this point. The Lorentz force $\mathbf{F} = \mathbf{J} \times \mathbf{B}$ can be decomposed in the directions parallel and perpendicular to the poloidal field: $F_\phi = (B_p/r)\nabla_\parallel(rB_\phi)$, $F_\parallel = -(B_\phi/r)\nabla_\parallel(rB_\phi)$, $F_\perp = -(B_\phi/r)\nabla_\perp(rB_\phi) + J_\phi B_p$, where $\nabla_\parallel$ and $\nabla_\perp$ indicate the derivatives parallel and perpendicular to the poloidal field respectively. These expressions show clearly that the poloidal component of the Lorentz force associated with the toroidal field is perpendicular to the isosurfaces $rB_\phi =$ constant which are the surfaces along which the poloidal electric current flows.

In addition, the component parallel to the poloidal field $F_\parallel$ and the toroidal F_ϕ are linked by the simple relation $F_\parallel = -(B_\phi/B_p)F_\phi$ (notice that our toroidal field $B_\phi < 0$).

Therefore a poloidal current–field configuration that accelerates the outflow along the field lines ($F_\parallel > 0$) accelerates the plasma also in the toroidal direction ($F_\phi > 0$) thus providing an additional centrifugal force. This represents the core of the so called magneto-centrifugal mechanism: the magnetic energy stored in the toroidal field at the base of the outflow both accelerates the plasma along the field lines and increases its angular momentum thus providing a centrifugal acceleration. As stated by the above relation the relative importance of the two mechanisms, poloidal and centrifugal acceleration, is given by the ratio $|B_\phi|/B_p$: when the toroidal field is stronger than the poloidal one, the gradient of B_ϕ along the field lines is the main accelerating mechanism while if $|B_\phi|/B_p < 1$ the plasma tends to corotate with the mainly poloidal magnetic field and the centrifugal force is dominant.

Inside the accretion disk conditions are reversed: as in a unipolar inductor, a radial electric current develops inside the conducting plasma generating a toroidal force $-J_r B_z$ which slows down the rotation. Since the Lorentz force now brakes the matter in the toroidal direction, a poloidal pinching occurs. As the disk is vertically pinched by gravity as well as by magnetic pressure, only the thermal pressure gradient can ensure a quasi-static vertical equilibrium gently lifting the accreting matter towards the surface of the disk where it can be accelerated to form the outflow. As pointed out in Ferreira (1997), the transition from accretion to ejection can be therefore achieved if the sign of the magnetic torque F_ϕ changes sign at the disk surface in order to provide a magnetic acceleration.

Examining the current circuits plotted in Fig. 2 we can understand how the toroidal field collimates the outflow against the centrifugal force of the rotating jet. In the outer part of the outflow, where the currents flow out from the disk with a positive J_z component, the Lorentz force associated with the toroidal field pushes the plasma outwards, thus uncollimating the jet: in fact, in this situation the outward directed pressure gradient due to the toroidal field is stronger than the collimating tension. On the other hand, in the inner part of the outflow, where the circuit closes back with a negative J_z component, the "hoop stress" exceeds the pressure gradient: as it is confirmed by the poloidal speed

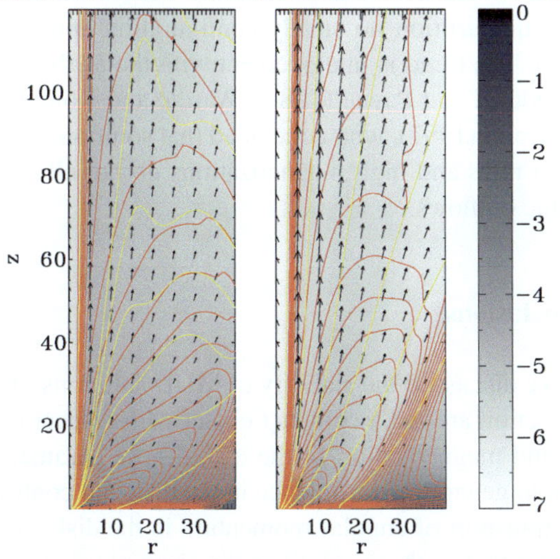

Fig. 2 Poloidal current circuits (*solid thin red lines*) at $t = 400$ for the simulations with $\alpha_m = 0.1$, $\chi_m = 1$ (*left panel*) and $\alpha_m = 1$, $\chi_m = 3$ (*right panel*). Plotted are also sample poloidal field lines (*solid thick yellow lines*) and the poloidal speed vectors (*arrows*). In the background, density maps in logarithmic *grayscale* are shown

vectors shown in Fig. 2, the outflow is therefore collimated only in its inner part, flows out of the computational domain at the outer boundary.

Finally we mention that we have explicitly calculated the constants of motion κ, λ in the Blandford and Payne self-similar solution showing that they are in fact constant in the numerical simulations when the outflows become quasi-steady. Our conclusion is that the Blandford and Payne solution for accelerating collimated jets from accreting disks is stable but for a limited range of relatively large resistivity only. These values require an enhanced resistivity in the disk, most likely due to turbulence developed by the action of MRI or shear instabilities.

4 MHD simulations and magnetization

In a second set of simulations we investigate the range of magnetization μ at the disk midplane that allows quasi-steady collimated jets. Analytic studies by Ferreira (1997) and Li (1995) indicate that this range should be $0.75 > \mu > 0.15$ to cross the slow-magnetosonic point. The four cases illustrated in Fig. 3 have been chosen to cover this range, while extending to higher and lower values of magnetization. In fact in all cases a strong outflow is generated. In the low magnetization limit an open magnetic field topology is achieved, even though, for $\mu = 0.1$, the outer field lines are poorly collimated. For strong magnetic fields, the comparatively smaller magnetic Reynolds number (due to smaller accretion rates and larger magnetic diffusion), combined with the magnetic tension, yields a field topology almost perpendicular to the meridional plane. Such an inclina-

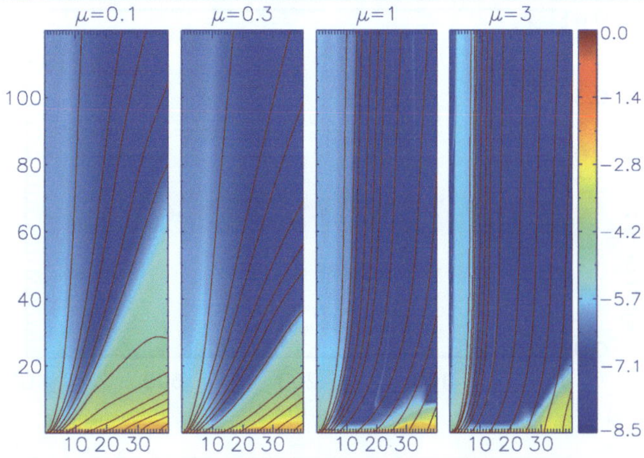

Fig. 3 Snapshots of the logarithmic density and poloidal magnetic lines at equal time $t = 400$ for increasing (from *left* to *right*) values of magnetization $\mu = 0.1, 0.3, 1.0, 3.0$; in all cases $\alpha_m = 1$, $\chi_m = 3$

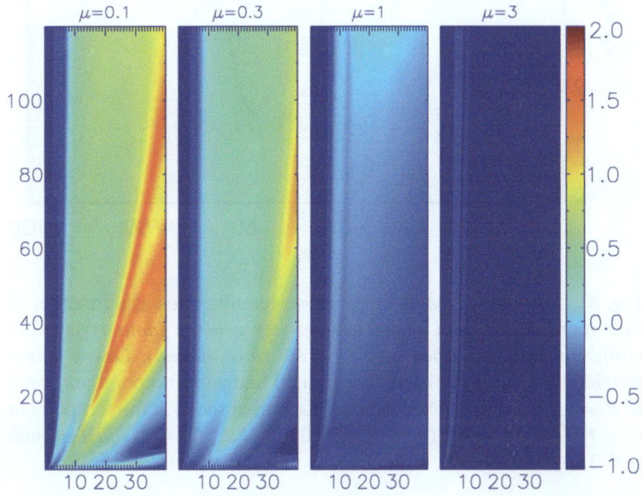

Fig. 4 The ratio of the absolute value of the toroidal magnetic field over the poloidal component at $t = 400$. Plots are in logarithmic scale. Regions where acceleration is due to the gradient of the magnetic pressure are in *yellow* and *red*, where it is due to centrifugal effect are in *green* and *blue*

tion of the magnetic surface does not favour an efficient acceleration by magneto-centrifugal mechanism, hence a sub Alfvenic (slightly super-slow) wind is produced. We therefore confirm that the best range for the magnetization parameter in order to generate a collimated jet is typically just below equipartition.

As discussed in the previous Sect. 3, the two components of the Lorentz force responsible for the magnetic acceleration are $F_\parallel$ parallel to the field lines (poloidal gradient of the magnetic pressure) and F_ϕ toroidal that gives rise to an effective centrifugal force. In Fig. 4 we represent the value of the ratio $|F_\parallel / F_\phi| = |B_\phi / B_p|$ in the integration domain for the four cases at increasing magnetization. Notice that the magnetization parameter plays a significant role on which

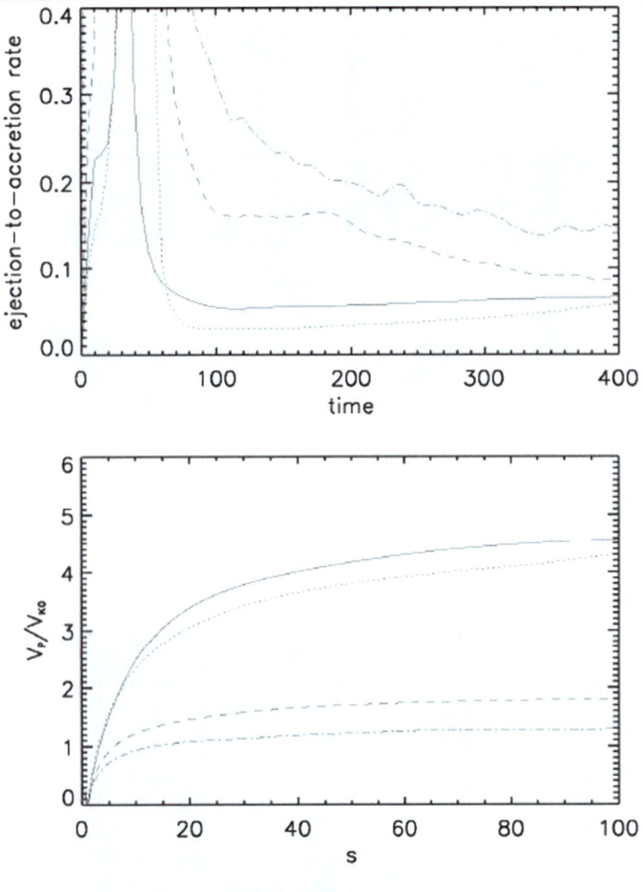

Fig. 5 *Upper panel.* Mass ejection vs accretion rates as a function of time (considering both jet and counter-jet): $\mu = 0.1$ (*dotted*), $\mu = 0.3$ (*solid*), $\mu = 1.0$ (*dashed*) and $\mu = 3.0$ (*dash-dotted*). *Lower panel.* Poloidal velocity calculated along a field line with footpoint anchored at $R = 1.5$ and normalized over the Keplerian speed at the footpoint at time $t = 400$: $\mu = 0.1$ (*dotted*), $\mu = 0.3$ (*solid*), $\mu = 1.0$ (*dashed*) and $\mu = 3.0$ (*dot-dashed*)

of the two forces is dominant. For weakly magnetized outflows it is clear that just above the disk surface the magnetic drive prevails, whereas for magnetically dominated configurations the centrifugal drive is more important. Nonetheless, in all cases most of the acceleration takes place in the sub-Alfvénic region, where the centrifugal drive dominates. In order to evaluate the efficiency of transfer of momentum and mass from the disk to the outflow we plot the poloidal velocity along the magnetic line whose footpoint is anchored in the inner part of the disk from which the collimated jet is exiting (lower panel of Fig. 5) and the ratio of mass ejection to mass accretion rate (upper panel of Fig. 5) as a function of time. Solutions with $\mu = 0.1$–0.3 reach large asymptotic poloidal velocities, larger than the Keplerian velocity at the footpoint ≈ 4.3–$4.5\ V_{K0}$; increasing the magnetization parameter this asymptotic velocity decreases and becomes $\approx 1.3\ V_{K0}$ for fields substantially above equipartition. Correspondingly the ejection rate $2\dot{M}_{ej}/\dot{M}_{acc}$, after an initial transient phase of the simulation, is of the order of 6–7 % for low magnetization, increasing slightly with μ; however above equipartition no steady configuration appears to be reached. The trend is in agreement with the analytic results of Blandford and Payne (1982), Wardle and Königl (1993) and Ferreira (1997), who concluded that solutions with high ejection rates and high magnetizations do not allow super-Alfvénic outflows.

5 Conclusions

We have discussed the stability of MHD solutions of accretion/ejection around a compact object with respect to resistivity and magnetization of the disk. These parameters are crucial to generating turbulence in the disk and contributing to the transport of angular momentum in the disk. However it is clear that in these solutions the same physical parameters determine also the coupling between the disk and the jet: and finally the transport of angular momentum in the disk is dominated by the process of jet acceleration.

Observations show that jets are produced for many different astrophysical conditions and they appear to be well behaved over relatively long time scales. Although in our simulations outflow solutions can be found for a large range of resistivity and magnetization, quasi-steady collimated jets can only be obtained for large resistivity and for magnetization around equipartition. Obviously this result has direct implications for comparison with observations.

Acknowledgements This work was supported by the U.S. DoE under grant No. B523820 to the Center of Astrophysical Thermonuclear Flashes at the University of Chicago and the EU Marie Curie Research Training Network JETSET under contract MRTN-CT-2004-005592.

References

Blandford, R.D.: Mon. Not. R. Astron. Soc. **176**, 465 (1976)
Blandford, R.D., Payne, D.G.: Mon. Not. R. Astron. Soc. **199**, 883 (1982)
Casse, F., Ferreira, J.: Astron. Astrophys. **353**, 1115 (2000)
Casse, F., Keppens, R.: Astrophys. J. **581**, 988 (2002)
Casse, F., Keppens, R.: Astrophys. J. **601**, 90 (2004)
Ferreira, J.: Astron. Astrophys. **319**, 340 (1997)
Ferreira, J., Pelletier, G.: Astron. Astrophys. **295**, 807 (1995)
Fryxell, B., Olson, K., Ricker, P., et al.: Astrophys. J. Suppl. Ser. **131**, 273 (2000)
Kato, S.X., Kudoh, T., Shibata, K.: Astrophys. J. **565**, 1035 (2002)
Königl, A.: Astrophys. J. **617**, 1267 (2004)
Krasnopolsky, R., Li, Z.Y., Blandford, R.D.: Astrophys. J. **526**, 63 (1999)
Li, Z.-Y.: Astrophys. J. **444**, 848 (1995)
Lovelace, R.V.E.: Nature **262**, 649 (1976)
Meliani, Z., Casse, F., Sauty, C.: Astron. Astrophys. **460**, 1 (2006)
Mignone, A., Bodo, G., Massaglia, S., Matsakos, T., Tesileanu, O., Zanni, C., Ferrari, A.: Astrophys. J. Supl. Ser. **170**, 228 (2007)
Ogilvie, G.I., Livio, M.: Astrophys. J. **553**, 158 (2001)
Ouyed, R., Pudritz, R.E.: Astrophys. J. **482**, 712 (1997a)
Ouyed, R., Pudritz, R.E.: Astrophys. J. **484**, 794 (1997b)
Shakura, N.I., Sunyaev, R.A.: Astron. Astrophys. **24**, 337 (1973)

Uchida, Y., Shibata, K.: Publ. Astron. Soc. Jpn. **37**, 515 (1985)
Ustyugova, G.V., Koldoba, A.V., Romanova, M.M., Chechetkin, V.M., Lovelace, R.V.E.: Astrophys. J. **516**, 221 (1999)
Wardle, M., Königl, A.: Astrophys. J. **410**, 218 (1993)
Zanni, C., Ferrari, A., Rosner, R., Bodo, G., Massaglia, S.: Astron. Astrophys. **469**, 811 (2007)

Astrophys Space Sci (2009) 322: 11–17
DOI 10.1007/s10509-008-9939-8

ORIGINAL ARTICLE

Supersonic plasma jet interaction with gases and plasmas

P. Nicolaï · C. Stenz · V. Tikhonchuk · X. Ribeyre · A. Kasperczuk · T. Pisarczyk · L. Juha · E. Krousky · K. Masek · M. Pfeifer · K. Rohlena · J. Skala · J. Ullschmied · M. Kalal · D. Klir · J. Kravarik · P. Kubes · P. Pisarczyk

Received: 3 June 2008 / Accepted: 30 October 2008 / Published online: 18 November 2008

Abstract The interaction of supersonic plasma jets with dense gases and plasmas has been studied experimentally and theoretically. Collimated plasma jets were generated from the laser pulse interaction with solid targets. The jet propagates with the velocity exceeding 400 km/s and transports the energy of a few kJ/cm^2. The interaction of such a jet with an Ar and He gases at various pressures has been studied by using optical and X-ray diagnostics. Qualitative estimates and numerical simulations with a radiative hydrodynamic code explain a sequence of physical processes during the interaction. Experimental and numerical results show that, by changing ambient material, the working surface structure changes from an adiabatic outflow to a radiative cooling jet. The applications of this phenomenon to astrophysical conditions and the inertial confinement fusion are discussed.

Keywords Supersonic plasma jet · Laser experiment · Shock

PACS 52.50.Jm · 52.72.+v · 52.35.Tc · 98.38.Fs

P. Nicolaï (✉) · C. Stenz · V. Tikhonchuk · X. Ribeyre
Centre Lasers Intenses et Applications,
Université Bordeaux 1-CNRS-CEA, 33405 Talence cedex, France
e-mail: nicolai@celia.u-bordeaux1.fr

A. Kasperczuk · T. Pisarczyk
Institute of Plasma Physics and Laser Microfusion, 00-908, Warsaw, Poland

L. Juha · E. Krousky · K. Masek · M. Pfeifer · K. Rohlena · J. Skala
Institute of Physics AS CR, Na Slovance 2, 182 21 Prague 8, Czech Republic

J. Ullschmied
Institute of Plasma Physics AS CR, Za Slovankou 3, 182 00 Prague 8, Czech Republic

M. Kalal
FNSPE, Czech Technical University in Prague, Brehova 7, 115 19 Prague 1, Czech Republic

D. Klir · J. Kravarik · P. Kubes
FEL, Czech Technical University in Prague, Technicka 2, 166 27 Prague 6, Czech Republic

P. Pisarczyk
ICS, Warsaw University of Technology, 00-665, Warsaw, Poland

1 Introduction

Laboratory studies can address issues relevant to astrophysical objects (Remington et al. 2006) and improve our understanding of the physical processes that occur in young stars and Herbig-Haro objects. In particular, the collimation of radiation cooling jets over propagation distance exceeding 10 jet diameters and their interactions with ambient medium remain an open question (Reipurth et al. 2001). The stellar jets can be detected through excitation of the line emission induced by shocks (Hartigan 1989). The structure and evolution of jets in their interaction with molecular gas were studied in numerical simulations (Blondin et al. 1990) and with a high resolution X-ray spectroscopy (Rosmej et al. 2006). Recent laboratory experiments using high power lasers (Farley et al. 1999; Blue et al. 2005; Foster et al. 2002) and magnetic pinch facilities (Lebedev et al. 2002) provide unique information in the jet structure, the interplay between radiation processes, bow, oblique or reverse shocks and their stability.

Laboratory plasmas and astrophysical objects have different length, time and density scales. However, the typical velocities are the same, of a few hundred km/s and the similarity criteria (Ryutov et al. 1999) can be applied to scale the laboratory jets to astrophysical conditions. Laser-created jets are also of interest for the Inertial Confinement Fusion (ICF). Jet formation and material mixing due to capsule defects at joints or surface bumps, or due to cryogenic fill tubes may perturb the target performance (Goldmann et al. 1999). The radiation cooling jet experiments provide also an excellent test for large multidimensional radiation hydrodynamic codes, leading to the possibility of improving the numerical algorithms and physical models (Blondin et al. 1990; Foster et al. 2002).

In the present work, we use a method of jet formation (Nicolaï et al. 2006; Kasperczuk et al. 2006), which allows to launch a very fast jet having a velocity around 400 km/s by using a relatively small laser energy, of the order of 100 J. The jet, launched from a heavy metal planar target, has a density above 10^{18} cm^{-3}, a Mach number greater than 10, a length of a few mm, and a radius of a few tenths of mm. Here, we present results of experimental, theoretical and numerical studies of interaction of a copper jet with a gas puff placed in front of the target. By varying the gas nature and pressure, we obtained different structures in the interaction zone covering a large range of parameters from quasi adiabatic to radiation cooled.

2 Experiment

The experiment was carried out at the Prague Asterix Laser System (PALS) iodine laser facility (Jungwirth et al. 2001). The laser beam of a diameter about 290 mm window was focused by an aspherical lens with the focal length of 600 mm at the wavelength of 0.438 μm. The laser energy was ~100 J, the focal spot radius on the target was 300 μm and the pulse duration ~250 ps (full width at half maximum). The laser was arriving at normal incidence on a planar solid copper target serving for the plasma jet production (Nicolaï et al. 2006). In front of it a secondary gaseous target was placed at the distance of 3.5 mm, see Fig. 1. A high pressure (2–40 bars) supersonic gas nozzle was used to create a cylindrical column of argon or helium. The corresponding density of neutral atoms was varying in the range $(0.2\text{–}8) \times 10^{18}$ cm^{-3}. The gas axis was parallel to the solid target surface and it was aligned in such a way that the plasma jet was incident perpendicularly to it. The gas density profile corresponds to a plateau of maximum density with the radius of 1.5 mm surrounded by a 1.5 mm smooth gradient. The axial gas velocity, ~10 km/s, has no influence on the interaction process in the time scale of 10 ns.

The main diagnostics used to study the jet propagation and its interaction with gas were the multi-frame images in the optical domain (interferometry and shadowgraphy) and in the soft X-rays. A three-frame interferometric/shadowgraphic system gives us access to the electron density distributions. It was illuminated by the third harmonic of the iodine laser and each channel was equipped with its own interferometer of the folding wave type (Borodziuk et al. 2005) with an objective focal length of 250 mm and a wedge angle of 3°. The images were recorded by high resolution 10 bits dynamics CCD cameras Pulnix TM-1300, with the matrix of 1300 × 1030 pixels. The exposure time of the interferometer was ~ 250 ps. The spatial resolution of 20 μm was deduced from the target edge washers. The delay between subsequent frames was set to 3 ns. Utilization of a four-frame X-ray pinhole camera gave us a possibility to discriminate the jet plasma and the ambient gas and to see the contribution of radiation effects. The exposure time was set to 4 ns, the line of vision was parallel to the target plane or 30° tilted, and the pinhole diameter was 80 μm. The soft X-ray in the energy range 10–1000 eV were transferred to the microchannel plate detector coupled with the CCD type digital camera Nikon D1X (4028 × 1324 pixels, 12 bits).

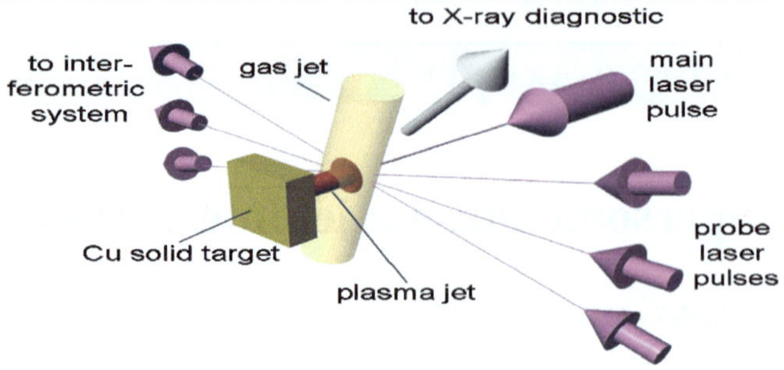

Fig. 1 Experimental setup and diagnostics alignment

3 Results from main diagnostics and preliminary analysis

Typical shadowgraphy/interferometry images for 10 bars argon gas and 20 bars helium gas are shown in Fig. 2. For the $t = 10$ ns images, the nozzle can be seen at the bottom. In panels a–c, one can see the converging traces delimiting the laser beam in the gas. These data allow us to situate the gas jet, which is ionized by the incoming laser pulse and by the thermal emission from the copper target seen in the left. The plasma jet is emerging from the expanding copper plasma. The jet has been characterized in Nicolaï et al. (2006): its velocity $v_j \sim 400$ km/s in vacuum, the temperature ~50 eV (extracted from simulations), the density $\sim 10^{18}$ cm^{-3}, and the mass density $\rho_j \sim 0.1$ mg/cc. The average kinetic energy of copper ions is ~50 keV, the jet radius $r_j \sim 0.3$ mm, and the total energy $\lesssim 15$ J. In vacuum the jet propagates

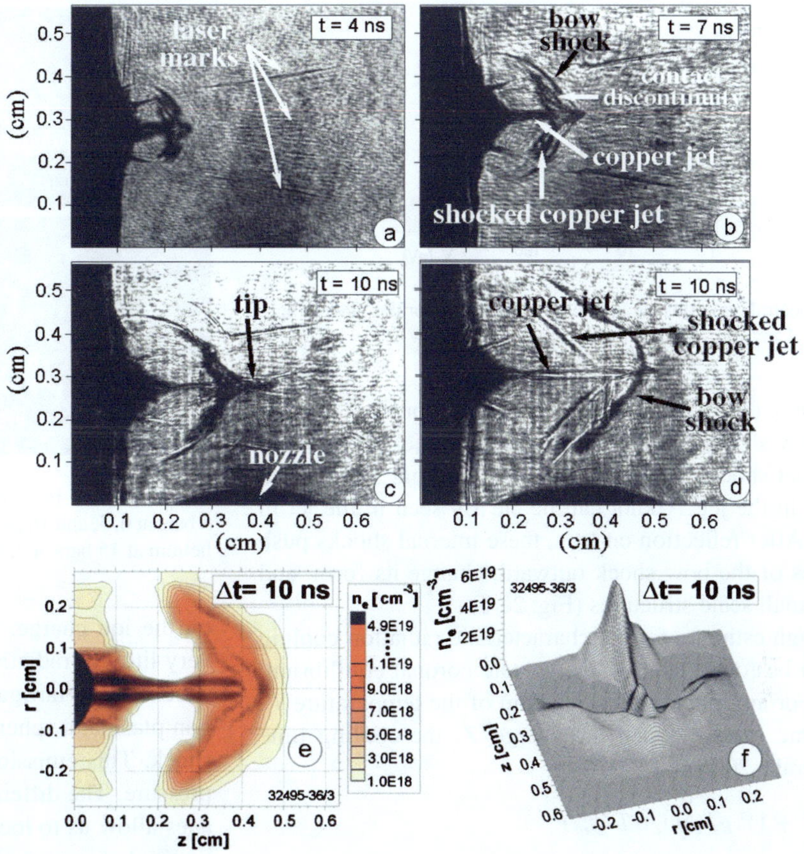

Fig. 2 Shadowgrams of the interaction of a plasma jet with argon gas at 10 bars at 4 (**a**), 7 (**b**), and 10 ns (**c**). *Panels* (**d–f**) correspond to helium gas at 20 bars at 10 ns. *Panel* (**d**) is a shadowgram, *panels* (**e**) and (**f**) show the density map obtained from interferograms unfolded with the Abel inversion

more than 20 times its radius without noticeable radial expansion. The jet is accompanied by a low density blow-off plasma. This latter has a very low ion density, below the resolution limit of the interferometry, that is a few 10^{16} cm^{-3}. The momentum flux carried with this halo is small and cannot perturb noticeably the gas puff.

The ambient plasma was created from argon or helium gas jet. The variation of gas allowed us to explore a large domain of density ratios, ρ_j/ρ_a from very low, $\sim$0.1, for helium at low pressure to rather large, $\sim$10, for a high pressure argon. As early as at 4 ns, the copper jet collides with the argon plasma. Figure 2a shows a density accumulation at the working interface of the copper and argon. It is an origin of the bow shock spreading radially and forming a nearly spherical surface. At later times of 7 and 10 ns the central part of the copper plasma jet overtakes the shock front and forms a narrow tip of the plasma plume. Specifically, at 7 ns, a double structure is visible in front of the interaction zone. The outer structure is a bow shock, which propagates in argon. The inner structure linked to the density accumulation could be a combination of the surface of contact and the reverse shock propagating in the copper plasma. The estimated bow shock velocity (from Figs. 2a and 2c, without the tip), is $v_{bs} \sim 200$ km/s, which agrees well with the estimate following from the pressure balance, $\rho_j(v_j - v_{bs})^2 \approx \rho_a v_{bs}^2$ (Blondin et al. 1990) where $\rho_j/\rho_a \simeq 1$ and ρ_a is the ambient gas density. At later times, the contact surface catches up to the bow shock. This is due to slowing down of the copper blow-off plasma, which lost a significant part of its kinetic energy supporting the blast wave.

Figure 2d presents at 10 ns, the shadowgram for the jet interaction with helium at 20 bars. The electron density distribution obtained from the interferograms unfolded with the Abel inversion is shown in panels e and f. The narrow traces seen in the shadowgram correspond to the steep density gradients. Here, the density ratio jet/plasma is much higher, $\sim$5. Consequently, the form and size of the bow shock are different. It propagates with the velocity very close to the initial jet velocity and the distance between the bow shock and the reverse shock is larger, $\sim$1 mm. This indicates a more adiabatic interaction with much smaller radiation losses. In this case, the region behind the bow shock contains a hot plasma and the tip of the shock has an obtuse shape. The density modulations in the bow shock are due to the gas pressure variation and to the hydrodynamic instability.

Theses observations agree qualitatively with the numerical analysis of radiation cooling effects presented in Blondin et al. (1990). In the case with radiation losses, the zone behind the shock is colder and denser and it is dominated by the accumulated inertia of the ambient plasma. The working surface takes a form of a pointed arrow, while the jet pre-

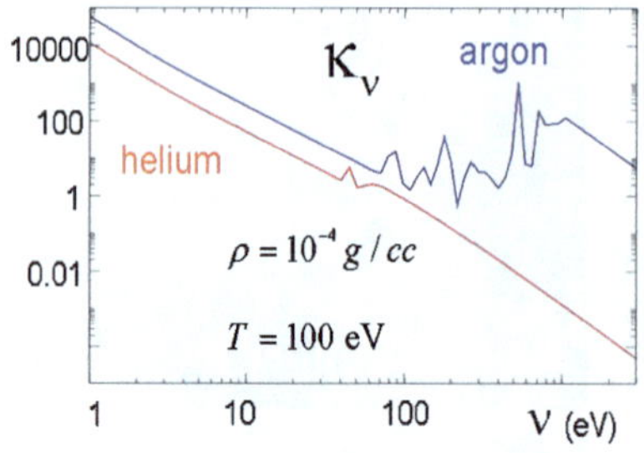

Fig. 3 Spectral opacities for the helium and argon for the plasma density $\rho = 0.1$ mg/cm^3 and the temperature $T = 100$ eV

serves more or less its initial radius. On the contrary, the adiabatic bow shock takes an obtuse form, the jet is perturbed by internal shocks emerging from the working surface and entering in the jet. A knot can be clearly seen in the jet in Fig. 2e. After reflection on axis, these internal shocks push the edges of the bow shock outward, change its form, and induce small scale structures (Fig. 2e, f).

A rough estimate for the characteristic radiation cooling time can be found from a steady state coronal equilibrium model. For an optically thin medium of the temperature T, the atomic mass m_a and the charge Z, the cooling time reads (Drake 2006):

$$\tau_{rc} \simeq (Z+1)k_B/(4m_a\sigma T^3 k_m^P).$$

Here, $k_m^P = 15\pi^{-4}\int du\, k_\nu^P u^3 e^{-u}$ is the mean Planck opacity, $u = h\nu/k_BT$, h, σ and k_B are the Planck, Stefan-Boltzmann and Boltzmann constants and $k_\nu^P(u)$ is the spectral Planck opacity (Tsakiris 1987). For the conditions at the working surface, $T \sim 100$ eV and $\rho \sim 0.1$ mg/cm^3, the spectral Planck opacities are shown in Fig. 3 for the helium and argon. The line emission in argon of photons with the energy above ~80 eV makes a strong difference of the factor of 1000 in the opacities. Indeed, after integration, one obtains $\kappa_m \simeq 80$ cm^2/g for argon and $\kappa_m \simeq 0.09$ cm^2/g for helium. The corresponding cooling times for a plasma of few hundreds microns thick are $\simeq$2 ns and $\simeq$2 μs, respectively. The Bremsstrahlung emission is a dominant process for the helium plasma. According to Huba (2007), the Bremsstrahlung cooling time for helium is 3.5 μs, which is comparable with our estimate above. For the argon case, the line emission dramatically decreases the cooling time from 450 ns to 2 ns as it is calculated above. The radiation cooling time has to be compared with the interaction time between the jet and plasma, which is ~6 ns. For the helium case, the radiation cooling is not important and the interaction is in the adiabatic regime. In the contrary, for argon, the radiation cooling plays a crucial role, and the interaction conditions can be controlled by choosing the plasma density.

Figure 4 shows two X-ray image sequences obtained for the jet interaction with 5 bars argon and 15 bars helium plasmas. The images are slightly diffused because of a long exposure time of 4 ns. The X-ray emission depends strongly

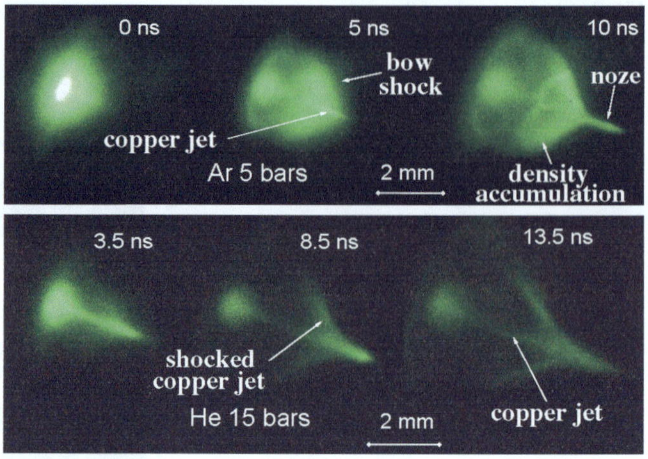

Fig. 4 (*Top*) X-ray images of the copper jet interaction with argon at 5 bars at 0, 5, and 10 ns. (*Bottom*) X-ray images for the interaction with helium at 15 bars at 3, 8, and 13 ns

on the ion charge. Therefore the helium plasma produces very little of radiation, and the images show the jet and the shocked jet (Hartigan 1989; Blondin et al. 1990). For the argon plasma, a spherical halo shows the position of the bow shock. This emission fades out as one decreases the argon pressure. The difference between the X-ray and optical images allow us to localize the position of the contact discontinuity between the jet and the plasma. An arrow shape associated with the copper jet emission appears in all images. For the helium plasma, the X-ray emission keeps the same shape whatever are the time and pressure. This confirms that the ambient plasma does not decelerate strongly the jet. For the argon plasma, the accumulation of matter is visible at the working surface, followed by an extended narrow region ahead of the spherical shape analogous to a nose cone.

4 Numerical simulations

For the conditions inferred in the experiment the mean free paths of electrons and ions, taking into account the kinetic energy, are of the order of a few microns to several tens of microns. This is much smaller than the characteristic dimensions of the jet and the bow shock. Therefore, the kinetic effects are not important and the hydrodynamic model applies. The experiment has been interpreted using the laser plasma interaction hydrodynamic code CHIC (Maire et al. 2007). The code includes two dimensional axially symmetric hydrodynamics, the ion and non-local electron heat conduction, the thermal coupling of electrons and ions and a detailed radiation transport. An Arbitrary Lagrangian-Eulerian (ALE) scheme is implemented to improve the geometrical quality of the grid elements (Hirt et al. 1997). The ionization and opacity data are tabulated (Tsakiris 1987), the equation of state (QEOS) is described in More et al. (1988). The laser

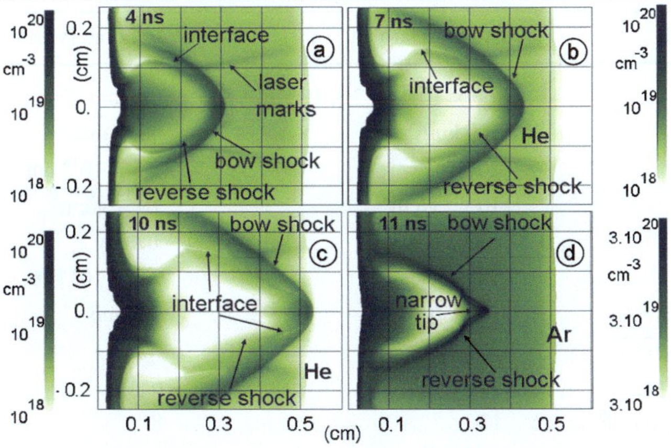

Fig. 5 Electron density map from simulations at 4 (**a**), 7 (**b**), and 10 ns (**c**) for the copper jet interaction with helium at 20 bars. (**d**) The same for the argon plasma at 11 ns, at 10 bars

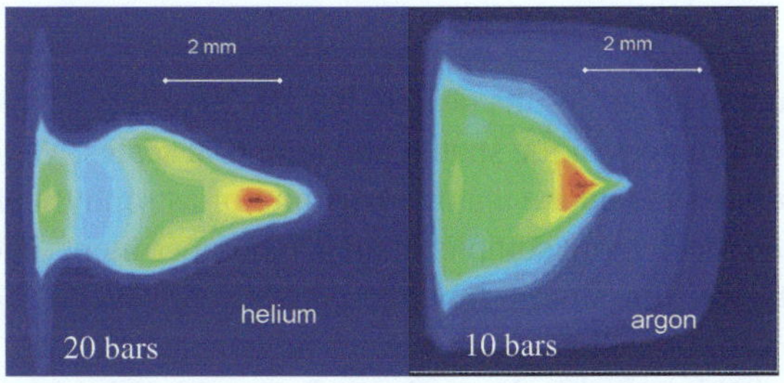

Fig. 6 Simulations of the X-ray emission about 10 ns for the jet interaction with the helium (*left*) and argon (*right*) plasmas.

propagation, refraction and collisional absorption are treated by a ray tracing algorithm.

Figure 5 shows the electron density map for the helium and argon plasmas. In the case of helium, one easily distinguishes the bow shock propagating through the plasma, the hot and denser zone just behind it, the surface discontinuity and the reverse shock propagating inside the copper blow-off plasma. The interaction zone thickness reaches the distance ~1 mm at 10 ns and agrees with experimental image shown in Fig. 2e. Figure 5d presents the density map at 11 ns for the argon plasma. The shape is more trapezoidal and the interaction zone, due to the radiation cooling is thinner and denser. The jet tip is observed, but it is less pronounced compared to the experiment. This difference is probably due to the fact, already observed in Nicolaï et al. (2006), that the jet in the simulations has a larger radius and it is more divergent. Nevertheless, the main features of copper plasma jet interaction with the ambient plasma are recovered and corroborate previous analysis.

From the calculated distributions of the plasma density and temperature, the X-ray pinhole camera images have been simulated. Figure 6 shows the side view X-ray images, integrated between 8 and 12 ns. Similarly to the observations, the X-ray image of the helium plasma is different from the electron density distribution obtained from the interferogram. This is explained by the lack of emission from helium. The emission zone corresponds to the copper density bump, formed between the interface and the reverse shock. For the argon plasma, one obtains a better correspondence between the optical and X-ray diagnostics, with a more spherical X-ray emission and a denser jet tip.

The morphology and X-ray emission of jets are well reproduced with our radiation hydrodynamic code. The latter simulates the whole process starting from the jet formation due to the laser-matter interaction to its collision with the secondary plasma. It allows to have access to various quantities not directly measured in the experiment. In that way, the temperature, estimated in the preliminary analysis of the radiation cooling time, is quantified with the simulations. In addition to the present correct modeling of measured quantities, confidence in CHIC code comes from good agreements between previous experiments and simulation results (Schurtz et al. 2007; Malka et al. 2008) and from direct comparisons with other hydro-codes (Atzeni et al. 2008).

5 Similarity conditions

The similarity criteria (Ryutov et al. 1999) for scaling of the experiment to astrophysical objects depend on the possibility to neglect the dissipative processes and the magnetic effects. In addition, radiation has to be not important or behaves identically in both systems. We assume in this study that magnetic fields effects are negligleable. The requirement for negligible dissipative processes demands that viscosity and thermal conduction are negligible. The measure of the magnitude of these processes can be done via dimensionless parameters: the Peclet number for thermal diffusion and the Reynolds number for viscosity. Systems in which both numbers are much greater than one, are those in which inertial forces and advection of heat dominate viscous drag and thermal conduction. The Peclet number: $Pe = v_j r_j/\kappa$, where v_j and r_j are the jet characteristic velocity and radius. The thermal diffusivity reads (Spitzer and Härm 1953)

$$\kappa = 2 \times 10^{21} T^{5/2}/\Lambda(Z+1)n_e \text{ cm}^2/\text{s},$$

where Λ is the Coulomb logarithm, the temperature is in eV and the electron density in cm^{-3}. The Reynolds number: $Re = v_j r_j/\nu$, where the kinematic viscosity reads:

$$\nu = 2 \times 10^{19} T^{5/2} A^{-1/2} Z^{-3}/\Lambda n_e \text{ cm}^2/\text{s}.$$

For the conditions presented in Table 1 and for the argon gas, one finds $\kappa \sim 2.5 \times 10^5$ cm²/s and $\nu \sim 4$ cm²/s. Note that, the density, the shock and fluid velocities and the X-ray emission are measured data, whereas the temperature used

Table 1 Astrophysical-laboratory jets comparison

	Symbol	Herbig-Haro jet	Laboratory jet
Jet radius, Jet length	r_j, L_j	7.5×10^{13} m, 3×10^{15} m	0.3 mm, 5 mm
Jet velocity	v_j	100 km/s	400 km/s
Ion density	n	20 cm^{-3}	1.5×10^{18} cm^{-3}
Density	ρ	10^{-22} g/cm^3	10^{-4} g/cm^3
Temperature	T	1 eV	100 eV
Pressure	P	10^{-16} bars	2.5×10^3 bars
Sound speed	$c_s = \sqrt{\gamma P/\rho}$	10 km/s	70 km/s
Time scale	$t_s = r_j\sqrt{\rho/P}$	280 yrs	6 ns
Mach number	$M = v_j/c_s$	9	6
Euler number	$Eu = v_j\sqrt{\rho/P}$	12	8
Peclet number	Pe	2.5×10^4	5
Reynolds number	Re	1.5×10^7	3×10^5
cooling time	t_{cool}	170 yrs	2 ns
cooling parameter	t_s/t_{cool}	1.5	3

in the previous formulae is extracted from simulations. The Reynolds numbers are sufficiently big for both systems ensuring the smallness of the viscosity effect even if there is a two orders of magnitude difference. The electron thermal conductivity is negligible in the astrophysical conditions but it is not so for the laboratory jets. The value $Pe \sim 5$ is greater than 1, as it should be, but the value is low enough such that heat conduction may play a role in smoothing small scale features. However, the fact that Pe is small is not of particular importance in our case, because the radiation losses dominate both in the present experiment and in the astrophysical conditions. As it can be seen in Table 1, the cooling time is of the order of observation time for both systems and, therefore, the ruling equations for our conditions are the equations of the radiation hydrodynamics, but not just the Euler equations. By choosing the appropriate working gas we are able to adjust the radiation characteristics in such a way the parameter of radiation transport in our experiment is approximately the same as in the astrophysical jets. As it is shown in previous section, the cooling time equals 2 ns for the argon gas, whereas the hydrodynamic time scale is 6 ns. Then, the cooling parameter, t_s/t_{cool}, is 3 for the argon gas. The Herbig-Haro jets evolve in a time scale of ~280 years and their cooling time is ~170 years. That is, the cooling parameter equals to 1.5, close to the argon case.

The last dimensionless scaling parameter is the density ratio between the plasma jet to the ambient material, $\eta = \rho_j/\rho_a$. This number defines the density contrast and is used in the hydrodynamic instability development studies. In the present experimental geometry, variation of the gas pressure and the gas atomic composition allows us to access to a large range of the density ratios $\eta = 0.1$–10.

From this analysis, we conclude that there are direct connections to astrophysics and that laboratory experiment may produce relevant results despite the 10–20 orders of magnitude differences in the spatial and temporal scales. We have nevertheless to notice that the structure of astrophysical jets is often much more complex than in our experiment. The astrophysical flows may show the presence of several shocks with variable velocities in the jet inducing collisions. The bow shock may breaks up into clumps, and the jet itself may appear to be clumpy both along and laterally to the jet. The morphology and especially the internal structures of astrophysical objects still appear too complicated to be completely reproduced by our laboratory jets. Multiple internal velocities in the jet induced by several laser pulses, combined with a non-homogeneous ambient gas could improve the comparison in a next laboratory experiment.

6 Conclusion

In conclusion, an experimental study of the interaction of supersonic jets with background gases was carried out at a 100 J single laser beam facility. By using a combination of two complementary diagnostics in the optical and X-ray spectral range we have characterized the main structures in the interaction zone: the bow shock propagating through gas, the shocked gas behind the shock, the contact discontinuity, the shocked part of the copper plasma jet and the plasma jet launched from a solid target. By choosing the parameters of ambient plasma, the plasma jet interaction can be studied in a broad range of the energy and momentum exchange conditions, from adiabatic to radiative cooling. Both by a theoretical analysis and by two dimensional radiation hydrodynamic simulation results, it is shown that the parameters satisfy the similarity criteria in order to scale to astrophysical conditions of interest to astronomical observations.

Acknowledgements We acknowledge the fruitful discussions and comments of M. Olazabal and T. Schlegel. This work is partly supported by the Aquitaine Region council and the European program LASERLAB.

References

Atzeni, S., et al.: Phys. Plasmas **15**, 056311 (2008)
Blondin, J., et al.: Astrophys. J. **360**, 370 (1990)
Blue, B., et al.: Phys. Rev. Lett. **94**, 095005 (2005)
Borodziuk, S., et al.: Opt. Appl. **35**, 242 (2005)
Drake, P.: High-Energy Density Physics. Springer, Berlin (2006)
Farley, D., et al.: Phys. Rev. Lett. **83**, 1982 (1999)
Foster, J.. et al.: Phys. Plasmas **9**, 2251 (2002)
Goldmann, S., et al.: Phys. Plasmas **6**, 3327 (1999)
Hartigan, P.: Astrophys. J. **339**, 987 (1989)
Hirt, C., et al.: J. Comput. Phys. **135**, 203 (1997)
Huba, J.D.: NRL Plasma Formulary. Washington, DC 20375 (2007)
Jungwirth, J., et al.: Phys. Plasmas **8**, 2495 (2001)
Kasperczuk, A., et al.: Phys. Plasmas **13**, 062704 (2006)
Lebedev, S., et al.: Astrophys. J. **564**, 113 (2002)
Maire, P.-H., et al.: SIAM J. Sci. Comput. **29**, 1781 (2007)
Malka, G., et al.: Phys. Rev. E **77**, 026408 (2008)
More, R., et al.: Phys. Fluids **31**, 3059 (1988)
Nicolaï, P., et al.: Phys. Plasmas **13**, 062701 (2006)
Reipurth, B., et al.: An. Rev. Astron. Astrophys. **39**, 403 (2001)
Remington, B., et al.: Rev. Mod. Phys. **78**, 755 (2006)
Rosmej, F.B., et al.: Europhys. Lett. **76**, 815 (2006)
Ryutov, D.D., et al.: Astrophys. J. **518**, 821 (1999)
Schurtz, G., et al.: Phys. Rev. Lett. **98**, 095002 (2007)
Spitzer, L., Härm, R.: Phys. Rev. **89**, 977 (1953)
Tsakiris, G., J. Quant. Spectrosc. Radiat. Transf. **38**, 353 (1987)

Astrophys Space Sci (2009) 322: 19–23
DOI 10.1007/s10509-009-9981-1

ORIGINAL ARTICLE

Formation of episodic magnetically driven radiatively cooled plasma jets in the laboratory

F. Suzuki-Vidal · S.V. Lebedev · A. Ciardi · S.N. Bland · J.P. Chittenden · G.N. Hall · A. Harvey-Thompson · A. Marocchino · C. Ning · C. Stehle · A. Frank · E.G. Blackman · S.C. Bott · T. Ray

Received: 2 July 2008 / Accepted: 16 December 2008 / Published online: 17 January 2009

Abstract We report on experiments in which magnetically driven radiatively cooled plasma jets were produced by a 1 MA, 250 ns current pulse on the MAGPIE pulsed power facility. The jets were driven by the pressure of a toroidal magnetic field in a "magnetic tower" jet configuration. This scenario is characterized by the formation of a magnetically collimated plasma jet on the axis of a magnetic "bubble", confined by the ambient medium. The use of a radial metallic foil instead of the radial wire arrays employed in our previous work allows for the generation of episodic magnetic tower outflows which emerge periodically on timescales of ~30 ns. The subsequent magnetic bubbles propagate with velocities reaching ~300 km/s and interact with previous eruptions leading to the formation of shocks.

F. Suzuki-Vidal (✉) · S.V. Lebedev · S.N. Bland · J.P. Chittenden · G.N. Hall · A. Harvey-Thompson · A. Marocchino · C. Ning
The Blackett Laboratory, Imperial College, London SW7 2BW, UK
e-mail: f.suzuki@imperial.ac.uk

A. Ciardi · C. Stehle
LERMA, Observatoire de Paris, CNRS and UPMC, Meudon 92195, France

A. Frank · E.G. Blackman
Department of Physics and Astronomy, University of Rochester, Rochester, NY, USA

S.C. Bott
Center for Energy Research, University of California, San Diego 92093-0417, USA

T. Ray
Dublin Institute for Advanced Studies, Dublin, Ireland

Keywords Jets · Protostars · Magnetic tower · Hydrodynamic jet · Young stellar object · YSO jet · Jet–ambient interaction · Radial foil · z-pinch · Magnetic field

1 Introduction

Highly collimated jets and outflows from protostars have been of increasing interest to observational and theoretical astrophysics as observational techniques and computer simulations continue to improve. Simultaneously a new branch of high-energy density physics—laboratory astrophysics—has been able to reproduce the dynamics shown on the stellar scale within the laboratory, maintaining the relevant dimensionless parameters (i.e. Mach number, Reynolds number, cooling parameter, etc.) given by the MHD scaling laws (Ryutov et al. 2000). Experimental facilities have shown to be capable of performing experiments which reproduce particular features of these objects by using high currents in a z-pinch machine (Lebedev et al. 2002, 2005b) and high power lasers (Farley et al. 1999; Foster et al. 2002; Blue et al. 2005).

Different models of the formation of protostellar jets have proposed that magnetic fields are responsible for driving and collimating outflows from a system composed by a star with an accretion disk (Blandford and Payne 1982). In a particular scenario the magnetic field topology evolves, due to differential rotation, into one with a predominantly toroidal magnetic field, which collimates ejected material from the system as a jet on the axis of a cavity confined by the external ambient pressure. This "magnetic tower" model (Lynden-Bell 1996) has been proposed as a mechanism of jet formation for different astrophysical objects ranging from protostars to neutron stars (Lynden-Bell 2006;

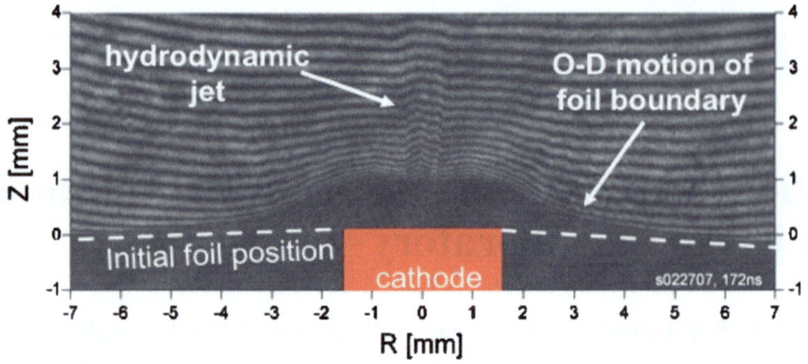

Fig. 2 Side-on laser interferometry of the radial foil at 172 ns. The pressure of the toroidal magnetic field below the foil starts displacing plasma axially and a hydrodynamic jet is formed on axis from plasma ablated from the foil

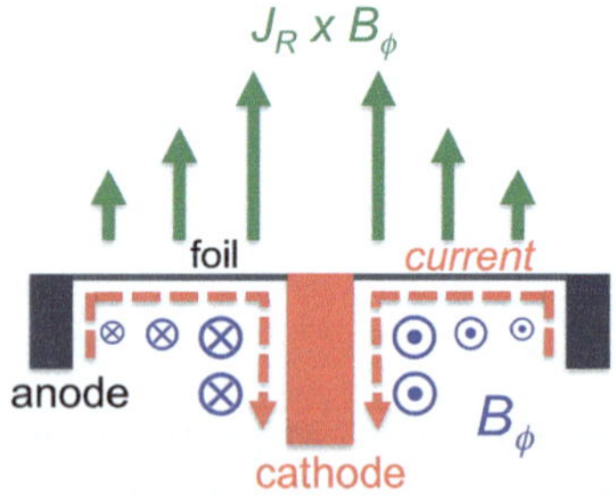

Fig. 1 Schematic of the radial foil setup showing the current path through the foil (J_R), toroidal magnetic field (B_ϕ) and the net $J_R \times B_\phi$ force acting on the plasma produced from the foil

Goodson et al. 1999; Uzdensky and MacFayden 2006; Kato et al. 2004). The experimental approach that reproduces some aspects of the plasma jet dynamics relevant to this model has been the radial wire array z-pinch configuration (Lebedev et al. 2005a), in which a plasma jet is collimated and driven onto the axis of a magnetic "bubble" by rising toroidal magnetic field loops. Magneto-hydrodynamic simulations have shown that dimensionless parameters in these experiments are relevant to jets from young stellar objects (Ciardi et al. 2007).

In this paper we present experiments in which episodic formation of magnetic tower jets was observed. The experimental set-up also allows us to vary controllably the density of the ambient medium through which the magnetic tower jets propagate. It is believed that the knots and shocks observed in the protostellar jets could originate from both the variability of the outflow at the jet formation stage, or could arise from the interaction with the ambient medium. The experimental capabilities developed in the present work can contribute to a better understanding of the issues related to variability of astrophysical jets.

2 Experimental setup

The experimental configuration is similar to the radial wire array z-pinch used in our previous experiments (Lebedev et al. 2005a). In the present experiments the current from the MAGPIE generator (peak current of 1 MA in 250 ns) (Mitchell et al. 1996) is driven into a 6–6.5 μm thick aluminum foil, which is held radially between two concentric electrodes (Fig. 1). The central electrode (cathode) is a hollow cylinder with a diameter of 3.1 mm, with the diameter of the outer electrode being 60 mm. Diagnostics included: laser probing ($\lambda = 532$ nm, $t \sim 0.4$ ns) providing 2-frame interferometry, shadowgraphy and schlieren imaging; time resolved ($\sim$2 ns exposure) pinhole cameras which recorded emission in the XUV region (>30 eV) providing up to 8 frames per experiment; magnetic "pick-up" probes to measure any trapped magnetic field inside the outflows; an inductive probe connected to the cathode to measure voltage and thus Poynting flux driving the outflow.

The imposed current path (Fig. 1) produces a toroidal magnetic field B_ϕ below the foil which is directly proportional to the current and decreases with the radial distance from the cathode ($B_\phi \alpha I(t)/r$). For peak current the toroidal magnetic field can reach magnitudes of $B_\phi \sim 100$ T (1 MG) at the cathode radius.

As the current increases in time the foil is ohmically heated by the current, showing evidence that is consistent with the foil being melted by the current together with the formation of plasma on the foil surface. The ablated plasma flows from the foil surface in the axial ($J_R \times B_\phi$) direction. Side-on laser probing imaging taken at 172 ns from the current start (Fig. 2), shows an axial displacement of the foil near the central electrode, which is reasonably well described by 0-D equations of motion, assuming that most of the foil mass is accelerated by the pressure of the toroidal magnetic field. The material ablated from the foil fills the region above it with a low-density background plasma with a typical electron density integrated along the laser line of sight measured by laser interferometry of $N_e \sim 10^{18}$ cm^{-2}. Axial density profiles reconstructed from interferometry show an exponential decay with the height above the foil, suggesting an isothermal expansion of the plasma with a typical sound speed of $C_s \sim 9$ km/s.

Figure 2 also shows an enhancement of density in the ablated plasma in the region near the axis. This hydrodynamic "precursor" jet is formed from plasma ablated from the surface of the foil, which is redirected towards the

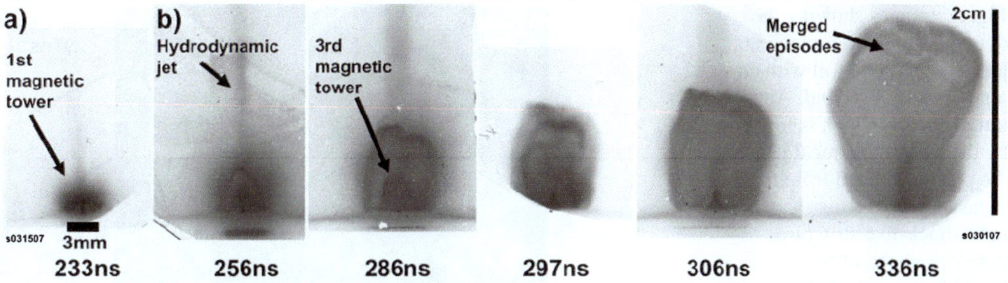

Fig. 3 Side-on XUV emission images showing the formation of episodic magnetic towers outflows which emerge with a periodicity of ~30 ns. The sequence shows results from a single experiment with the exception of the image at 233 ns

axis by radial pressure gradients. The formation of such jets by converging plasma flows has been also reported in previous laboratory astrophysics experiments using conical wire arrays and radial wire arrays (Ampleford et al. 2008; Lebedev et al. 2002, 2005b). When an ambient gas was injected into the region above the foil in the present experiments, the presence of this hydrodynamic jet affected the early time dynamics of the interaction of the outflow with the ambient, as will be discussed later in this paper.

3 Formation of episodic magnetic tower jets

The formation of magnetically driven jets starts later in time, when the Lorentz $J_R \times B_\phi$ force (which is strongest at the cathode radius) leads to ablation of all of the foil mass near the cathode and to the formation of a small radial gap between the cathode and the remainder of the foil. From this moment the Poynting flux can be injected through this gap into the region above the foil. The toroidal magnetic field pushes the ablated plasma axially and radially outwards and also pinches the plasma on axis, forming a magnetic tower jet configuration. At this stage the current flows along the jet on the axis of the magnetic cavity and along the walls of the cavity, in the same way as in our previous experiments (Lebedev et al. 2005a; Ciardi et al. 2007). The magnetic pressure from these rising toroidal loops inside the cavity inflates it both radially and axially, with measured velocities of $V_R \sim$ 50–60 km/s and $V_Z \sim$ 130–200 km/s respectively. Experimental results showing such dynamics are shown in Fig. 3. It can be seen that the initial diameter of the bubble is given by the diameter of the cathode. The most prominent feature of this new experimental set-up is that we now observe several subsequent outflows formed in the same experiment. It is possible to follow the axial positions of the subsequent episodes of the outflows shown in Fig. 3, with Fig. 4 presenting the measurements that allowed the determination of their velocities along the axis. It is seen that each outflow is expanding with approximately constant velocity, and the extrapolation of the trajectories back in time allows determining the starting time for each episode. Each subsequent bubble expands with a faster velocity, reaching $V_Z = 325$ km/s for the third observed magnetic cavity. This increase in

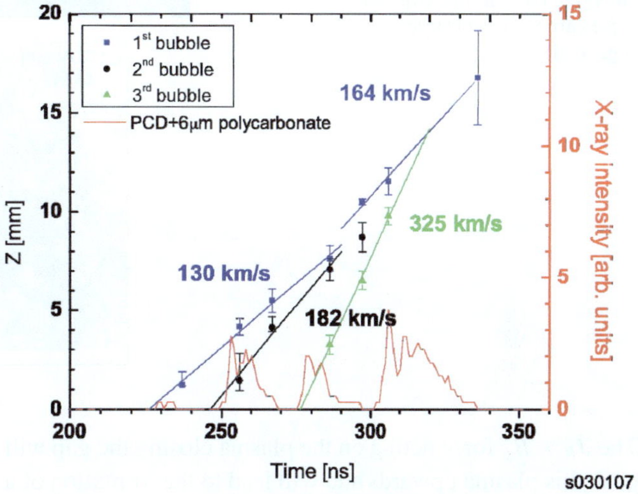

Fig. 4 Measurements of the axial extent of the episodic magnetic bubbles for the sequence in Fig. 3. Each episode is correlated with soft X-ray emission produced by the pinching of a plasma jet on axis

velocity is consistent with sweeping of the ambient plasma by the earlier episodes, thus allowing the subsequent magnetic bubbles to propagate through a lower ambient density. Figure 4 also shows that the episodic outflows are accompanied by episodic outbursts of soft X-rays (photon energy between 200–300 eV and above 800 eV), which can be well correlated with the formation of each new magnetic tower jet. This is an indication that each new episode starts from the pinching of plasma on the axis of the magnetic cavity and that pinched plasma is the source of the X-ray emission. Both the axial expansion dynamics and the periodicity of X-ray emission show a timescale of ~30 ns for the formation of subsequent magnetic tower outflows.

The formation of episodic magnetic tower outflows occurs due to reconnection of current at the base of the cavity, as the gap formed between the foil and the central electrode is closed by plasma. The mass distribution in the radial foil set-up is different from that in radial wire arrays used in our previous experiments ($\sim r$ for a foil and constant when using wires). This could lead to a smaller gap formed in the foil at the start of the first outflow and for a faster closure of the gap by the plasma expanding from the cathode and the remaining foil. The typical width of this gap estimated from the shape of the magnetic cavity walls is $\Delta r \sim$ 0.3–0.7 mm.

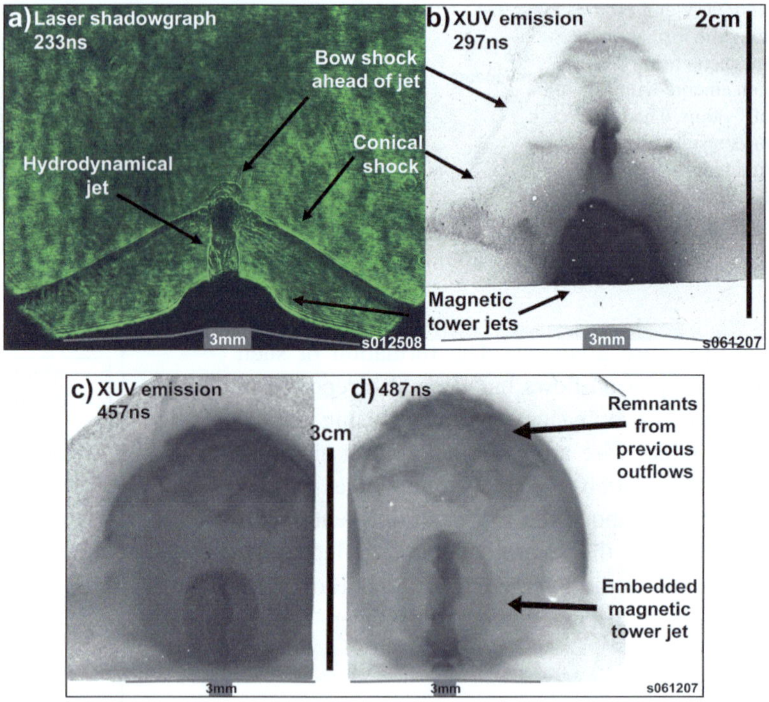

Fig. 5 Images showing the dynamics of the interaction of jets from a radial foil with an argon gas background. An initial conical shock is formed by the hydrodynamical jet and the plasma ablated from the foil. A faster propagating bow-shock-like structure develops ahead of the hydrodynamical jet (figs. **a**, **b**). Magnetically driven jets emerging later in time overtake these earlier formed structures (figs. **c**, **d**)

The $J_R \times B_\phi$ force acting on the plasma closing the gap will push this plasma upwards and will lead to the formation of a new magnetic tower outflow. The process of gap closure and formation of new outflows will continue for the duration of the current pulse from the generator, allowing to obtain 4–5 episodes for each single experiment. We should note that preliminary results show that episodic jets can also be obtained by using radial wire arrays with a high wire number.

4 Jet propagation in an ambient gas

The radial foil setup readily allows for the addition of an ambient neutral gas, which is injected via a supersonic gas nozzle into the space above the foil before the start of the current pulse, in a similar way to previous experiments with conical wire arrays (Ampleford et al. 2005). Argon was used in most of the experiments although other gases (e.g. Xe, He) were also tested. In the experiments reported here an estimated initial number density of the argon gas was $N \sim 10^{17}$–10^{18} cm^{-3}.

The presence of the ambient gas above the foil led to several new features in the jet formation and propagation. At early times the interaction of the hydrodynamic jet and the plasma ablated from the foil with the background gas led to the formation of a conical shock with an opening angle (measured from the jet axis to the conical shock boundary) of $\sim 60°$ moving with an axial velocity of $V_Z \sim 60$ km/s (Fig. 5a). It was observed that later in time (at ~260 ns) this conical shock splits into two oblique shock structures. The reasons for this phenomenon are still under investigation. It is also observed that ahead of the tip of the hydrodynamical jet a second shock feature is formed at ~230 ns. This bow shock, best seen in XUV images, is a spherical front moving at a faster velocity of $V_Z \sim 110$ km/s (Fig. 5a,b). The interaction of this bow shock with the initial conical shock forms a contact boundary ("Mach stem") which can be seen in Fig. 5b as a horizontal dark (emitting) line. It is possible that radiation from the working surface at the end of the hydrodynamic precursor jet is playing a role in the formation of this fast bow-shock like structure, but a more detailed investigation is needed here. The addition of an ambient gas shows no significant effect on the periodicity of magnetic tower jet formation. Figure 5c,d show the evolution of an embedded magnetic tower jet formed inside the remnants from the earlier outflow episodes. For this embedded magnetic tower both the envelope and the central jet are clearly seen. As the subsequent magnetic bubbles reach axial expansion velocities of ~300 km/s they can catch-up and interact with the initial shock features produced by the earlier episodes. This interaction could be responsible for the complex shock structure seen at the top of the cavity in Fig. 5d.

5 Future directions

The results of radial foil experiments show for the first time a way of producing episodic magnetically driven jets in the laboratory. The similarities of the dynamics of these episodic jets with the single-episode magnetic tower jets studied in our previous experiments with radial wire arrays (Lebedev et al. 2005a; Ciardi et al. 2007) indicate that, although some of the plasma parameters in the radial foil experiments are yet to be accurately measured, the dimensionless numbers are expected to be similar and therefore these new experiments are relevant to the physics of young stellar objects-jet launching. Preliminary measurements of trapped toroidal magnetic field inside the magnetic towers ($B_\phi \sim 0.5$ T) indicate a sufficiently high magnetic Reynolds number, which is consistent with a temperature of the magnetically driven jet of $\sim$200 eV, estimated using X-ray spectroscopy. Voltage measured with an inductive probe across the gap of the magnetic towers have given first estimates of Poynting flux and magnetic energy available to drive these outflows. Detailed analysis of these results will be subject of future publications. The process of current reconnection responsible for the formation of the episodic jets in the experiments can be compared to the phenomenon of mass accretion and magnetic reconnection that could be responsible for episodic outflows in young stellar objects (Goodson et al. 1999).

Acknowledgements This research was supported by the European Community Marie Curie JETSET network (contract MRTN-CT-2004 005592) and the SSAA program of the NNSA (DOE Cooperative Agreement DE-FC03-02NA00057).

References

Ampleford, D.J., et al.: Formation of working surfaces in radiatively cooled laboratory jets. Astrophys. Space Sci. **298**, 241–246 (2005)

Ampleford, D.J., et al.: Supersonic radiatively cooled rotating flows and jets in the laboratory. Phys. Rev. Lett. **100**, 035001 (2008)

Blandford, R.D., Payne, D.G.: Hydromagnetic flows from accretion discs and the production of radio jets. Mon. Not. R. Astron. Soc. **199**, 883–903 (1982)

Blue, B.E., et al.: Experimental investigation of high-mach-number 3D hydrodynamic jets at the National Ignition Facility. Phys. Rev. Lett. **94**, 095005 (2005)

Ciardi, A., et al.: The evolution of magnetic tower jets in the laboratory. Phys. Plasmas **14**, 056501 (2007)

Farley, D.R., et al.: Radiative jet experiments of astrophysical interest using intense lasers. Phys. Rev. Lett. **83**, 1982–1985 (1999)

Foster, J.M., et al.: Supersonic jet and shock interactions. Phys. Plasmas **9**, 2251 (2002)

Goodson, A.P., et al.: Jets from accreting magnetic young stellar objects. I. Comparison of observations and high-resolution simulation results. Astrophys. J. **524**, 142–158 (1999)

Kato, Y., et al.: Formation of semirelativistic jets from magnetoshperes of accreting neutron stars: injection of hot bubbles into a magnetic tower. Astrophys. J. **600**, 338–342 (2004)

Lynden-Bell, D.: Magnetic collimation by accretion discs of quasars and stars. Mon. Not. R. Astron. Soc. **279**, 389–401 (1996)

Lynden-Bell, D.: Magnetic jets from swirling discs. Mon. Not. R. Astron. Soc. **369**, 1167–1188 (2006)

Lebedev, S.V., et al.: Laboratory astrophysics and collimated stellar outflows: the production of radiatively cooled hypersonic plasma jets. Astrophys. J. **564**, 113–119 (2002)

Lebedev, S.V., et al.: Magnetic tower outflows from a radial wire array z-pinch. Mon. Not. R. Astron. Soc. **361**, 97–108 (2005a)

Lebedev, S.V., et al.: Production of radiatively cooled hypersonic plasma jets and links to astrophysical jets. Plasma Phys. Control Fusion **47**, B465–B479 (2005b)

Mitchell, I.H., et al.: A high impedance mega-ampere generator for fiber z-pinch experiments. Rev. Sci. Instrum. **67**, 1533–1541 (1996)

Ryutov, D., et al.: Criteria for scaled laboratory simulations of astrophysical MHD phenomena. Astrophys. J. Suppl. **127**, 465–468 (2000)

Uzdensky, D.A., MacFayden, A.I.: Stellar explosions by magnetic towers. Astrophys. J. **647**, 1192–1212 (2006)

Astrophys Space Sci (2009) 322: 25–29
DOI 10.1007/s10509-009-0025-7

ORIGINAL ARTICLE

Experimental results to study astrophysical plasma jets using Intense Lasers

B. Loupias · C.D. Gregory · E. Falize · J. Waugh · D. Seiichi · S. Pikuz · Y. Kuramitsu · A. Ravasio · S. Bouquet · C. Michaut · P. Barroso · M. Rabec le Gloahec · W. Nazarov · H. Takabe · Y. Sakawa · N. Woolsey · M. Koenig

Received: 17 June 2008 / Accepted: 9 March 2009 / Published online: 19 March 2009

Abstract We present experimental results of plasma jet, interacted with an ambient medium, using intense lasers to investigate the complex features of astrophysical jets. This experiment was performed in France at the LULI facility, Ecole Polytechnique, using one long pulse laser to generate the jet and a short pulse laser to probe it by proton radiography. A foam filled cone target was used to generate high velocity plasma jet, and a gas jet nozzle produced the well known ambient medium. Using visible pyrometry and interferometry, we were able to measure the jet velocity and electronic density. We get a panel of measurements at various gas density and time delay. From these measurements, we could underline the growth of a perturbed shape of the jet interaction with the ambient medium. The reason of this last observation is still in debate and will be presented in the article.

Keywords Plasma · Laser · Astrophysical jet · Laboratory astrophysics · Young stellar objects

B. Loupias (✉) · C.D. Gregory · A. Ravasio · M. Rabec le Gloahec · M. Koenig
LULI, Ecole Polytechnique, CNRS, CEA, UPMC, Route de Saclay, 91128 Palaiseau, France
e-mail: berenice.loupias@polytechnique.fr

E. Falize · S. Bouquet
CEA-DAM-DIF, BP 12 91680 Arpagon, France

E. Falize · S. Bouquet · C. Michaut
Laboratoire de l'Univers et de ses Théories, UMR8102, Observatoire de Paris, 92195 Meudon, France

P. Barroso
GEPI, Observatoire de Paris, CNRS, Universite Paris Diderot, Place Jules Janssen, 92190 Meudon, France

J. Waugh · N. Woolsey
Department of Physics, University of York, York, YO10 5DD, UK

D. Seiichi · Y. Kuramitsu · H. Takabe · Y. Sakawa
Institute of Laser Engineering, Osaka University, 2-6 Yamadaoka, Suita 565-0871, Japan

S. Pikuz
Joint Institute for High Temperatures of RAS, Izhorskaya 13/19, Moscow 125412, Russia

W. Nazarov
School of Chemistry, University of St. Andrews, North Haugh, St. Andrews, UK

1 Introduction

Experimental research using high-energy density facilities, like intense laser or Z-pinch facilities, is a recent promising area to investigate astrophysical phenomena reviewed in Remington et al. (2006). Several approaches are possible such as comparison or similarity to answer astrophysical questions. In the term "comparison", we arise experiments which can create the exact conditions of matter observed in astrophysical objects, for instance in the core of giant planets (Chabrier et al. 2007). The experimental measurements provide crucial input for astrophysical theoretical or numerical models. Whereas, the meaning for "similarity" between two area (experiment and astrophysics) highlights the possibility to simulate the whole evolution of the system and to keep it invariant compare to its astrophysical counterpart. The measurements of these evolution, when properly diagnosed, allow theoretical models and computer simulation codes to be tested. This last branch of experiments required the existence of scaling laws for the read up system. Scaling laws has been already demonstrated for phenomena which follow a set of equations, like pure hydrodynamic (Ryutov et al. 1999), magnetohydrodynamic (Ryutov et al. 2001) or radiative hydrodynamic systems (optically thin and thick medium) (Falize et al. 2008). The scaling law ensures the complete similarity between the astrophysical object and the experiment assuming some conditions. These conditions are the dimensionless numbers which define, by their order of magnitude, the limits of the scaling laws application. Strong shock driven turbulent dynamics relevant to supernovae explosions was investigated following this method in Ryutov et al. (2001) and can be applied to study astrophysical jets.

The collimated supersonic jets observed around stars in formation (YSO) can be studied by similarity in experiments and has been already studied following this way using intense lasers (Blue et al. 2005; Gregory et al. 2008) and Z-pinch facilities (Lebedev et al. 2005). The jets associated with YSO are often seen to have a chain of high visible emission knots and to terminate with a bow shock. The whole structure is known as Herbig-Haro (HH) objects (Blondin et al. 1989). The nature and complexity of astrophysical jets is such that it is very complex for an experiment to probe the whole of their dynamics. In this experiment, we decided to select and examine part of the physics follow by the astrophysical jet: its interaction with the interstellar medium (ISM). The bow shock structure observed in astronomical data appears with a very perturbed and fragmented shape (Hartigan 1989). We hence have decided in the presented experiment to pursue studies of astrophysical jets using intense laser (Loupias et al. 2007) to study the interaction of the jet with the ambient medium.

To check the similarity via the dimensionless parameters we have to measure all the jet parameters such as velocity, temperature and density to calculate the dimensionless parameters. This last point is crucial to experimentally check the similarity of the experimental plasma jet to its astrophysical counter part. The analyses are still in progress so we present in this article only the transverse diagnostics results and the related estimation concerning the dimensionless parameter η corresponding to the ratio of the jet density to the ambient medium density.

2 Experimental setup

For this experiment we used a long pulse to generate the plasma jet and a short pulse beam to produce protons for radiography. The lasers characteristics used was 1 kJ in 1.5 ns for the shock driven beams, and 100 J in 1 ps for the backlight beam. We used to generate plasma jet a brominate doped foam filled cone. This target was well characterized in previous experiments and the jet parameters were measured for its propagation in vacuum (Loupias et al. 2007). The 1.5 ns pulse was converted to the second harmonic, and

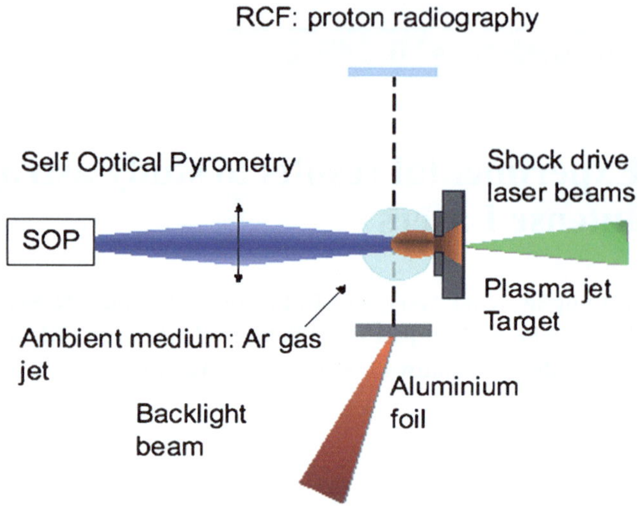

Fig. 1 Experimental setup from the top view. Using the rear side SOP, we measured the jet radius evolution as well as its equivalent black body temperature using absolute calibration

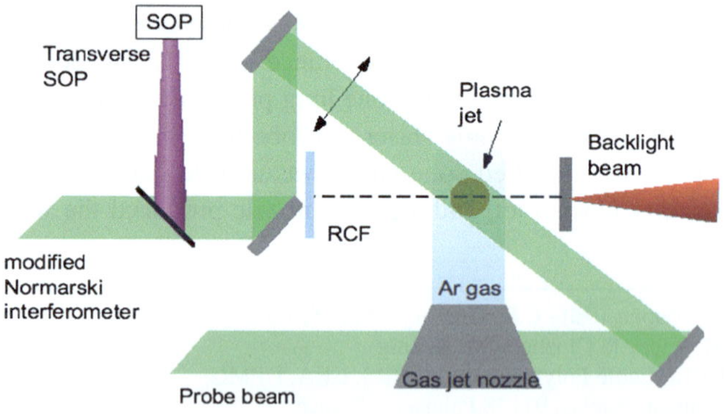

Fig. 2 Experimental setup from the transverse side

focused through a phase plate resulting in around 300 J in a 400 µm focal spot, with a pulse duration of 1.5 ns. The target is comprised of a conical former with entrance hole of diameter 500 µm, an exit hole of diameter 100 µm, and length 250 µm. A solid target as a pusher, from laser side 9 µm of plastic and 2 µm of titan, is placed over the entrance hole to drive a strong shock through the target. This shock is guided by the cone walls and results in the expulsion of a plasma jet from the exit hole. A washer of length 100 µm, and hole diameter 100 µm, is attached to the rear side of the target to increase the collimation of the flow for the initial stage of the expansion. The foam density was 50 $\mathrm{mg{\cdot}cm^{-3}}$ or 100 $\mathrm{mg{\cdot}cm^{-3}}$ 30% brominated in mass. The experimental setup is presented in Fig. 1 for the top view of the diagnostics implementation and 2 for the transverse view. Thanks to the cylindrical symmetry of the jet, the tilted probe beam allowed to have simultaneously a visible transverse view of the jet and its proton radiography.

The experiment was diagnosed through time resolved optical emission at a wavelength of 450 ± 10 nm, both transverse and end-on to the jet propagation direction, a transverse 532 nm probe, analysed with a modified Normarski interferometer, and transverse proton radiography using radiochromic film stack as detector (RCF). In this case, the probe beam delivered an 8 ns pulse, and the interferograms were recorded with a gated optical imager (GOI), with a temporal resolution of 250 ps. The ambient medium was characterized in the gas jet facility experimental room at LULI. A Mach-Zehnder interferometer was used to measure the neutral density at various initial pressure of the argon gas used to simulate the ISM. The nozzle pressure was varied between 5 bar and 80 bar, resulting in an Argon (Ar) ambient number density of between 6×10^{17} cm^{-3} and 1×10^{19} cm^{-3}. The ambient medium density is proportional to the initial input pressure to the nozzle. The plasma jet generation at the rear side of the target is a very important point, to limit any jet or ambient medium interaction with the laser beam. In fact the laser beam itself can induce spurious effects like in Edens et al. (2004) modifying the jet evolution and disturbing the shock launching from the jet propagation into the ambient medium (bow shock). That are the reasons why we privilege jet generation from the rear side of the target.

3 Experimental results

3.1 Jet and shock velocity

The jet velocity was measured from the transverse Self Optical Pyrometer (SOP). In the Fig. 3 we present the result for a 50 mg/cm^{-3} 30% brominated foam density target and 5 bar input pressure of argon gas. We observe the emission length evolution by time. Without ambient medium the velocity is the jet velocity, whereas with an ambient medium it supposes to be the gas emission, heated by a shock. In fact, we notice different emission profile with and without gas. The profile from Fig. 3, show an intensity increasing along the length. For instance at ~17 ns, the intensity increases from 0 μm (the target side) to ~1300 μm to reach its maximum. In comparison, for a shot without gas, we observe roughly constant emission level all along the jet. Furthermore, by increasing the density, the measured emission velocity decreases. This observation is consistent with the dependence of the shock velocity with the ambient medium density. Thanks to these observations we conclude that the velocity measured with an ambient medium comes from the shock front generated by the jet interaction with the gas: the bow shock. Its velocity (V_{BS}) varies according to the jet density (n_j), the ambient medium density (n_a) and the jet velocity (V_j) (Hartigan 1989):

$$\beta n_j (V_j - V_{BS})^2 = n_a V_{BS}^2, \tag{1}$$

where β correspond to the momentum transfer efficiency (from 0.4 to 0.8). By measuring the bow shock velocity at different ambient medium density (Fig. 4) and using the best fit following equation 1 (red curve) we can estimate the dimensionless parameters at the early stage of the phenomenon: $\eta = n_j/n_a$. Astrophysical conditions correspond to $\eta = 1$–10. In the presented experimental conditions we obtained: $2.5 \leqslant \eta \leqslant 22$ (using limited values of β).

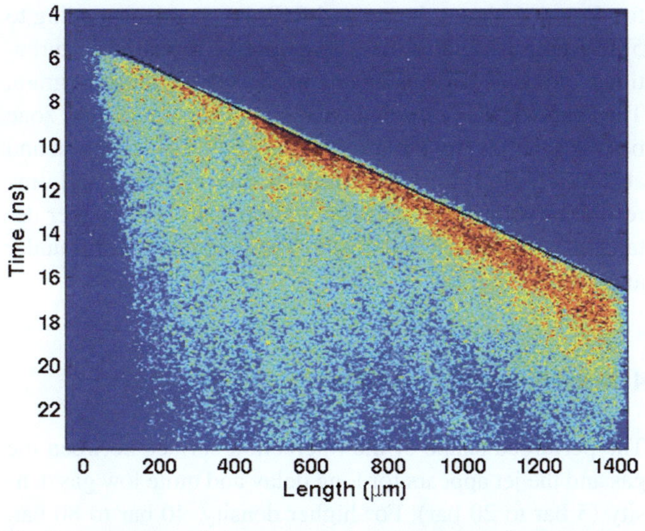

Fig. 3 Jet length evolution and velocity measurement from the transverse SOP diagnostic. The shot is with 6×10^{17} cm^{-3} of argon neutral density and 50 mg/cm^{-3} 30% brominated foam density. The *black line* represents the linear fit of the length evolution and corresponds to 123 km·s^{-1}

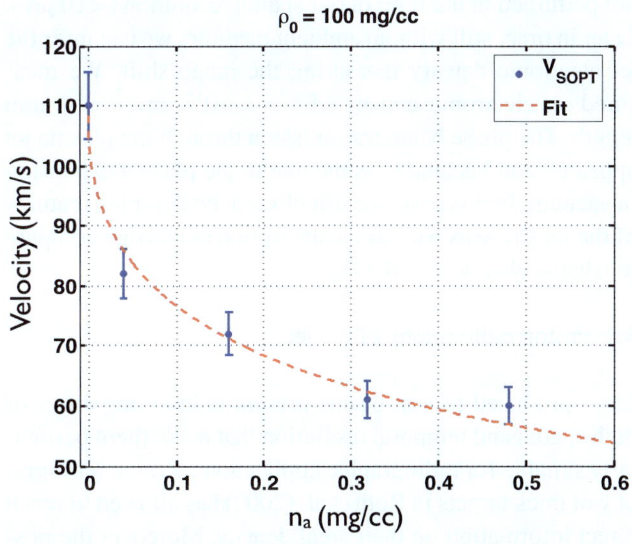

Fig. 4 Velocity measurements using transverse SOP (initial foam density 100 mg/cc) at different ambient gas density (*blue points*). The *red curve* corresponds to the best fit using (1)

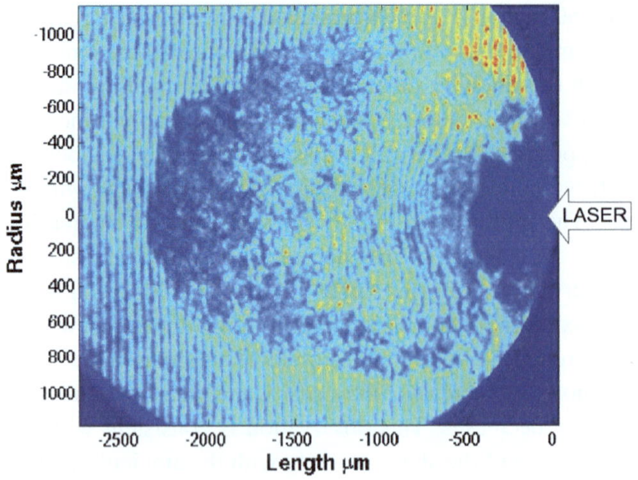

Fig. 5 Interferogram for a 50 mg·cm^{-3} 30% brominated foam density target (30 ns delay, 5 bar gas pressure)

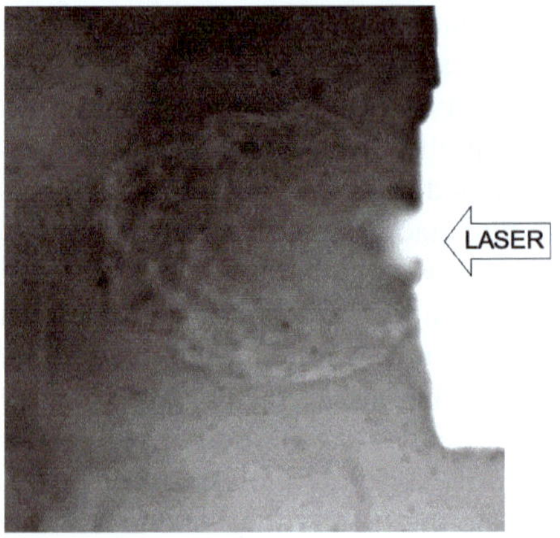

Fig. 6 Proton radiography of the jet for a 50 mg·cm^{-3} density 30% brominated foam filled cone target (30 ns delay, 5 bar gas pressure). The proton energy for this radiograph is ≈5 MeV

3.2 Interferometry results

In the Fig. 5, the probe delay is 30 ns from the arrival of the drive laser beam, and we observe the effect of introducing Ar gas. Differences are seen from the vacuum case: a high density feature is seen to propagate in front of the jet, as the flow acts as a piston driving a shock through the ambient gas. The shock front is not uniform, and small scale perturbations are seen around the leading edge. Due to the higher density in this shock front, the optical probe is not able to provide any quantitative measurement of the electronic density for this case. Whereas in the case without ambient medium, we observe a regular shape of the plasma jet. Earlier in time, the jet highly compressed by the cone guiding effect absorb and refract the probe beam. The shadow profile of the jet seems not perturbed in the limit of our spatial resolution (<10 µm). Later in time, still without ambient medium, we can infer the jet electronic density measuring the range shift. We measured an electronic density of $\sim 5 \times 10^{18}$ cm^{-3} at 1 mm length. The probe beam transmission through the plasma jet appeared and became possible due to the plasma expansion in vacuum. In this case, we still observe no complex features of the jet shape as well as no absorption at its edge compare to what is observed in the Fig. 5.

3.3 Proton radiography of the jet

Laser-produced proton beams present unique properties of high spatial and temporal resolution that make them particularly suitable for radiographic applications. Proton radiography of thick targets in Roth et al. (2002) has allowed to reach direct information on their areal density. Moreover the possibility to radiograph thin objects, i.e. with an areal density smaller than the proton range, has also been demonstrated, Borghesi et al. (2004). Recently, this diagnostic has been used to characterize the density profile in shock compressed targets (Ravasio 2007).

Broadly speaking, proton propagation into matter is affected by two important processes: multiple Coulomb scattering and energy losses. The mutual dependence of these processes makes it difficult to establish a functional relation between the target density and the radiograph. That's the reason why density measurements from proton radiography are obtained with the aide of statistical simulations, such as Monte-Carlo methods, which describe the proton propagation through matter. In our case the aim was to measure the jet density structure and the influence of an ambient medium on it. After propagation through the jet, the proton beam was detected by a Radiochromic Film (RCF) multilayer stalk. Each film selects a narrow band of proton energies, leading to a spectral capability of the multilayer detector. Figure 6 shows the second RCF layer, corresponding to 5MeV protons. In this image we observe the same perturbations and same dimensions as in the visible interferogram. The shape of the perturbed envelope of the interaction zone between the jet and the gas is delimited by a deflection and absorption of protons. The high spatial resolution and time resolution allowed to perform a radiograph with higher details than the visible diagnostic. These results are still under analysis.

4 Discussion and conclusion

The perturbed shape of the interaction surface between the gas and the jet appears for long delay and quite low gas density (5 bar to 20 bar). For higher density, 40 bar to 80 bar, delay longer than 30 ns needs to be applied. Several hypothesis can be cited to explain this kind of shape. Some of them can be shock instability, gas clusters or ambient medium perturbed profile. For the moment, the reasons are not completely found and are still in debate. To conclude, we have performed plasma jet propagation through an ambient medium at various jet and ambient medium densities. For all this shots we were able, thanks to a panel of diagnostics to measure electronic density by visible interferometry, propagation and radial velocity thanks to the SOP diagnostics and jet shape with proton radiography. The propagation of the supersonic jet through the ambient medium at the target rear side as well as the SOP observation suggests that the observed highly perturbed region is the bow shock launching by the plasma jet.

Acknowledgements This work was supported by grants from Région Ile-de-France.

References

Blondin, J.M., Konigl, A., Fryxell, B.A.: Astrophys. J. Lett. **337**, L37 (1989)

Blue, B.E., Weber, S.V., Glendinning, S.G., Lanier, N.E., Woods, D.T., Bono, M.J., Dixit, S.N., Haynam, C.A., Holder, J.P., Kalantar, D.H., MacGowan, B.J., Nikitin, A.J., Rekow, V.V., van Wonterghem, B.M., Moses, E.I., Stry, P.E., Wilde, B.H., Hsing, W.W., Robey, H.F.: Phys. Rev. Lett. **94**(9), 095005 (2005)

Borghesi, M., MacKinnon, A.J., Campbell, D.H., Hicks, D.G., Kar, S., Patel, P.K., Price, D., Romagnani, L., Schiavi, A., Willi, O.: Phys. Rev. Lett. **92**(5), 055003 (2004)

Chabrier, G., Saumon, D., Winisdoerffer, C.: Astrophys. Space Sci. **307**, 263 (2007)

Edens, A.D., Ditmire, T., Hansen, J.F., Edwards, M.J., Adams, R.G., Rambo, P., Ruggles, L., Smith, I.C., Porter, J.L.: Phys. Plasmas **11**, 4968 (2004)

Falize, É., Bouquet, S., Michaut, C.: J. Phys. Conf. Ser. **112**(4), 042016 (2008)

Gregory, C.D., Howe, J., Loupias, B., Myers, S., Notley, M.M., Sakawa, Y., Oya, A., Kodama, R., Koenig, M., Woolsey, N.C.: Astrophys. J. **676**, 420 (2008)

Hartigan, P.: Astrophys. J. **339**, 987 (1989)

Lebedev, S.V., Ciardi, A., Ampleford, D.J., Bland, S.N., Bott, S.C., Chittenden, J.P., Hall, G.N., Rapley, J., Jennings, C., Sherlock, M., Frank, A., Blackman, E.G.: Plasma Phys. Control. Fusion **47**, 465 (2005)

Loupias, B., Koenig, M., Falize, E., Bouquet, S., Ozaki, N., Benuzzi-Mounaix, A., Vinci, T., Michaut, C., Rabec Le Goahec, M., Nazarov, W., Courtois, C., Aglitskiy, Y., Faenov, A.Y., Pikuz, T.: Phys. Rev. Lett. **99**(26), 265001 (2007)

Ravasio, A.: Novel diagnostics for Warm Dense Matter : Application to shock compressed target. PhD thesis, Ecole Polytechnique (2007)

Remington, B.A., Drake, R.P., Ryutov, D.D.: Rev. Mod. Phys. **78**, 755 (2006)

Roth, M., Blazevic, A., Geissel, M., Schlegel, T., Cowan, T.E., Allen, M., Gauthier, J.C., Audebert, P., Fuchs, J., Meyer-Ter-Vehn, J., Hegelich, M., Karsch, S., Pukhov, A.: Phys. Rev. Special Topics Accelerators Beams **5**(6), 061301 (2002)

Ryutov, D., Drake, R.P., Kane, J., Liang, E., Remington, B.A., Wood-Vasey, W.M.: Astrophys. J. **518**, 821 (1999)

Ryutov, D.D., Remington, B.A., Robey, H.F., Drake, R.P.: Phys. Plasmas **8**, 1804 (2001)

Astrophys Space Sci (2009) 322: 31–35
DOI 10.1007/s10509-009-0036-4

ORIGINAL ARTICLE

A jet production experiment using the high-repetition rate Astra laser

J.N. Waugh · C.D. Gregory · L.A. Wilson · B. Loupias · E. Brambrink · M. Koenig · Y. Sakawa · Y. Kuramitsu · H. Takabe · R. Kodama · N.C. Woolsey

Received: 21 June 2008 / Accepted: 30 March 2009 / Published online: 29 April 2009

Abstract Plasma jets were produced using a high repetition rate laser by laser-ablation of coatings on the surface of conical impressions machined into solid blocks of an aluminium alloy. The ablating plasmas emerged into background gases generating shock waves. The jet-shock system was diagnosed using interferometry. The use of a high repetition rate laser allowed examination of a large number of combinations of jet materials, background gases and gas pressures.

Keywords Jet · Shock · Laser-plasma · Laboratory astrophysics

1 Introduction

The plasma jet is one of many interesting astrophysical objects seen with telescopes such as the Hubble Space Telescope (e.g. Coffey et al. 2007), which makes observations at optical wavelengths, and the Chandra X-ray Observatory (e.g. Kataoka et al. 2003), which makes observations at X-ray wavelengths. Jets vary (de Gouveia Dal Pino 2005a) from the exceedingly large (over 1 MParsec), highly relativistic jets seen emerging from active galactic nuclei (AGN) down to much smaller ($\sim$10–100 lightdays), non-relativistic jets associated with "Young Stellar Objects" (YSOs)—systems in which stars are in the process of turning on. Common to AGN and YSOs is the following structure: a massive object—be it a supermassive black hole in the case of AGN or a protostar in the case of YSOs—is surrounded by an accretion disc, and perpendicular to this disc highly-collimated jets of plasma material are expelled at high velocity. These systems are believed to be threaded with large-scale magnetic fields.

The experiment described in this report is part of an ongoing campaign (e.g. Gregory et al. 2008; Loupias et al. 2008) to develop scaled laboratory models of astrophysical plasma jets, specifically those associated with YSOs. Laboratory modelling experiments have a number of useful features which mean that they can potentially provide useful information about jets to complement that coming from telescope observations and computer simulations. Jets produced in the laboratory can be observed over their entire lifetimes and, by changing experimental parameters such as laser pulselengths and energies and target geometries and materials, it is possible to vary some of their physical properties. Also, being real physical systems, they can exhibit non-linear, multi-scale behaviour. In addition, the experiments are repeatable and a wide variety of instruments are available to probe and measure important plasma properties.

In order to gather a large amount of data, our experiment was performed on a high-repetition rate titanium-sapphire laser system. The ability to explore many configurations is

J.N. Waugh (✉) · C.D. Gregory · L.A. Wilson · N.C. Woolsey
Department of Physics, University of York, York, YO10 5DD, UK
e-mail: jnw501@york.ac.uk

B. Loupias · E. Brambrink · M. Koenig
LULI, Ecole Polytechnique, CNRS, CEA, UPMC, Route de Saclay, 91128 Palaiseau, France

Y. Sakawa · Y. Kuramitsu · H. Takabe
Institute of Laser Engineering, Osaka University, Yamada Oka 2-6 Suita, Yamada Oka, Osaka 565-0871, Japan

R. Kodama
Graduate School of Engineering, Osaka University, 2-1 Yamada Oka, Suita, Osaka 565-0871, Japan

Present address:
C.D. Gregory
LULI, Ecole Polytechnique, CNRS, CEA, UPMC, Route de Saclay, 91128 Palaiseau, France

one of the principal reasons for pursuing laboratory-based experiments but, due to low shot rate and limited availability, this is restricted on large systems. The very large number of configurations we were able to investigate is one of the main advantages over work done on larger systems, for example that of Shigemori et al. (2000). Unlike the Shigemori work, where plasmas were produced in vacuo, we were able to produce plasmas both in vacuo and within a wide range of different types and pressures of gas.

Astrophysical and laboratory plasma jets typically have very different underlying microphysics. For example, the dominant processes by which radiative energy losses occur are different in the laboratory experiment and an astrophysical jet. However, under certain circumstances, the bulk motions of the two kinds of jet can be modelled with the same set of (magneto-)hydrodynamic equations. In this case, a direct scaling between the two systems may be possible via either the Euler similarity (Ryutov et al. 1999) or ideal MHD (Ryutov et al. 2000). Strictly, these scaling relationships are only valid for ideal plasmas when dissipation may be neglected. Such plasmas have negligible thermal conduction and negligible viscous forces, emit negligible amounts of energy via radiation etc. In practice, this means one has to create plasmas for which both the Reynold's number, the ratio of the strengths of the inertial forces and the viscous forces within the plasma, and the Peclet number, a ratio indicating the relative importance of heat convection with respect to heat conduction, are large. It could in addition be expected that one needs to create plasmas with low radiative emission. However, astrophysical jets are highly radiative, and using the scaling relations to scale between a laboratory experiment and the astrophysical situation is then only approximate. The validity of the approximation is judged by how closely the values of a set of scaling parameters calculated for the laboratory plasma match the corresponding values inferred from observations of the astrophysical plasma of interest. For astrophysical jets there are three important similarity parameters as discussed by Blondin et al. (1990) and Lebedev et al. (2002). The first parameter is the jet internal Mach number, which is the ratio of the bulk flow speed of the plasma inside the jet to the sound speed of the plasma inside the jet. The second parameter is defined as the ratio of the mass density of the plasma in the jet to the mass density of the surrounding medium. The third parameter is the ratio of the radiative cooling lengthscale of the jet plasma to the jet radius. By a two-step method, first inferring whether a plasma is dissipationless and second inferring the values of the three scaling parameters, it can be determined whether a laboratory plasma is in a regime where the scaling relationships are applicable, and further whether the plasma accesses the relevant physics to scale to a YSO jet.

One observed feature of astrophysical jets is that the jet radius is much less than the jet length. Inertial confinement, radiative emission and magnetic fields are all thought to contribute to this tight collimation. It is possible for experiments to address some aspects of this.

An astrophysical jet moves in a background plasma. Within a YSO jet, a typical electron number density is between 10 and 10^5 cm^{-3}, and the mass density ratio between the YSO jet and the medium it propagates into is about 1–20 (de Gouveia Dal Pino 2005b). It is expected that having a background plasma surrounding a jet source improves the jet collimation; this is what is here meant by inertial confinement.

As mentioned above, astrophysical jets are observed to be radiative. This means that the timescale for a jet plasma to dissipate its energy by emitting radiation is less than or comparable to the hydrodynamic timescale, the timescale on which the bulk flow of the plasma evolves. The emission of radiation results in a plasma cooling and losing pressure. For a jet propagating in a vacuum the effect of radiative emission on the jet collimation is straightforward. The pressure of the plasma inside the jet promotes its lateral expansion leading to decollimation. Emission of radiation results in a decrease in the internal energy of the jet, reducing the pressure that drives decollimation. A jet propagating in an ambient medium rather than in a vacuum is a complex structure. A detailed discussion of astrophysical jet morphology in this case can be found in Blondin et al. (1990), and one would anticipate the observation of similar structures in laboratory jets if the effect of radiation on jet dynamics is similar in an astrophysical and a laboratory jet.

Also, astrophysical jets are associated with accretion disk systems and their large-scale magnetic fields. It is thought that these fields have a significant role to play in jet collimation.

2 Experimental method

In the experiment, jets were made under a variety of circumstances to explore the effects of inertial confinement and radiation both individually and in combination. The material the jets were made from was varied, as was the type and pressure of the background gas into which the jets emerged. Four target materials were used, n-type polypropylene (C_8H_8), Al, Cu and Au, and three gases were used, He, N_2 and Xe. The motivation for using these materials and gases was to explore the effects of changing radiative cooling rates and inertial masses. Higher atomic number materials radiate more efficiently than lower atomic number materials. This enables the comparative study of radiative cooling. An additional important effect of changing background gas conditions was alteration of the gas opacity.

The jet creation method used in this experiment (described below) is similar to that used by Shigemori et al.

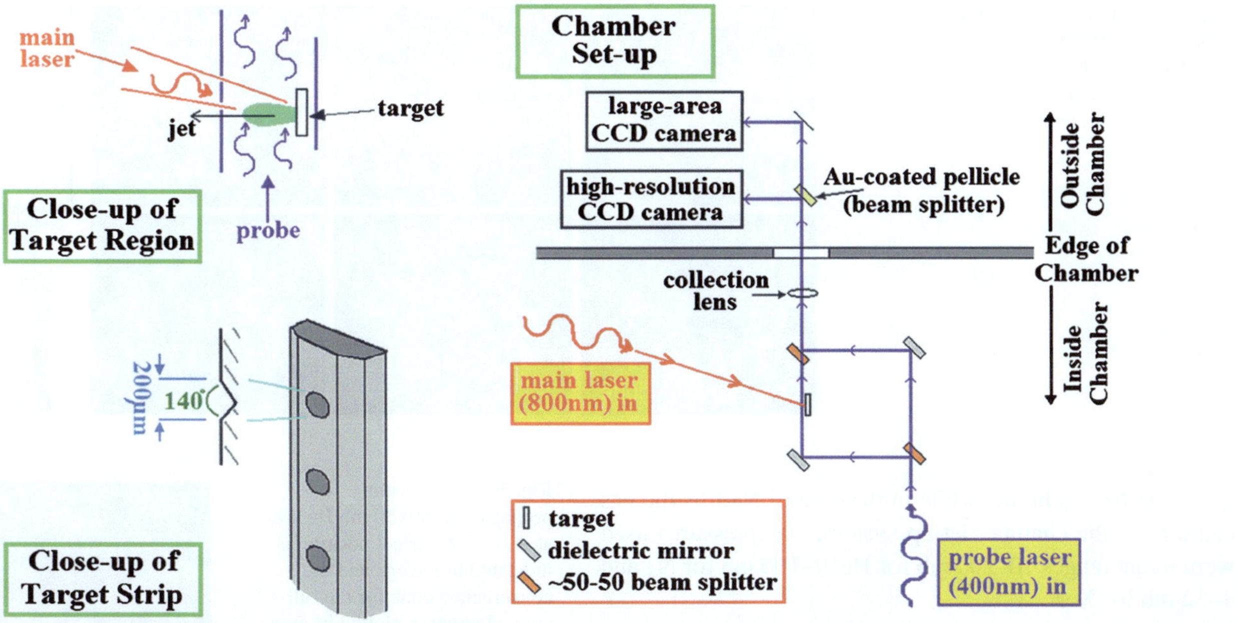

Fig. 1 Experimental set-up. The main laser pulse comes in from the left at 8° to target normal, and the probe pulse comes in from the bottom and passes the target perpendicular to target normal

(2000), although as mentioned earlier we used a small, high-repetition rate laser to allow us to study a wide range of experimental configurations. The set-up of the experiment is shown in Fig. 1. Laboratory jets were created by firing the Astra laser into shaped targets within a vacuum chamber. The Astra laser is a high-repetition rate titanium-sapphire system delivering approximately 0.5 J of energy at a wavelength of 800 nm at 2 Hz. We stretched the approximately 40 fs duration of the pulses to $\sim$2 ps and focused the beam to a 100 μm FWHM spot using an $f/6.5$, off-axis parabola. This gave a peak laser intensity at the target of around 3×10^{15} W cm^{-2}. The central axis of the parabola was displaced by 8° from target normal. The target cones were coated conical depressions, each of diameter 200 μm and full opening angle 140°, machined into lapped solid blocks of an aluminium alloy. The inter-cone spacing was 2 mm and there were 49 cones on each target strip, which allowed us to take advantage of the high-repetition rate of the laser.

Each laser pulse ablated the plastic, Al, Cu or Au material coating the surface of a cone resulting in a flow of plasma away from the cone surface. It was expected that this ablating plasma would stagnate on the cone axis and move away from the target as a jet. In addition to conical targets, coated flat targets were also shot thus enabling us to establish whether using a conical target geometry is beneficial for jet creation.

Several diagnostics were used to observe the jets created. The main diagnostic was optical interferometry. The interferometry used a probe beam split from the main Astra beam and frequency-doubled to 400 nm. This beam was sent through a Mach-Zehnder interferometer, with the target placed in one arm of the interferometer, and the beam was collected and imaged using an $\sim f/3$ lens. After exiting the vacuum chamber through a window anti-reflection coated for 400 nm, the probe beam was split using a gold-coated pellicle and both the reflected and transmitted light were imaged onto charge-coupled device (CCD) cameras filtered with 400 nm interference filters. One CCD was a large-area, 16-bit detector with 13.5 μm square pixels, and the other CCD was a smaller-area, 10-bit camera with 6.45 μm square pixels. This second camera gave magnified, but smaller field of view, images of the central regions of our plasmas. The large-area camera enabled the imaging of a 2.9×3.0 mm region albeit at a lower spatial resolution.

In addition, a pinhole camera and a gated optical imager were fielded. The optical imager had a $\sim f/3$ collecting lens placed $\sim$16 cm from the target at an angle of 9° with respect to target normal. The collecting lens imaged the self-emission from the plasma onto a gated CCD camera filtered with a 632 nm interference filter and a neutral density filter of optical density 2.0. The pinhole camera recorded the time-integrated self-emission of X-rays with energies above 1.5 keV. The pinhole camera and optical imager are not shown in Fig. 1 and no data from them is presented in this paper.

The laser-target interaction occurred inside a vacuum chamber. In addition to producing jets that moved into a vacuum with pressures below 10^{-2} mb, for many of the shots the chamber was pumped out to vacuum and then re-filled with a single gas, either He, N_2 or Xe. For He and Xe fills the

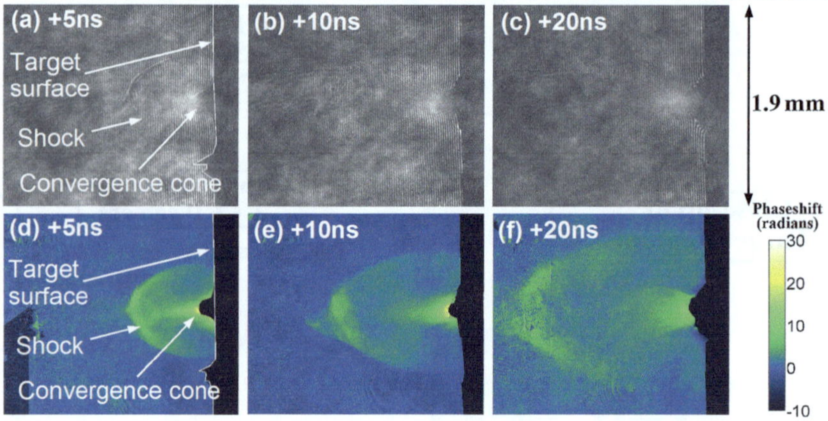

Fig. 2 Partial time series for a copper plasma emerging into 50 mb He. (**a**)–(**c**) are interferograms and (**d**)–(**f**) are the corresponding phasemaps

gas came from a bottle, while in the case of N_2 fills the gas came from the chamber let-up system. Gas pressures used were in the ranges 10–200 mb for He, 9–100 mb for N_2 and 4–32 mb for Xe.

3 Results

Examples of our raw interferometry data for a copper-coated target in a 50 mb He background gas are shown in Fig. 2, the images of which come from the large-area CCD camera. The images of Fig. 2(a)–(c) are taken at times of 5, 10 and 20 ns after the laser pulse struck the target and together form part of a time series showing the plasma evolution. In all the images the target is in profile and is a fringe-free region on the right hand side. The Astra laser pulse comes in from the left through the gas back-fill, ablates the target and produces a plasma that propagates to the left. The time series was built up over a sequence of laser-target shots. It is not a sequence of images from the same shot. Between shots the target strip was moved to align a new cone, and the same laser, target and gas conditions were replicated in all shots in the sequence. The only difference between shots was the time delay of the probe beam with respect to the main Astra pulse. This was altered by changing the physical distance the probe beam traveled before it entered the chamber.

The interferometer was aligned such that, in the absence of a plasma in front of the target, a pattern of narrow, parallel fringes was seen by the cameras in the probe line. As the phase speed of a probe beam photon is higher when passing through a plasma than through a vacuum, the probe photons incurred shifts in phase (relative to the vacuum case) when there was a plasma in front of the target. This resulted in the bending of the otherwise straight fringes seen on the CCD cameras, and in Fig. 2(a)–(c) both shocks and the plasmas coming from the targets show up as departures from the parallel fringe pattern. Using the IDEA computer program (Hipp et al. 2004) to compare images taken both with and without a plasma in front of the target, the raw

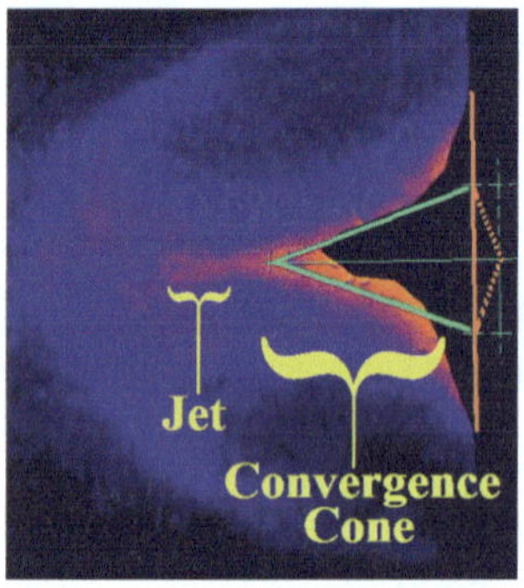

Fig. 3 A Cu plasma propagating into 50 mb He seen at +3ns. The target position is indicated in *orange* and the convergence cone for the full conical target is plotted in *green*

interferometry data can be processed to produce 2-D maps of the shift in phase (relative to the vacuum case) incurred by probe photons in traveling through our plasmas to the CCDs. Figure 2(d)–(e) shows the phasemaps extracted from Fig. 2(a)–(c).

At early times, Fig. 2(a, d), the plasma flow emerging from the target surface has a triangular appearance. This is a side-on view of the convergence of the plasma coming from the target, the conical shape of which is a result of plasma leaving the target along the normal to the cone surface. The converging plasma was collisional, and we anticipated a collisional interaction on-axis between plasma coming from opposite sides of the target. It was expected that this would result in the plasma stagnating on the cone axis but continuing to move away from the target. Figure 3 is from the same time series as the images in Fig. 2 and shows a plasma after 3 ns. In the figure, the target strip surface is indicated in orange and the approximate position of the conical target is indicated by the orange dotted line. In green is indicated the convergence cone that would have resulted if, over its entire surface, plasma had come away from the conical target perfectly along the surface normal. It can be seen that the plasma ablating from the target maintains a conical convergence shape right up to the axis of the system. Beyond the tip of the convergence cone a collimated outflow is seen. This is the stagnated plasma and is our jet.

As the plasmas created in our experiment propagated into a background gas, the question arises as to whether the plasmas seen to converge and form jets are from the target or the background gas. Comparative measurements in vacuum, lower density background gases and with other target materials indicate that these features are due to flow of the target materials. In summary, the evolution of the Cu plasma propagating into the 50 mb He gas was observed to be as follows. At early times, a jet is clearly seen extending from the tip of the convergence cone in the direction of the shock front. By 10 ns, Fig. 2(e), the conical shape of the converging plasma has been lost but a broad axial feature is still visible. By the late time of 20 ns, the convergence cone has disassembled and no axial feature can be seen.

Shocks are clearly seen in Fig. 2, both at early times and at late times, lasting beyond 20 ns. The shocks arise as the ablation plasmas leaves the targets with bulk speeds above 100 $\mathrm{km\,s^{-1}}$, well above the sound speed in the background gas. The shock front in Fig. 2(e) is seen to be smooth except near to target normal. This is a typical feature of our results and is believed to be due to the interaction of the shock and the plasma behind it with ionised background gas within the laser focusing cone.

4 Conclusions

We have created plasma jets by the laser ablation of the inner surface of conical targets. Using the high-repetition rate Astra laser, we have explored the plasmas formed from a number of different materials in different background gases and pressures for time delays up to 20 ns. As an example, we have presented a partial time series of interferometric measurements of plasma coming from a Cu-coated target in 50 mb He gas. The measurements show a convergent flow of plasma from the target surface and, at the tip of this flow, a jet-like region formed by stagnation. Different targets do show significant differences, and some part of these differences is expected to be down to differences in radiative cooling rate. The large amount of data taken, possible because of the high-repetition rate of the laser, means that there is a sizable database for comparisons between plasmas made of different materials and plasmas propagating into different gases, and there is still plenty of analysis work to be done. In order to ascertain whether and to what extent the jets we've created can be scaled to YSO jets, the values of the scaling parameters need to be inferred for our measurement.

We note that there are aspects of YSO systems that are not included in the experiment. Two of these are large-scale magnetic fields and angular momentum, which will always be present in YSO jets.

Acknowledgements We thank the staff of the CLF, and acknowledge provision of funding from EPSRC, STFC, the Royal Society and JSPS. In addition, C.D. Gregory acknowledges funding from Région Ile-de-France.

References

Blondin, J.M., Fryxell, B.A., Königl, A.: Astrophys. J. **360**, 370–386 (1990)

Coffey, D., Bacciotti, F., Ray, T.P., Eislöffel, J., Woitas, J.: Astrophys. J. **663**, 350–364 (2007)

de Gouveia Dal Pino, E.M.: AIP Conf. Proc. **784**, 183–194 (2005a)

de Gouveia Dal Pino, E.M.: Adv. Space Res. **35**(5), 908–924 (2005b)

Gregory, C.D., Howe, J., Loupias, B., Myers, S., Notley, M.M., Sakawa, Y., Oya, A., Kodama, R., Koenig, M., Woolsey, N.C.: Astrophys. J. **676**, 420–426 (2008)

Hipp, M., Woisetschläger, J., Reiterer, P., Neger, T.: Measurement **36**, 53–66 (2004)

Kataoka, J., Edwards, P., Georganopoulos, M., Takahara, F., Wagner, S.: Astron. Astrophys. **399**, 91–97 (2003)

Lebedev, S.V., Chittenden, J.P., Beg, F.N., Bland, S.N., Ciardi, A., Ampleford, D., Hughes, S., Haines, M.G., Frank, A., Blackman, E.G., Gardiner, T.: Astrophys. J. **564**, 113–119 (2002)

Loupias, B., Falize, E., Gregory, C.D., Akira, O., Vinci, T., Howe, J., Koenig, M., Woolsey, N.C., Ozaki, N., Benuzzi-Mounaix, A., Bouquet, S., Michaut, C., Rabec le Goahec, M., Nazarov, W., Pikuz, T., Faenov, A.Ya., Aglitskiy, Y., Atzeni, S., Schiavi, A., Sakawa, Y., Takabe, H., Kodama, R.: J. Phys. Conf. Ser. **112**, 042022 (2008)

Ryutov, D., Drake, R.P., Kane, J., Liang, E., Remington, B.A., Wood-Vasey, W.M.: Astrophys. J. **518**, 821–832 (1999)

Ryutov, D.D., Drake, R.P., Remington, B.A.: Astrophys. J., Suppl. Ser. **127**, 465–468 (2000)

Shigemori, K., Kodama, R., Farley, D.R., Koase, T., Estabrook, K.G., Remington, B.A., Ryutov, D.D., Ochi, Y., Azechi, H., Stone, J., Turner, N.: Phys. Rev. E **62**(6), 8838–8841 (2000)

Astrophys Space Sci (2009) 322: 37–41
DOI 10.1007/s10509-009-0035-5

ORIGINAL PAPER

Colliding plasma experiments to study astrophysical-jet relevant physics

C.D. Gregory · J. Howe · B. Loupias · S. Myers · M.M. Notley · Y. Sakawa · A. Oya · R. Kodama · M. Koenig · E. Falize · S. Bouquet · C. Michaut · N.C. Woolsey

Received: 18 June 2008 / Accepted: 30 March 2009 / Published online: 15 April 2009

Abstract We present the results of experiments in which jets are created through the collision of two laser-produced plasmas. These experiments use a simple 'v-foil' target design: two thin foils are placed at an angle of 140° to each other, and irradiated with a high-energy laser. The plasmas from the rear face of these foils collide and drive plasma jets moving with a velocity of $\sim$300 km s^{-1}. By choosing the foil thickness and material to suit the laser conditions available, it has proven possible to create plasma jets for which the relevant scaling parameters show significant overlap with those of outflows associated with young stellar objects (YSOs). Preliminary results are also shown from experiments to study the effect of an ambient gas on jet propagation. Nominally identical experiments are conducted either in vacuum or in an ambient medium of 5 mbar of nitrogen gas. The gas is seen to increase the jet collimation, and to introduce shock structures at the head of the outflow.

Keywords Young stellar objects · Jets and outflows · Laboratory astrophysics

1 Introduction

Highly collimated, supersonic plasma jets are a common astrophysical phenomenon, and are observed from a wide range of sources including active galactic nuclei (Begelman et al. 1984), x-ray binary systems (Mirabel and Rodríguez 1999) and young stellar objects (YSOs) (Reipurth and Bally 2001). The jets span a large range of sizes and powers, but share many similarities (de Gouveia Dal Pino 2005): they are highly collimated, and in many cases two sided; origi-

C.D. Gregory (✉) · J. Howe · S. Myers · N.C. Woolsey
Department of Physics, University of York, Heslington
YO10 5DD, UK
e-mail: chris.gregory@polytechnique.edu

Present address:
C.D. Gregory
Laboratoire pour l'Utilisation des Lasers Intenses, UMR7605, CNRS - CEA - Université, Paris VI - Ecole Polytechnique, 91128 Palaiseau Cedex, France

B. Loupias · M. Koenig
Laboratoire pour l'Utilisation des Lasers Intenses, UMR7605, CNRS - CEA - Université, Paris VI - Ecole Polytechnique, 91128 Palaiseau Cedex, France

M.M. Notley
Central Laser Facility, Science and Technology Facilities Council, Rutherford Appleton Laboratory, Harwell Science and Innovation Campus, Dicot OX11 0QX, UK

Y. Sakawa
Institute of Laser Engineering, Osaka University, Suita, Osaka 565-0871, Japan

A. Oya · R. Kodama
Graduate School of Engineering, Osaka University, Suita, Osaka 565-0871, Japan

E. Falize · S. Bouquet
CEA/DIF/DPTA, BP 12, 91680 Bruyers-le-Chatel, France

E. Falize · S. Bouquet · C. Michaut
LUTH, Observatoire de Paris, CNRS, Université Paris Diderot, Place Jules Janssen, 92190 Meudon, France

Present address:
E. Falize · S. Bouquet
Laboratoire pour l'Utilisation des Lasers Intenses, UMR7605, CNRS - CEA - Université, Paris VI - Ecole Polytechnique, 91128 Palaiseau Cedex, France

nate in compact objects; show a chain of regularly spaced emission knots; and are often seen to terminate in emission lobes—labelled as 'working surfaces'—as the jet material shocks against the ambient medium. In the case of jets associated with YSOs, the regions of emission from the knots and working surface are know as Herbig-Haro (HH) objects, and these HH objects radiate into spectral lines in the optical and infra-red. YSO jets represent the closest examples of these jets to Earth, and are relatively abundant. This, coupled with the fact that the jets radiate, means that a large amount of high-quality observational data have been collected from these systems, and as such they represent excellent laboratories for the investigation of astrophysical jets. Progress in the understanding of YSO jets has been made through observational, theoretical and numerical studies. In addition there exists a number of areas where well-designed laboratory experiments could lead to further insight into the physics of these jets (Hartigan 2005).

Laboratory simulations of YSO jets can be broadly divided into two categories, jet launch and jet propagation. A comparatively small number of experimental studies have been conducted into jet launch mechanisms, but experiments have succeeded in launching jets from geometries analogous to that of a YSO in both z-pinch (Lebedev et al. 2005) and spheromak (Hsu and Bellan 2002) systems. In contrast, the physics of jet propagation and collimation has been the focus of a number of experiments. These experiments have propagated jets into an ambient medium (Logory et al. 2000; Foster et al. 2005), studied the interaction between the jet and a plasma cross wind (Lebedev et al. 2004), and demonstrated the effects of radiative cooling on the collimation of the flow (Farley et al. 1999; Shigemori et al. 2000; Lebedev et al. 2002).

In what follows, an experimental design to study physics relevant to jet propagation is presented. The experiment uses a simple, flexible target to generate high-velocity plasma jets through irradiation of thin foils with a high energy laser. These jets have been propagated into both vacuum and into 5 mbar of nitrogen gas, in order to investigate the effect of an ambient medium on the flows properties. Scaling analysis shows that these jets are in a regime where scaling to astrophysics may be possible, and as such represent a potential future avenue for a controlled laboratory studies of YSO jets.

2 Experimental design

The target design is shown in Fig. 1. Two 0.8 μm aluminium foils were placed at an angle of 70° to the mirror symmetry axis. Each of these foils was irradiated with one beam of the VULCAN Nd:glass laser, at the Rutherford-Appleton laboratory, UK. The spacing between the two focal spots was 1 mm. The beams were converted to the second harmonic, and delivered 80 J of laser light at 527 nm in 1 ns, with a focal spot of 260 μm achieved using random phase plates. The thickness of the aluminium is chosen such that irradiation by this laser results in an explosion of the foil, and the plasma flow from the rear face is supersonic. As the two plasmas reach the mirror symmety axis and collide, there is a stagnation of the flow in the transverse direction, while the axial kinetic energy is conserved. This leads to a heating of the plasma and a bulk flow along axial direction.

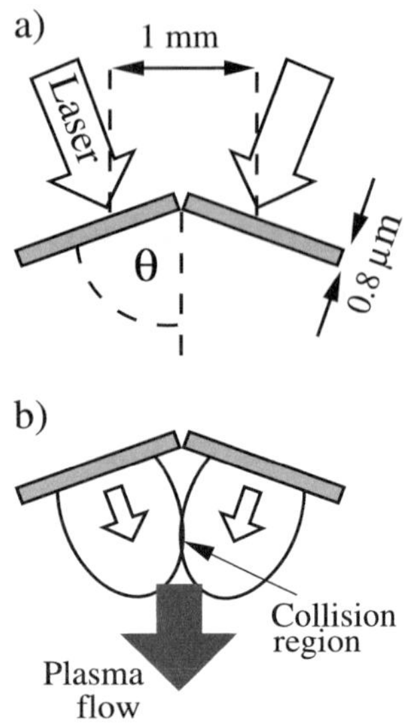

Fig. 1 The target design: (**a**) two laser beams irradiate thin aluminium foils placed at an angle of $\theta = 70°$ to the vertical. The foils explode and collide, (**b**) driving a narrow plasma flow

The experiment was diagnosed with a 1 ps, 263 nm probe laser coupled to a modified Normarski interferometer (Benattar et al. 1979), allowing the electronic density to be inferred from the resulting interfergrams. The probe laser could be delayed with respect to the drive laser by up to 12 ns. By taking 1 ps snapshots of the plasma expansion at different times, over a series of nominally identical shots, a time history of the jet expansion is obtained.

3 Results

Figure 2 shows raw interferometric data, taken with a probe delay of 10 ns. The original positions of the foils are shown with white lines, and the jet propagates in the direction from the top to the bottom of the image. In the region adjacent to the mirror symmetry axis the fringes are no longer well defined. This may be due to turbulent mixing of the two flows, and as a result no quantitative electron density information can be extracted here. In the analysed data presented below, this region is masked for clarity.

Figure 3 shows the extracted phase shift map for delay times between 4 and 10 ns. After 4 ns the foils have begun to collide. At 6 ns a narrower outflow has formed in front of the collision region, and by 10 ns, this narrow outflow is seen at distances of greater than 3 mm from the original

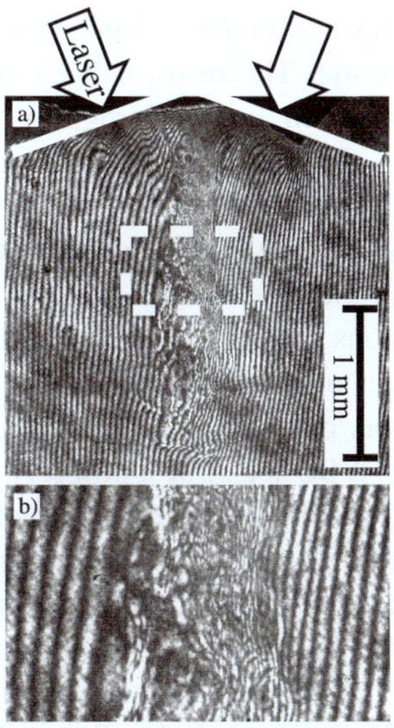

Fig. 2 Raw intereferometric data, taken with a probe delay of 10 ns is shown in (**a**). The initial foil positions are shown by white lines. The fringes in the collision region are scrambled, and the zone enclosed by a white dashed rectangle in show magnified in panel (**b**)

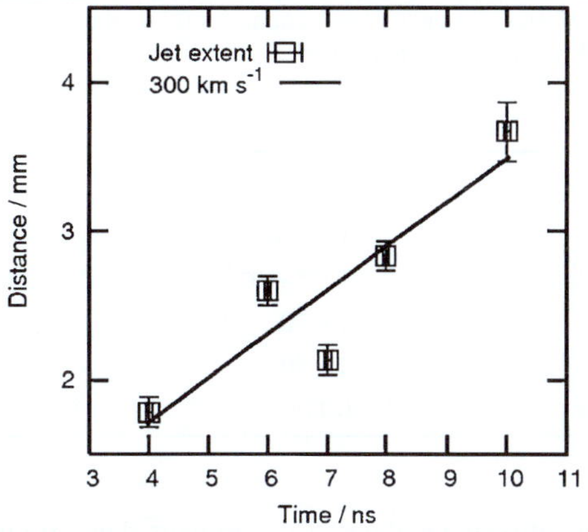

Fig. 4 The jet length as a function of time. A least squares fit to the data gives a velocity of 300 ± 60 km s^{-1}

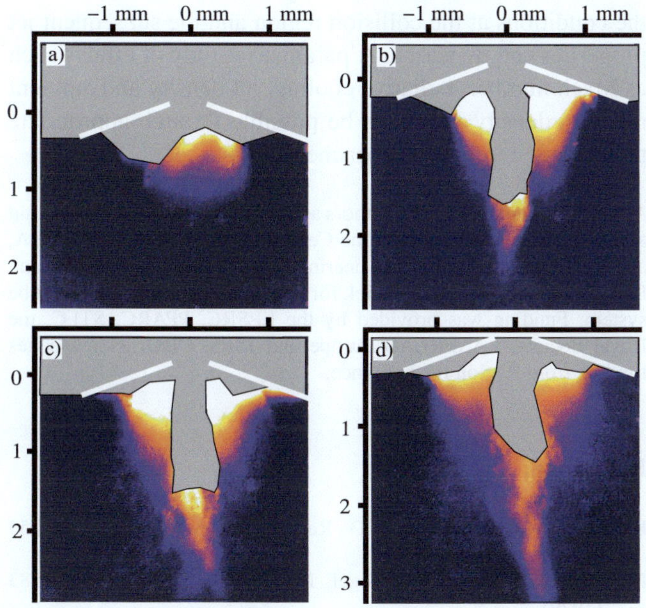

Fig. 3 Phase shift maps extracted from interferograms taken with a probe delay of (**a**) 4 ns, (**b**) 6 ns, (**c**) 8 ns and (**d**) 10 ns

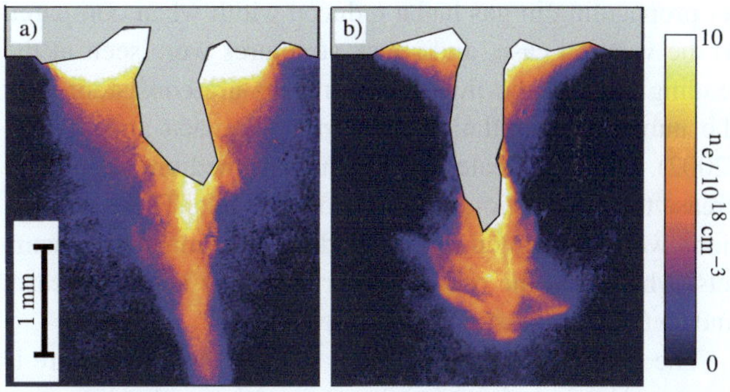

Fig. 5 Inferred electron density maps from (**a**) a jet propagating in vacuum, and (**b**) and jet propagating in 5 mbar N_2 gas. The probe delay is 10 ns in both cases

laser-foil interaction. The extension of the leading edge of the expansion as a function of time is plotted in Fig. 4, and a least squares fit to this data gives a velocity of 300 ± 60 km s^{-1}.

Experiments were also performed with the vacuum chamber back-filled with 5 mbar of nitrogen gas. A comparison of nominally identical shots propagating into vacuum and into nitrogen gas, with a probe delay of 10 ns, are shown in Fig. 5. The electron density averaged along the probe path through the plasma is shown here. The extent of the plasma in the probing direction is measured from an interferometric image taken in the orthogonal direction for a different laser shot, with the same probe delay. The addition of the ambient gas results in a region of higher density at the head of the jet, as the supersonic outflow shocks against the ambient medium. Behind this zone of increased electron density, the width of the jet propagating in gas is reduced in comparison with the vacuum case. The reduction ratio is between 0.6–0.7.

4 Discussion

The probe results presented above show that the v-foil targets produce jets when irradiated with a suitable laser. Probe measurements of experiments using a single beam on a plane foil (not presented) show that the initial expansion velocity from each foil is 425 ± 40 km s^{-1}. It is noted that the axial component of this initial foil velocity is 400 km s^{-1}, as compared to measured value of around 300 km s^{-1} for the jet velocity. This suggests that a large amount of the initial axial velocity is conserved in the jet.

The effect of an ambient medium of 5 mbar of nitrogen gas was investigated. This gas pressure was chosen to recreate astrophysical conditions as closely as possible, and

Table 1 Inferred dimensionless parameters for a laboratory jet propagating in 5 mbar N_2, and a YSO jet

Parameter	V-foil	YSO jet
Re	10^5–10^7	10^7
Pe	10^2–10^3	10^6
ζ	10^{-8}–10^{-6}	10^{-4}
η	10–50	1–20
M	20–40	20–40
χ	0.1–1	0.1–10

to give a ratio of jet density to ambient density similar to that found in YSO jets. Nominally identical jets were propagated into vacuum and into a gas, and probe images taken after 10 ns showed differences between the two cases. The jet propagating in gas had a reduced width when compared to the vacuum case, and shock structures were seen at the leading edge of the flow. These results are consistent with the numerical simulation of Mizuta, Yamada and Takabe (2002), which indicate that an ambient medium should increase the collimation of laser produced jets. These simulations were performed with cylindrically symmetric plasma jets, whereas in the present experiment this is not the case, and so the comparison is qualitative only.

The goal of experiments such as those presented here is to study physics relevant to the propagation and collimation of the jets associated with YSOs. In order for this to be the possible, the two systems should first satisfy a set of scaling relations (Ryutov et al. 1999). The approach taken is that these relations can be applied for plasmas having a Reynolds number, Re and a Péclet number, Pe both $\gg$1, indicating that viscous effects and thermal diffusion are unimportant, respectively. In addition, the particles must localized on a scale length much smaller than the system size, leading to a constraint on the collsionality, $\zeta \ll 1$. Here ζ is defined as the ion-ion mean-free-path λ_{ii} divided by the width of the jet. If these strong inequalities are satisfied, then the similarity of the two systems is measured by three dimensionless parameters: the internal Mach number of the jet, M; the ratio of the jet-to-ambient density, η; and the ratio of the jet width to the characteristic cooling length, χ. The results of evaluating these parameters for a YSO jet (Gregory et al. 2008), and for a laboratory jet propagating in an ambient gas (Fig. 5b) are shown in Table 1.

When calculating λ_{ii} for the YSO jets, a characteristic particle density of 10^5 cm^{-3} and a temperature of 1 eV are used (de Gouveia Dal Pino 2005). From these values λ_{ii} is estimated to be 3×10^{11} cm, compared with a typical YSO radius of 3×10^{15} cm. For the laboratory jets, the density is taken to be 5×10^{18} cm^{-3}, as measured by interferometry. In similar experiments performed at the LULI laboratory, France, a temperature of $\sim$10 eV was measured for these flows after the collision, using an optical pyrometry diagnostic. The result is a λ_{ii} of $\sim 3 \times 10^{-8}$ cm, while the jet width in the interferograms is around 5×10^{-2} cm.

The range of values for the dimensionless parameters in the case of the laboratory jet represent upper and lower limits given by the accuracy of the experimental measurements. In the case of both the laboratory and YSO jets, the conditions of $Re \gg 1$, $Pe \gg 1$ and $\zeta \ll 1$ are satisfied, and the values of M, η and χ show substantial overlap. These laboratory experiments are thus in a regime which may be scalable to YSO jets. For a properly scaled experiment, the two systems should also be geometrically similar. The astrophysical jets are cylindrically symmetric, while this is not yet the case in the laboratory. Future experiments will correct for this with a cylindrically symmetric target design.

A large advantage presented by this target design is it flexibility and simplicity. Through changing the foil material and thickness, and the angle between the two foils, the conditions at the collision region and the subsequent jet properties can by varied. A parametric study of effects such as Mach number, radiative cooling, jet density and ambient medium density may then be possible through appropriate target design in future experiments.

Acknowledgements The authors acknowledge the assistance of, and thank, the many colleagues at the Central Laser Facility, LULI, CEA, and the Institute of Laser Engineering. They would also like to thank Phil Nilson and Karl Krushelnick for help with the design of the probe system. Funding was provided by the EPSRC, PPARC, STFC, the Royal Society, LASERLAB-Europe and JSPS. CDG acknowledges funding from Région Ile de France.

References

Begelman, M.C., Blandford, R.D., Rees, M.J.: Rev. Mod. Phys. **56**, 255 (1984)

Benattar, R., Popovics, C., Sigel, R.: Rev. Sci. Instrum. **50**, 1583 (1979)

de Gouveia Dal Pino, E.M.: Adv. Space Res. **35**, 908 (2005)

Farley, D.R., Estabrook, K.G., Glendinning, S.G., Glenzer, S.H., Remington, B.A., Shigemori, K., Stone, J.M., Wallace, R.J., Zimmerman, G.B., Harte, J.A.: Phys. Rev. Lett. **83**, 1982 (1999)

Foster, J.M., Wilde, B.H., Rosen, P.A., Williams, R.J.R., Blue, B.E., Coker, R.F., Drake, R.P., Frank, A., Keiter, P.A., Khokhlov, A.M., Knauer, J.P., Perry, T.S.: Astrophys. J. Lett. **634**, L77 (2005)

Gregory, C.D., Howe, J., Loupias, B., Myers, S., Notley, M.M., Sakawa, Y., Oya, A., Kodama, R., Koenig, M., Woolsey, N.C.: Astrophys. J. **676**, 420 (2008)

Hartigan, P.: Astrophys. Space Sci. **298**, 99 (2005)

Hsu, S.C., Bellan, P.M.: Mon. Not. R. Astron. Soc. **334**, 257 (2002)

Lebedev, S.V., Chittenden, J.P., Beg, F.N., Bland, S.N., Ciardi, A., Ampleford, D., Hughes, S., Haines, M.G., Frank, A., Blackman, E.G., Gardiner, T.: Astrophys. J. **564**, 113 (2002)

Lebedev, S.V., Ampleford, D., Ciardi, A., Bland, S.N., Chittenden, J.P., Haines, M.G., Frank, A., Blackman, E.G., Cunningham, A.: Astrophys. J. **616**, 988 (2004)

Lebedev, S.V., Ciardi, A., Ampleford, D.J., Bland, S.N., Bott, S.C., Chittenden, J.P., Hall, G.N., Rapley, J., Frank, C.A.J.A., Blackman, E.G., Lery, T.: Mon. Not. R. Astron. Soc. **361**, 97 (2005)

Logory, L.M., Miller, P.L., Stry, P.E.: Astrophys. J. Suppl. Ser. **127**, 423 (2000)
Mirabel, I.F., Rodríguez, L.F.: Annu. Review Astron. Astrophys. **37**, 409 (1999)
Mizuta, A., Yamada, S., Takabe, H.: Astrophys. J. **57**, 635 (2002)
Reipurth, B., Bally, J.: Annu. Rev. Astron. Astrophys. **39**, 403 (2001)
Ryutov, D., Drake, R.P., Kane, J., Liang, E., Remington, B.A., Wood-Vasey, W.M.: Astrophys. J. **518**, 821 (1999)
Shigemori, K., Kodama, R., Farley, D.R., Koase, T., Estbrook, K.G., Remington, B.A., Ryutov, D.D., Ochi, Y., Azechi, H., Stone, J., Turner, N.: Phys. Rev. E **62**(6), 8838 (2000)

Astrophys Space Sci (2009) 322: 43–47
DOI 10.1007/s10509-008-9957-6

ORIGINAL ARTICLE

Spiral mode of standing accretion shock instability in core-collapse supernovae

Wakana Iwakami · Naofumi Ohnishi · Kei Kotake · Shoichi Yamada · Keisuke Sawada

Received: 15 June 2008 / Accepted: 24 November 2008 / Published online: 3 December 2008

Abstract The study of standing accretion shock instability (SASI) in core-collapse supernova cores has been done with three-dimensional (3D) computer simulations. Rotations with various perturbations were introduced from outer boundary of an initial steady accreting flow. We found that one or two armed spiral accreting flow onto the proto-neutron star (PNS) is formed inside the shock wave depending on perturbations. The linear growth of spiral modes are clearly diagnosed by the mode analysis of the shock surface, and the lower m modes grow quickly in the linear regime.

Keywords Supernovae · Hydrodynamics · Instability

1 Introduction

A core-collapse supernova is a majestic explosion of a massive star at the end of its life and is also the origin of a neutron star. That is one of the scientifically significant phenomena since extremely high energy and density conditions around the center of the supernova core would contribute to the development of promising disciplines such as neutrino astronomy, gravitational astronomy, and high energy density physics. In the field of astrophysics, core-collapse supernovae are associated with various phenomena, e.g., gamma-ray bursts, r-process nucleosynthesis, pulsar spin/kick, and so on (Burrows 1998; Woosley and Janka 2005). Although this spectacular phenomenon is a key issue for astrophysics, the explosion mechanism has not been entirely clear.

Recently, SASI has been noticed since it may perform a crucial function for the large asymmetric explosion of a core-collapse supernova (Blondin and Mezzacappa 2006; Ohnishi et al. 2006; Burrows et al. 2006; Foglizzo et al. 2007; Laming 2007, 2008; Scheck et al. 2008). SASI is a hydrodynamic instability of the spherical shock wave in an accretion flow onto the PNS. The key feature of SASI is that the lower ℓ modes dominate the flow dynamics, where ℓ is the degree of the spherical harmonics. For axisymmetric flows, the spherical shock wave, therefore, oscillates upward and downward repeatedly, and the turbulent flow is generated inside the shock wave due to the non-linear saturation (Blondin et al. 2003; Scheck et al. 2004; Ohnishi et al. 2006). The growth of non-axisymmetric ($m \neq 0$) modes has also been confirmed with recent 3D simulations (Blondin and Mezzacappa 2007; Iwakami et al. 2008), where m is the order of the spherical harmonics.

The growing modes can be classified into axisymmetric modes and non-axisymmetric modes called *sloshing* modes and *spiral* modes, respectively. If axisymmetry is broken by giving non-axisymmetric perturbations or rotation into the flow, spiral modes appear. The spiral modes developing shock wave deformation lead to a spiral accreting flow, which transports the angular momentum onto the PNS,

W. Iwakami (✉) · N. Ohnishi · K. Sawada
Department of Aerospace Engineering, Tohoku University, 6-6-01 Aramaki-Aza-Aoba, Aoba-ku, Sendai, 980-8579, Japan
e-mail: iwakami@rhd.mech.tohoku.ac.jp

K. Kotake
Division of Theoretical Astronomy, National Astronomical Observatory of Japan, 2-21-1, Osawa, Mitaka, Tokyo, 181-8588, Japan

S. Yamada
Science and Engineering, Waseda University, 3-4-1 Okubo, Shinjuku, Tokyo, 169-8555, Japan

S. Yamada
Advanced Research Institute for Science and Engineering, Waseda University, 3-4-1 Okubo, Shinjuku, Tokyo, 169-8555, Japan

inside the shock wave (Blondin and Mezzacappa 2007). Blondin and Shaw (2007) have shown the linear growth of spiral modes by the two-dimensional non-axisymmetric simulations on the equatorial plane. They dropped multiple density bars on the non-rotating initial steady accretion flow as the perturbation. The single ($m = 1$), double ($m = 2$), and triple ($m = 3$) armed spiral flows appear without any explicit rotation, and the lower m modes are predominant in the perturbation growth. On the other hand, Yamasaki and Foglizzo (2008) have found in their perturbative analysis for cylindrical shock wave that the spiral $\pm m$ modes are degenerate without rotation and that the modes rotating in the same direction as the flow increase more than the modes rotating in the opposite direction (also pointed out by Laming 2007, 2008). We have to confirm it in the actual situation that both of them may be applicable for growing spiral modes in 3D rotating flow. In this paper, we report on the linear growth of spiral modes by numerical simulations with imposing rotation perturbed with spherical harmonics in 3D flow.

2 Numerical procedure

Our numerical code is an Eulerian hydrodynamics code based on ZEUS-MP/2 (Hayes et al. 2006). ZEUS-MP/2 code was developed with the finite-difference method using a staggered mesh. We modified the original code according to our preceding two-dimensional work (Ohnishi et al. 2006) and with the tensor type artificial viscosity (Stone and Norman 1992) for the numerical stability on the spherical shock wave.

We performed 3D numerical simulations of the accreting flow surrounding a central object termed PNS in the iron core with mass accretion rate $\dot{M} = 1.0\ M_\odot$/s. The flow is gravitationally attracted by the PNS having the mass of $1.4\ M_\odot$ and is irradiated by neutrinos emitted from the neutrino sphere near the PNS. The basic evolution equations are the compressible Euler equations. We added to them the terms associated with the neutrino heating and cooling and the gravity due to the central mass while ignoring the self-gravity. Moreover, we assumed that the falling matters outside the PNS are optically thin, and the electron and anti-electron neutrinos are emitted isotropically from the neutrino sphere with the luminosity of $L_\nu = 6.0 \times 10^{52}$ erg/s. A realistic tabulated equation of state formulated by Shen et al. (1998) was utilized.

The spherical coordinates (r, θ, ϕ) were adopted in the code with the origin at the center of the PNS. We set 300 radial grid points to cover $r_{\rm in} \le r \le r_{\rm out}$, where $r_{\rm in} \sim 50$ km is the radius of the inner boundary located roughly at the neutrino sphere and $r_{\rm out}$=2000 km is the radius of the outer boundary at which the supersonic flow is falling inward. The radial grid width Δr increased with r to resolve the turbulent motion inside the shock wave. 30 polar and 60 azimuthal grid points were set to cover the whole spherical surface.

The rigid rotation was given from the outer boundary as follows,

$$v_\phi^{2D}(r, \theta) = k v_r^{1D}(r) \sin\theta, \tag{1}$$

where v_ϕ^{2D}, $v_r^{1D}(r) = \dot{M}/4\pi r^2 \rho(r)$, and k denote the unperturbed ϕ component of velocity, the r component of velocity, and a parameter for the rotating flow, respectively. The density ρ is determined by solving the one-dimensional steady equations following Yamasaki and Yamada (2005) under the condition of $\rho_{\rm in} = 10^{11}$ g cm^{-3}. The radial distribution of $\rho(r)$ can be found in Iwakami et al. (2008). In this paper, we fixed to set $k = 0.01$ at $r = r_{\rm out} = 2000$ km, which corresponds to the specific angular momentum $L \sim 4.6 \times 10^{15}$ cm^2 s^{-1} on the equatorial plane. The L is slightly lower as 10^{16}–10^{17} cm^2 s^{-1} obtained from the results of the presupernova evolution of rotating stars performed by Heger et al. (2000).

In order to induce non-spherical instability, we added the non-axisymmetric ϕ velocity perturbations to the rotational flow as follows,

$$v_\phi(r_{\rm out}, \theta, \phi) = v_\phi^{2D}(r_{\rm out}, \theta) + \delta v_\phi(r_{\rm out}, \theta, \phi), \tag{2}$$

where $v_\phi(r, \theta, \phi)$ is the ϕ component of velocity. Three types of velocity perturbations were considered,

(1) a $\ell = 1$, $m = 1$ single-mode perturbation:

$$\delta v_\phi(r_{\rm out}, \theta, \phi) \propto -\sqrt{\frac{3}{8\pi}} \sin\theta \cos\phi \cdot v_\phi^{2D}(r_{\rm out}, \theta), \tag{3}$$

(2) a $\ell = 2$, $m = 2$ single-mode perturbation:

$$\delta v_\phi(r_{\rm out}, \theta, \phi) \propto \sqrt{\frac{15}{32\pi}} \sin^2\theta \cos 2\phi \cdot v_\phi^{2D}(r_{\rm out}, \theta), \tag{4}$$

(3) a random multimode perturbation:

$$\delta v_\phi(r_{\rm out}, \theta, \phi) \propto {\rm rand} \cdot v_r^{2D}(r_{\rm out}, \theta), \ (0 \le {\rm rand} < 1), \tag{5}$$

where "rand" denotes a pseudo-random number. These perturbation amplitudes are set to be less than 1% of the unperturbed velocity. The perturbed rotation introduced only on the outer boundary is advected inward by accreting flow and reach the shock wave around 100 ms.

3 Results

Figure 1 shows the velocity vector fields on the equatorial plane inside the shock wave. If we impose $\ell = 1$, $m = 1$ perturbation without rotation, the flow field maintains axisymmetry with respect to x-axis in the linear phase (Fig. 1(a)).

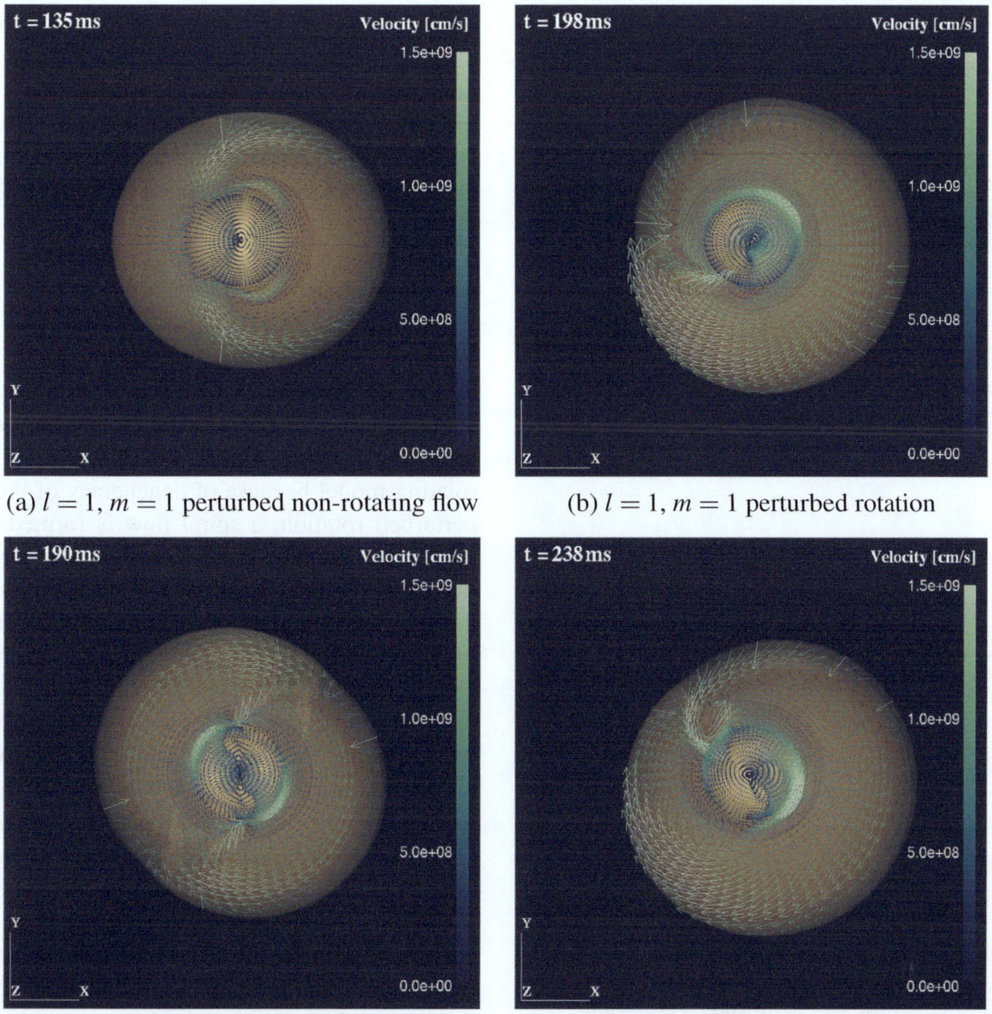

(a) $l=1$, $m=1$ perturbed non-rotating flow

(b) $l=1$, $m=1$ perturbed rotation

(c) $l=2$, $m=2$ perturbed rotation

(d) Random perturbed rotation

Fig. 1 Velocity fields on the equatorial plane inside the shock wave. The direction of *arrow* denotes that of velocity, and the length of *arrow* and the color contour express the magnitude of velocity

On the other hand, with rotation, a single-armed accretion flow appears (Fig. 1(b)). One can find a triple line on the shock wave in the left of Fig. 1(b). A triple line is the meeting place for the matters which flow in the direction of clockwise and counter-clockwise along the shock surface. Moreover, with $\ell=2$, $m=2$ perturbation, a double-armed accretion flow is formed (Fig. 1(c)), which have two triple lines in the upper right and the lower left of Fig. 1(c). For the case of random perturbation, a single-armed spiral flow with a triple line is developed (Fig. 1(d)). It should be noted that the matters accrete from the triple line to the PNS with high velocity. For all perturbations with rotation, a clockwise spiral flow makes the matters accrete onto the PNS with the velocity in the direction of the counter-clockwise.

Figure 2 shows the time evolution of normalized amplitude obtained by mode analysis of the deformation of the shock wave as expanded with the spherical harmonics. The perturbations does not grow until about 100 ms because the rotating flow perturbed on the outer boundary at 0 ms cannot reach the shock surface in this duration. After 100 ms, the perturbations grow exponentially until 200 ms: this phase is the so-called linear phase. After 200 ms, the normalized mode amplitudes are saturated, i.e., the non-linear phase begins. For the case of $\ell=1$, $m=1$ single-mode perturbation, the fundamental $\ell=1$, $m=1$ mode grows dominantly and the other modes also grow by mode coupling (Fig. 2(a)). On the other hand, with $\ell=2$, $m=2$ single-mode perturbation, the fundamental $\ell=2$, $m=2$ mode grows dominantly and the other even number modes grow, but the odd number modes do not grow (Fig. 2(b)). For both single-mode perturbations, the growth rates are nearly same in the linear phase, and the lower modes are distinguished even in the non-linear phase. According to Fig. 1 in the paper of Yamasaki and Foglizzo (2008), the growth rates between

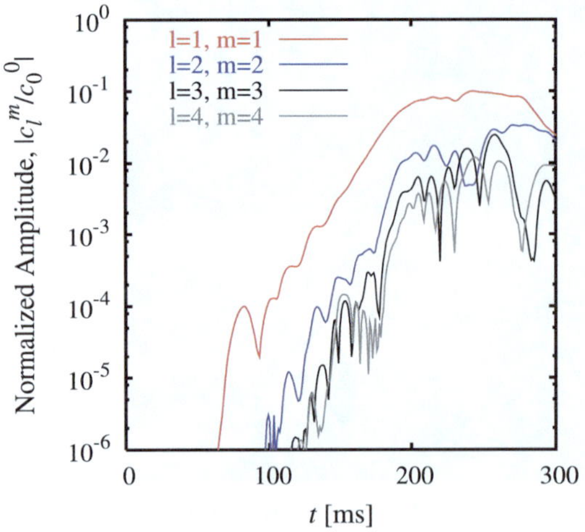

(a) $l = 1$, $m = 1$ perturbed rotation

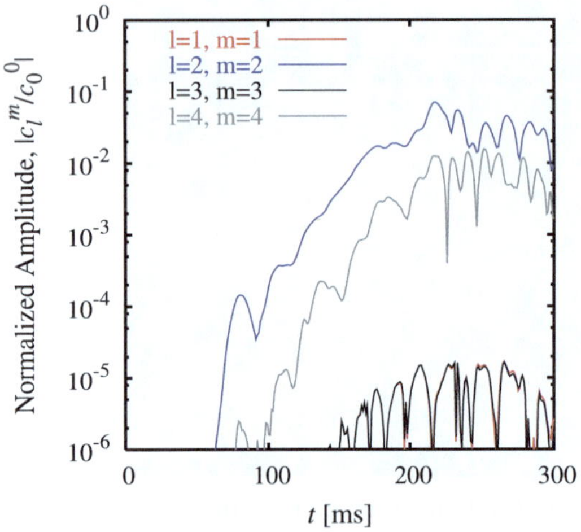

(b) $l = 2$, $m = 2$ perturbed rotation

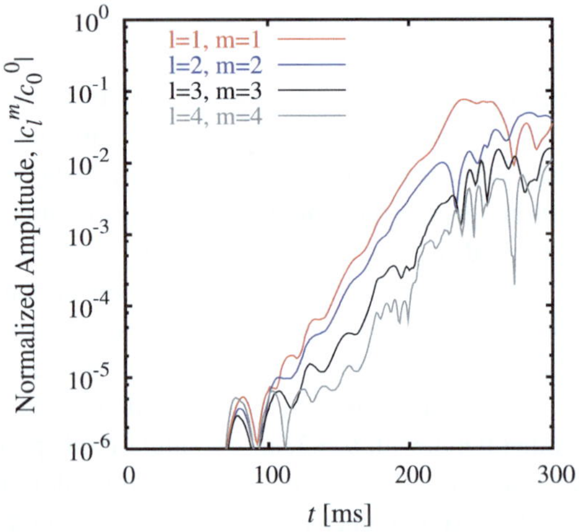

(c) Random perturbed rotation

Fig. 2 Time development of normalized amplitude obtained by mode analysis of the deformation of the shock wave. Only spiral modes are plotted

$m = 1$ and $m = 2$ are approximately same for $L \sim 730$ normalized by $2\pi \times 10^{12}\ \mathrm{cm}^2\,\mathrm{s}^{-1}$. For the case of multi-mode random perturbation, the ℓ=1 modes grow most quickly in the linear phase while the other modes also develop with smaller growth rate, and the lower modes may be sustained at higher level in the non-linear phase (Fig. 2(c)).

4 Conclusions

We have investigated the growth of spiral modes of SASI by the 3D simulations with imposing a perturbed rotation. When we impose $\ell = 1$, $m = 1$ perturbation, spiral modes grow for rotating model, but they never grow for non-rotating model because of degeneracy of them. Imposing a perturbed rotation, a spiral flow is formed and makes the matters accrete onto the PNS with high velocity in the direction of the counter-rotation. The perturbation amplitudes of the spiral modes grow exponentially in the linear phase and the lower m modes are predominant even in the non-linear phase. These results show in part the same characteristics of the results given by Blondin and Shaw (2007). However, they have used multiple density bars for non-rotating flow instead of introducing perturbation based on the spherical harmonics into rotating flow and have examined the flow field limited to the equatorial plane. Comparing the growth rates of $m = 1$ and $m = 2$, our models are more applicable to the linear analysis of Yamasaki and Foglizzo (2008) and consistent with the results of them though they performed the analysis for the cylindrical flow. In order to investigate the effect of rotation on the linear growth of SASI in detail, however, we need to impose a perturbation on the initial steady rotational flow.

Acknowledgements We have conducted numerical computations on Altix3700Bx2 at the Institute of Fluid Science, Tohoku University as well as on VPP5000 and the general common use computer system at the center for Computational Astrophysics, CfCA, the National Astronomical Observatory of Japan.

References

Burrows, A.: Nature **393**, 121 (1998)
Burrows, A., Livne, E., Dessart, L., Ott, C.D., Murphy, J.: Astrophys. J. **640**, 878 (2006)
Blondin, J.M., Mezzacappa, A.: Astrophys. J. **642**, 401 (2006)
Blondin, J.M., Mezzacappa, A.: Nature **445**, 58 (2007)
Blondin, J.M., Shaw, S.: Astrophys. J. **656**, 366 (2007)
Blondin, J.M., Mezzacappa, A., DeMarino, C.: Astrophys. J. **584**, 971 (2003)
Foglizzo, T., Galletti, P., Sheck, L., Janka, H.-T.: Astrophys. J. **654**, 1006 (2007)
Heger, A., Langer, N., Woosley, S.E.: Astrophys. J. **528**, 368 (2000)
Hayes, J.C., Norman, M.L., Fiedler, R.A., Bordner, J.O., Li, P.S., Clark, S.E., ud-Doula, A., Mac Low, M.-M.: Astrophys. J. Suppl. Ser. **165**, 188 (2006)

Iwakami, W., Kotake, K., Ohnishi, N., Yamada, S., Sawada, K.: Astrophys. J. **678**, 1207 (2008)
Laming, J.M.: Astrophys. J. **659**, 1449 (2007)
Laming, J.M.: Erratum. Astrophys. J. **687**, 1461 (2008)
Ohnishi, N., Kotake, K., Yamada, S.: Astrophys. J. **641**, 1018 (2006)
Stone, J.M., Norman, M.L.: Astrophys. J. Suppl. **80**, 753 (1992)
Shen, H., Toki, H., Oyamatsu, K., Sumiyoshi, K.: Nucl. Phys. A **637**, 43 (1998). 109, 301
Scheck, L., Plewa, T., Janka, H.-T., Kifonidis, K., Müller, E.: Phys. Rev. Lett. **92**, 011103 (2004)
Scheck, L., Janka, H.-T., Foglizzo, T., Kifonidis, K.: Astron. Astrophys. **477**, 931 (2008)
Yamasaki, T., Foglizzo, T.: Astrophys. J. **679**, 607 (2008)
Yamasaki, T., Yamada, S.: Astrophys. J. **623**, 1000 (2005)
Woosley, S., Janka, H.-T.: Nature Phys. **1**, 147 (2005)

Astrophys Space Sci (2009) 322: 49–55
DOI 10.1007/s10509-008-9967-4

ORIGINAL ARTICLE

Image processing of radiographs in 3D Rayleigh-Taylor decelerating interface experiments

C.C. Kuranz · R.P. Drake · M.J. Grosskopf · H.F. Robey · B.A. Remington · J.F. Hansen · B.E. Blue · J. Knauer

Received: 20 June 2008 / Accepted: 8 December 2008 / Published online: 5 February 2009

Abstract This paper discusses high-energy-density laboratory astrophysics experiments exploring the Rayleigh-Taylor instability under conditions similar to the blast wave driven, outermost layer in a core-collapse supernova. The planar blast wave is created in an experimental target using the Omega laser. The blast wave crosses an unstable interface with a seed perturbation machined onto it. The perturbation consists of a 3D "egg crate" pattern and, in some cases, an additional longer wavelength mode is added to this 3D, single-mode pattern. The main diagnostic of this experiment is x-ray radiography. This paper explores an image processing technique to improve the identification and characterization of structure in the radiographic data.

Keywords Supernovae · Hydrodynamic instabilities · Laboratory astrophysics

1 Introduction

The experiments described in this paper are motivated by the mixing that occurs in a core-collapse supernova (SN). Of specific interest, is the core-collapse supernova SN1987A because there was a difference between the light curve data from the SN and the models at the time (Muller et al. 1991). Data collected from this SN showed higher velocities and earlier than expected x-ray emission of dense elements originating from the center of the star. This occurrence led to interest in further understanding the mixing between different elements and layers of the star.

Prior to the core-collapse and explosion, this type of supernova has a layered density structure with a dense iron core surrounded by layers of decreasing density with increasing radius (Arnett 1996). This results in a density gradient in the inward direction. Once the star explodes, a shock wave moves outward through the layers of decreasing density. The pressure is greatest just behind the shock front, which creates a pressure gradient in the radially outward direction. The opposition of the density and pressure gradients initiates the Rayleigh-Taylor (RT) instability (Rayleigh 1900; Taylor 1950). This instability grows following the initial, Richtmyer-Meshkov (Meshkov et al. 1997; Richtmyer 1960) response of the interface to the shock passage. At the times that this experiment is observed, it is dominated by the RT instability (Robey et al. 2003).

These experiments are designed to study the RT instability induced at the interface between the helium and hydrogen layers, the outermost layers of the star. Conditions similar to the SN are created using the Omega Laser (Boehly et al. 1995). The experiment is scaled so that it will have similar hydrodynamic evolution to that of the SN (Remington et al. 2006; Ryutov et al. 1999). These experiments build on past work and the fundamental dynamics and the specifics of the experimental scaling have been previously discussed (Drake et al. 2004; Kane et al. 1997; Kuranz et al. 2005, 2007; Robey et al. 2001). The primary diagnostic data produced in the experiments is radiographic images, from

C.C. Kuranz (✉) · R.P. Drake · M.J. Grosskopf
University of Michigan, Ann Arbor, MI, USA
e-mail: ckuranz@umich.edu

H.F. Robey · B.A. Remington · J.F. Hansen
Lawrence Livermore National Laboratory, Livermore, CA, USA

B.E. Blue
General Atomics, San Diego, CA, USA

J. Knauer
University of Rochester, Rochester, NY, USA

which one desires to extract quantitative measures of instability development, such as the location of the RT bubble tips and spike tips as well as to calculate the mass of features produced by the RT instability. The present paper is concerned with the analysis of data from a recent sequence of experiments, in which an improved diagnostic technique enabled us to obtain better diagnostic resolution, but introduced a challenging background feature into the experimental images. This background feature increases the difficultly in accurately locating and characterizing the features related to the RT instability. We describe and demonstrate here a method to remove enough of this background feature that improved analysis of the images is possible. We also illustrate a change in experimental technique that eliminates the background feature for future experiments, however, there is still a large amount of data that requires the image analysis technique described in this paper.

2 Experiment and experimental setup

The experimental target has two key components, high-density plastic layer and low-density foam. This establishes a density drop where the RT instability develops. The plastic layer is 150 µm thick and is composed mostly of polyimide ($C_{22}H_{10}O_5N_2$). It is followed by a few mm of carbonized resorcinol formaldehyde (CRF) foam. These main components of the target are shown in an exploded view in Fig. 1a. The densities of the materials are 1.41 g/cc and 50 mg/cc, respectively. On the rear surface of the polyimide a slot 75 µm deep, 230 µm wide and the length of the polyimide disk, 905 µm, is milled out of the piece. A strip of plastic that has been doped with bromine is glued into the slot. This strip is shown as the dark region of Fig. 1b. The density of the brominated plastic (CHBr) is 1.42 g/cc and the chemical composition is $C_{500}H_{457}Br_{43}$. Since the polyimide and CHBr have such similar densities the hydrodynamics of the experiment should not be affected substantially by the different materials. The purpose of doping the plastic with bromine is to provide contrast in the resulting side-on x-ray radiographs used to diagnose the experiment.

On the rear surface of the plastic piece, a 3D seed perturbation is machined. The basic pattern resembles an "egg crate" and is composed of two sine waves in orthogonal directions. The amplitude of the perturbation is 2.5 µm and the wavelength is 71 µm. In some experiments a multimode pattern is used to explore how the initial conditions affect the growth of the Rayliegh-Taylor instability. There are two multimode patterns that have a long wavelength sinusoid of either 212 µm or 424 µm in one direction. A mathematical model of one wavelength of the 2-mode perturbation with the additional long wavelength mode having a wavelength of 424 µm is shown in Fig. 1c.

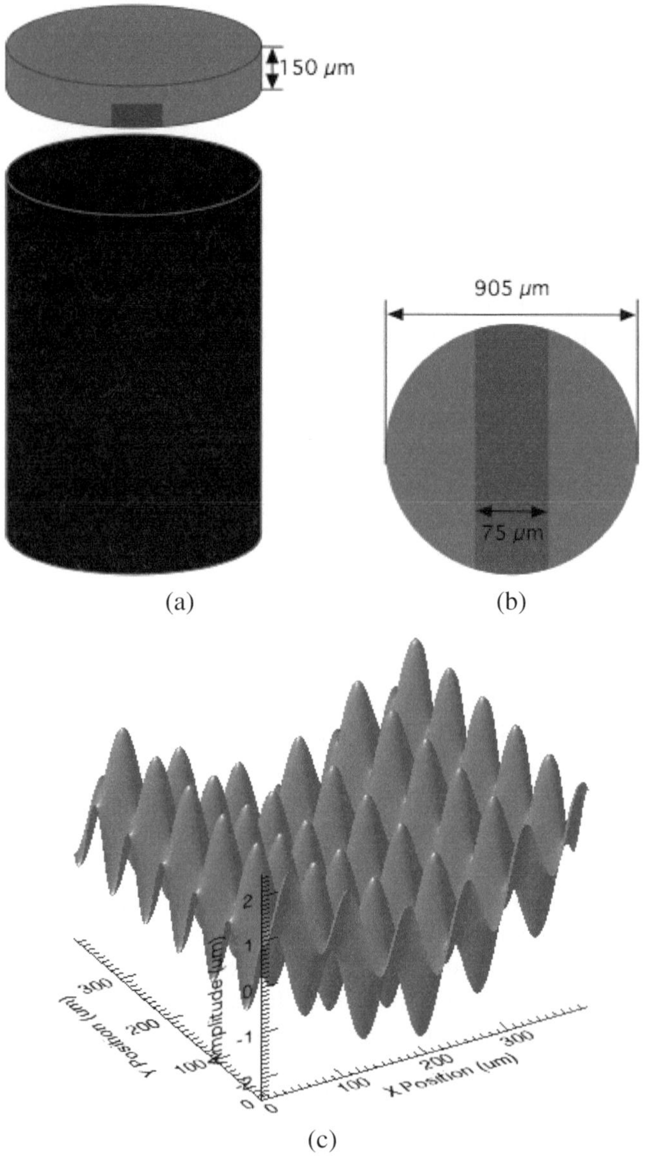

Fig. 1 (**a**) Exploded view of the key target components. The *gray* cylinder is the 150 µm thick, 905 µm diameter plastic piece and the black cylinder is the CRF foam that is 2–3 mm long. (**b**) The rear surface of the plastic piece where the *dark gray area* represents the CHBr. A slot 230 µm wide, 75 µm deep and the length of the component is milled out of the polyimide and the CHBr is glued into the slot. (**c**) This is an example of one of the perturbation types. This is a multimode pattern that has amplitude of 2.5 µm with orthogonal sine waves of 71 µm in wavelength and an additional mode of 424 µm in wavelength added in one direction. This image shows one full wavelength of the pattern that covers the entire disk

The plastic component, with the machined interface, and the foam are inserted into a polyimide tube with a nominal inner and outer diameter of 905 µm and 955 µm, respectively. The tube is attached to a conical, acrylic shield ~21 mm at the opening and decreasing to ~4 mm at the base. A gold wedge is attached to the shield on the side nearest the diagnostic. The purpose of the acrylic shield and gold

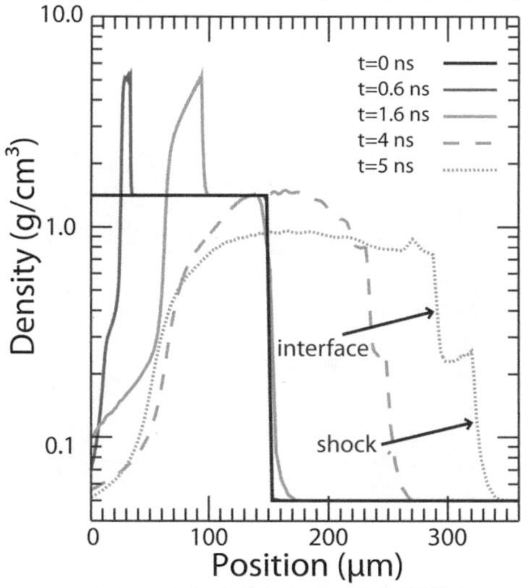

Fig. 2 A 1D Hyades plot of density vs. position for the experimental setup at several times. The laser irradiates the high-density material (plastic) and the interface of the low-density material (foam) is at 150 μm. This image shows the initial shock wave that is created by the ablation pressure from the laser and the blast wave that forms after the material begins to rarify. The blast wave crosses the interface at ∼2 ns and the evolution of the shock and interface is observed throughout the experiment

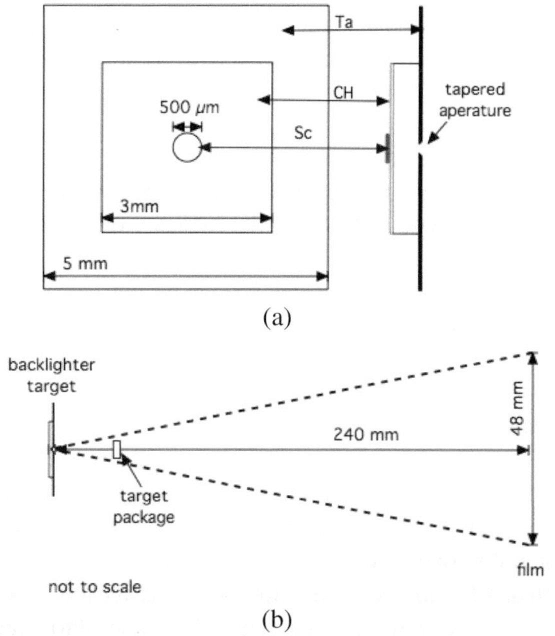

Fig. 3 (**a**) The backlighter target consists of a 500 μm, 12.5 μm thick Sc foil attached to a thin layer of plastic and mounted 500 μm from a 20 μm opening tapered to 10 μm in a piece of 5 mm × 5 mm Ta foil. (**b**) A schematic of the backlighter and diagnostic positions (not to scale) showing that, with the aperture specified above, there should be a uniform intensity from the x-rays over about ∼90% of the film

wedge are to prevent the ungated diagnostic used in the experiment from detecting emission from the hot plasma created when laser beams create the initial shock wave. There is also a gold washer at the base of the shield and underneath the edges of the plastic disk to prevent plasma from leaking out of the base of the shield. The gold washer has an outer diameter of 2.5 mm and an inner diameter of 1 mm, and it is covered with a thin layer of plastic and is not irradiated by the laser.

Ten Omega laser beams irradiate the plastic piece of the target with a 1 ns square pulse. The laser beams have an energy of ∼450 J/beam and pass through a distributed phase plate producing spot size of 820 μm FWHM. The beam is smooth overall with speckles on the 5-μm scale. The laser beams create an intensity of $\sim 10^{15}$ W/cm^2 and an ablation pressure of ∼50 Mbars or ∼50 million atmospheres. This ablation pressure creates a shock wave in the plastic layer. The propagation of this shock wave is shown at several times in Fig. 2. This plot shows density vs. position from a 1D Hyades (Larsen and Lane 1994) simulation.

The black line, which represents $t = 0$ ns, shows the initial condition of the target; 150 μm of dense material followed by the less dense foam (50 mg/cm^3). At 0.5 ns, the sharp increase in density at a position of 50 μm indicates that a shock wave has formed in the plastic layer. The laser pulse ends at 1.0 ns. By 1.5 ns, the shocked material has started to expand (decompress) which leads to a rarefaction wave moving to the right. This rarefaction quickly overtakes the shock turning the shock into a blast wave moving towards the interface between the plastic and foam. The blast wave crosses the density drop at the interface at ∼3 ns, which causes a forward shock to launch into the foam, seen clearly at 5 ns in the simulation. This experiment observes the evolution of the shock and interface.

The primary diagnostic is side-on, x-ray radiography. A second target, seen in Fig. 3a, is a point projection pinhole backlighter used to create 4.32 keV x-rays for the purpose of radiography. The target consists of a 50 μm, 5 mm square Ta foil with a pinhole aperture of 20 μm tapered to 10 μm. A 12.5 μm thick, 500 μm diameter disk of Sc is mounted on a 50 μm thick piece of CH. The CH mounted with Sc is positioned 500 μm from the Ta substrate. Three or 4 additional laser beams irradiate the Sc foil in a 1 ns pulse. The beams have an energy of ∼400 J/beam and a converging spot size of 1000 μm. The purpose of the large spot size is to overfill the Sc foil in order to tamp the metal plasma and prevent it from expanding laterally beyond the edge of the Ta foil (Kuranz et al. 2006). This backlighter target creates a directed flow of 4.32 keV x-rays to radiograph the target. The x-rays from the backlighter pass through the target and are recorded on x-ray film loaded in an ungated static pinhole camera array (SPCA).

The SPCA is loaded with a 12.5 μm thick piece of Ti to filter out high energy x-rays above the cold K-edge at 4.493 keV that may be created in the experiment. It is also loaded with a 500 μm piece of Be to filter out low energy x-rays to reduce the amount of noise in our images. In these experiments the SPCA was loaded with Agfa D7 x-ray film. After the experiment the film is digitized with a resolution of 22 μm per pixel on the film. The film is 240 mm from the backlighter target as shown in Fig. 3b. From a geometric-optics perspective, this distance and the aperture of the pinhole should create a relatively flat intensity over an area of $\sim$18 cm^2 and area of the entire film is $\sim$20 cm^2. Thus, the intensity from the backlighter should be flat over roughly 90% of the film.

3 Results and discussion

Figure 4a shows a radiographic image taken at 21 ns after the laser beams irradiated the target. This particular target had a multimode seed perturbation with an additional wavelength of 424 μm in one direction. In this image the interface and shock are moving to the right. The long, dark fingers are spikes due to the Rayliegh-Taylor instability. Bubbles of foam are in between the spikes of brominated plastic. The gold grid, with a wire period of 63 μm, can be seen in the upper right hand corner. The dark portion of the image on the far right is the edge of the gold covered acrylic shield. The various thin, white streaks on the image are scratches on the film.

For this experimental series several targets are imaged at different times after the initial laser pulse. The resulting radiographs are similar to the one seen in Fig. 4a. It is important to fully characterize the features described above in each image. Some examples of characterization are finding the location of the tip of the spike and the tip of the bubble as well as finding the amount of mass in a spike, bubble or the entire interface region. This information can be taken from several radiographic images at several different times, which allows the study of the evolution of the RT instability.

Figure 4b is an image of the entire piece of film, which is about 50.8 mm in diameter. An interesting feature in this image is the large spot, outlined by a black circle, that is a little over 1 mm in diameter in the target plane and $\sim$20 mm on the film. This spot appears on all of our data images under similar conditions although its location, brightness and shape vary. Since the spot is circular it is thought that it is coming through the pinhole. The spot is apparently produced by very-high-energy photons or electrons, absorbed very weakly by the target, because it reduces the contrast of the observed structures in the image. The size of the pinhole and its position in relation to the experiment should provide a mostly flat illumination as discussed earlier in this paper and shown in Fig. 3b. One can look at the unshocked foam, where the intensity in the image should be relatively uniform to see that this is not the case in this image. A 1D

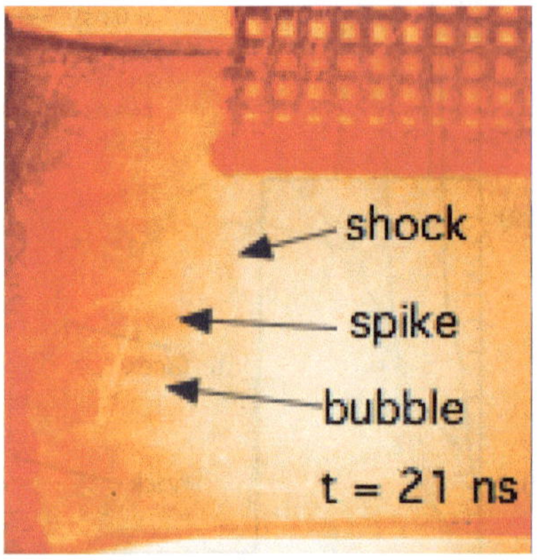

(a)

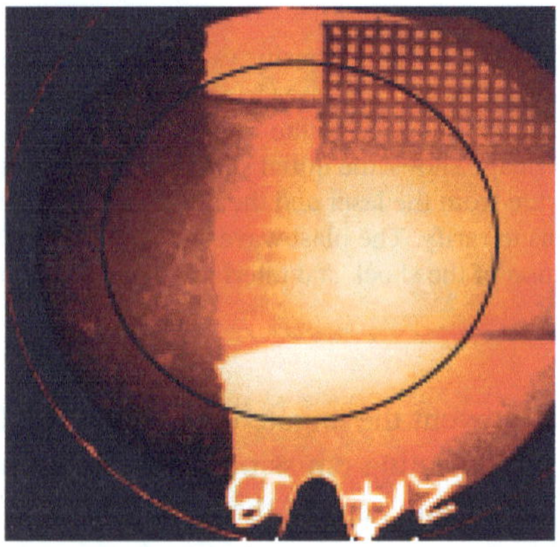

(b)

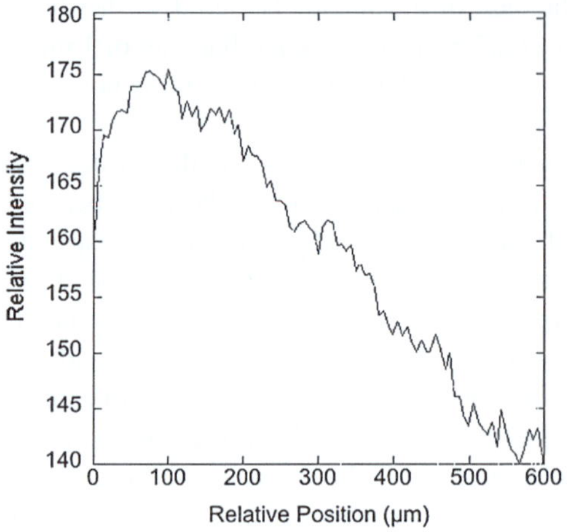

(c)

Fig. 4 (**a**) An x-ray radiograph of an experiment that had a multimode perturbation machined onto the interface of the target. This multimode pattern had an additional wavelength of 424 μm. The shock wave and interface are moving to the right in this image. (**b**) This is the entire piece of film from this experiment. Note the circular spot that is roughly in the center of the image. (**c**) This is a 1D profile through the unshocked foam region. This plot of relative intensity vs. relative position should be uniform, but there is a prominent drop off due to the bright, circular emission on the film profile from the unshocked region in the foam is shown in Fig. 4c. This plot shows the relative pixel intensity vs. relative position in the unshocked foam. The fluctuations in the profile probably result from Poisson noise due to the limited number of x-ray photons. In similar, previous data we have shown this to be the primary source of noise (Kuranz et al. 2006).

This spot is of unknown origins and diminishes the contrast in the image. Eliminating the bright area from the image would greatly improve the image contrast and make the spike and bubble tips as well location of the shock front less obscured by the bright area on the radiographic image. Also, when calculating the mass of a spike, another way of analyzing this data, the background spot would lead to erroneous measures of mass. One possible way to remove much of the spot is to invert the image to k-space using a Fast Fourier transform (FFT), subtract the modes that are related to the spot in the image, and then invert the image back to optical density space. These operations were performed using the Interactive Data Language (IDL). When an FFT is taken of a 2D array in IDL there are 4 elements in the power spectrum that corresponds to each 2D mode in the image, the real and imaginary parts of k_x and k_y. To remove the spot, one must identify the modes that correspond to it and subtract them in Fourier space.

The image shown in Fig. 4a contains numerous modes of varying amplitudes. The spot is roughly 1 mm in diameter in the target plane or covers about half of the image. As an approximation, assume this relates to a single low-order mode with a wavelength the entire length of the image. The four elements for that mode are then set to zero in the array of the power spectrum. Then the power spectrum is inverted to optical density space and then image should be of better quality. The image with the long wavelength mode removed is shown in Fig. 5a and a magnified image of the film is seen in Fig. 5b. Even in this approximation the quality has improved from the initial image both in contrast and intensity uniformity. There is still a variation in the unshocked foam region, however, it is smaller and of less intensity than the original spot. This can be seen by the 1D profile taken from the image of Fig. 5b, shown in Fig. 5c. Similar to Fig. 4c, this plot is the relative pixel intensity vs. the relative position in unshocked foam. There also remains a variation outside of the shock tube, which can be seen in the image of the entire piece of film, Fig. 5a. The fact that the spot was not completely removed means that the spot is not entirely sinusoidal in nature. The majority of the spot can be modeled as a 2D sinusoid; however, when that is subtracted some smaller wavelength components of the spot still remain and can be seen on the image. Largely, the image with the long wavelength component of the bright feature removed is much improved.

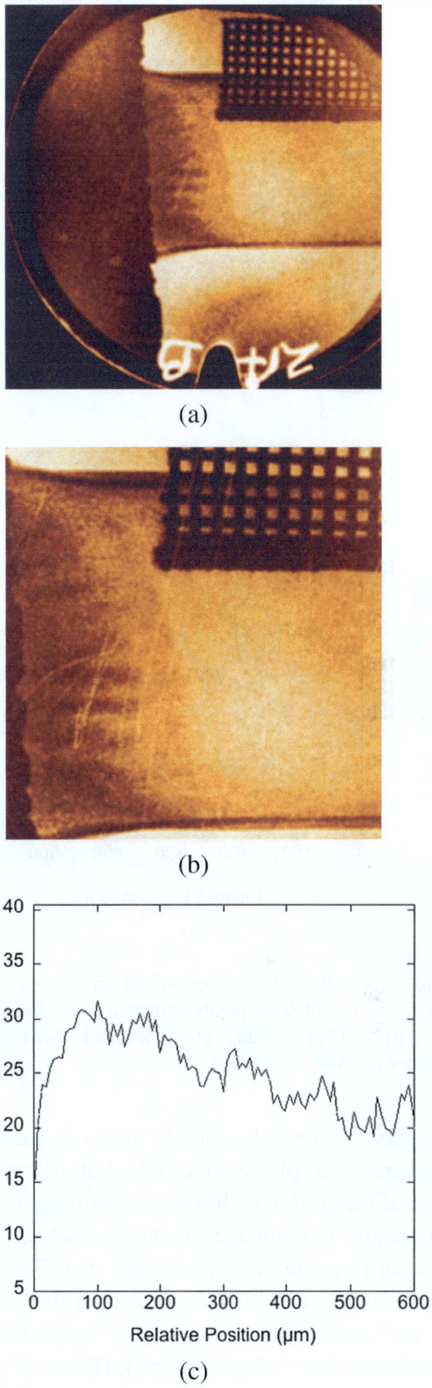

Fig. 5 (**a**) This is the x-ray radiograph shown in Fig. 4b with the bright spot mostly removed using the FFT. (**b**) This expanded image shows that the bright area is greatly diminished. (**c**) A 1D profile through the unshocked foam region shows a much flatter intensity that in Fig. 4c

Another way to remove the spot is to use the 1D profile in the unshocked foam from Fig. 4c in order to model the structure of the spot in the film. Then this modeled spot can be subtracted from the image. It is assumed that spot is circular and the 1D profile is essentially rotated 360° to create this model. The center of the spot is estimated by looking at the

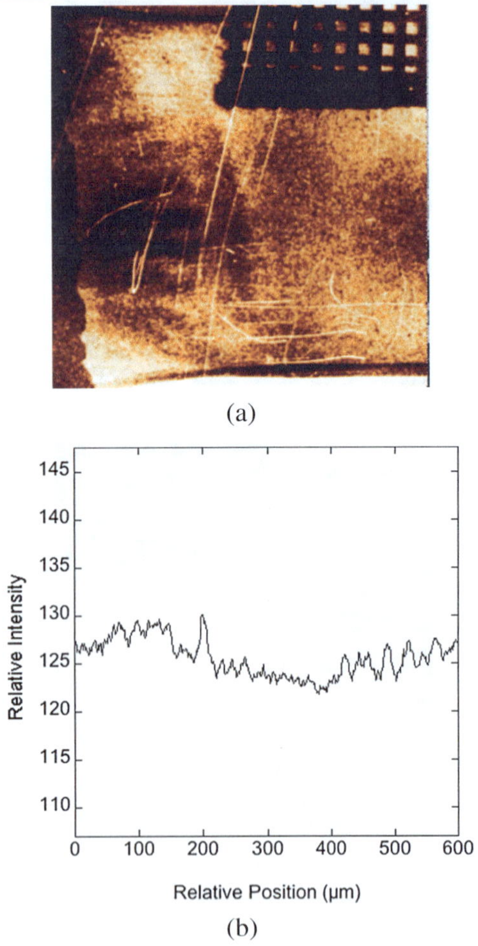

Fig. 6 (**a**) This is the x-ray radiograph similar to the one shown in Fig. 4a with the bright spot mostly removed using the spot modeling technique. (**b**) A 1D profile through the unshocked foam region shows a much flatter intensity that in Fig. 4c

brightest area on the radiographic image as well as measuring the overall size of the spot and taking the midpoints. If the center of the spot, which we are assuming to be the maximum intensity, lies outside of the unshocked foam region it is difficult to accurately model the spot. This is because there is variation in the shocked material due to the RT instability and one cannot ascertain how much of the pixel intensity is due to the background spot. However, this method captures the general size and variation in the background feature.

The result from subtracting the modeled spot from the radiographic image is shown in Fig. 6a. As with the FFT method, the important features in the radiograph, the spike and bubble tips as well as the shock front, are clearer than in original radiograph. Also, a 1D profile in the unshocked foam section from the resulting image is shown in Fig. 6b. Again, this is relatively flat compared to Fig. 4c.

It is important to compare the results of both methods as they can both be utilized for different purposes. Important features can be seen more clearly in Fig. 5b than in Fig. 6. For this image it would be best to use the FFT method to find the location of the spike tip, bubble tip and shock front. Also, when the center of the background spot is located far from the unshocked foam region the spot subtraction method would become less accurate, as discussed previously, and it would also be best to use the FFT method. However, the spot subtraction method proves to be more useful in certain cases. In observing the RT instability, it is of great interest to be able to study the flow of mass due to the instability. For instance, one can calculate the amount of mass in the spikes using method described by Hansen et al. (2007). The method described uses the pixel intensity as one parameter in calculating the mass from a radiographic image. If an image had the background feature shown in this paper the pixel intensity would not truly reflect the mass of the material of that pixel. In this case, modeling the spot and subtracting it from the image would yield more accurate pixel intensities and therefore masses. Since the method of "removing" the mode includes setting the value of that mode to zero this method changes the pixel intensity as can be seen by comparing the 1D profiles in Figs. 4b and 5b. Therefore, when calculating the amount of mass of a specific region in the image, the spot subtraction method is more precise.

As mentioned previously, the cause of this spot is unknown, although it plausibly may result from laser-plasma-interaction processes that can produce very energetic electrons. Accordingly, it was suggested that the spot would not appear if the beams irradiating the pinhole backlighter target came to the backlighter foil as a divergent beam, having passed through its focus, as opposed to a converging focus. (The diverging beam is more uniform than the converging one.) This hypothesis was tested on another experimental day using the same experimental setup except that the backlighter beams were diverging and the film was D8, another type of Agfa film, instead of D7. The resulting image shown in Fig. 7a does not show a bright spot. A 1D profile of the relative intensity vs. relative position in the unshocked foam is shown in Fig. 7b. The profile is flatter overall. The use of D8 film rather than D7 cannot explain the disappearance of the spot, since D8 is more sensitive than D7. If the spot existed it would only appear brighter on D8. That same experimental day D7 film was used and it did not show a bright spot either, however, the overall quality of the film was poor due to low signal from our backlighter.

4 Conclusions

Recent experiments, motivated by supernovae hydrodynamics, have observed growth due to the RT instability from 3D multimode initial conditions. The quality of the resulting radiographic images is variable. This is due to a bright spot

(a)

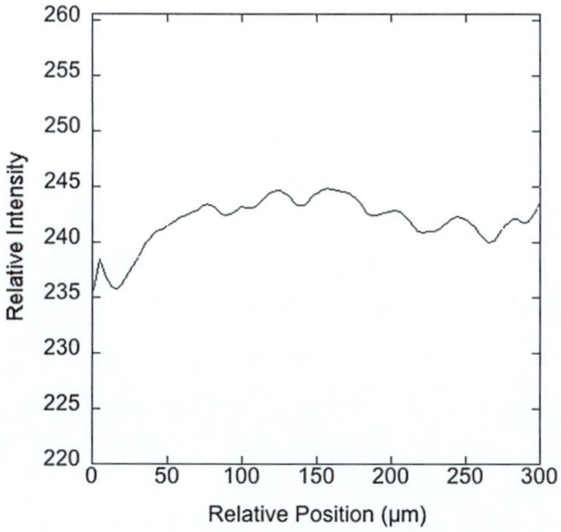

(b)

Fig. 7 (**a**) This x-ray radiograph was taken under conditions similar to the previous radiograph except that the beams that irradiated the backlighter target came to a diverging focus whereas in the previous image the backlighter beams came to a converging focus. (**b**) A 1D profile in the unshocked foam is more uniform than in Fig. 4c as is predicted by out backlighter geometry shown in Fig. 3

of unknown origin that appears on the film in some images, reducing the contrast of the data. However, the data can be made more suitable for quantitative analysis by taking the FFT of the image and removing the dominant modes that correspond to the global scale of the spot. An alternative method is to model the structure of the spot and subtract it from the image. Both of these methods improve quality of the radiographic image. Which method should be used depends on what type of further analysis will be implemented on the image pertaining to the RT instability. In limited tests, the spot has not appeared when the backlighter beams are diverging rather than converging. The data analysis approach discussed here will enable more accurate determination of the structure produced by the instabilities. The change in experimental technique should lead to improved data in future experiments.

Acknowledgements The authors would like to acknowledge Nick Lanier and Chuck Sorce for useful technical discussions. Also the authors would like to recognize the support of the Omega operations team and the Michigan target fabrication team and General Atomics component fabrication group for their work on the experimental targets. Financial support for this work included funding from the Stewardship Science Academic Alliances program through DOE Research Grant DE FG03-99DP00284, and through DE-FG03–00SF22021 and other grants and contracts.

References

Arnett, W.D.: Supernova and Nucleosynthesis. Princeton University Press, Princeton (1996)
Boehly, T.R., et al.: Rev. Sci. Instrum. **66**, 508 (1995)
Drake, R.P., et al.: Phys. Plasmas **11**, 2829 (2004)
Hansen, J.F., Robey, H.F., Klein, R.I., Miles, A.R.: Astrophys. J. **662**, 379 (2007)
Kane, J., et al.: Astrophys. J. **478**, L75 (1997)
Kuranz, C.C., et al.: Astrophys. Space Sci. **298**, 9 (2005)
Kuranz, C.C., et al.: Rev. Sci. Instrum. **77**, 10E327 1 (2006)
Kuranz, C.C., et al.: Astrophys. Space Sci. **307**, 115 (2007)
Larsen, J.T., Lane, S.M.: J. Quant. Spectrosc. Radiat. Transfer **51**, 179 (1994)
Meshkov, E., et al.: In: 6th International Workshop on the Physics of Compressible Turbulent Mixing, Marseille, France (1997)
Muller, E., Fryxell, B., Arnett, D.: Annu. Rev. Astron. Astrophys. **251**, 505 (1991)
Rayleigh, L.: Scientific Papers II. Cambridge University Press, Cambridge (1900)
Remington, B.A., Drake, R.P., Ryutov, D.D.: Rev. Mod. Phys. **78**, 755 (2006)
Richtmyer, D.H.: Commun. Pure Appl. Math. **13**, 297 (1960)
Robey, H.F., et al.: Phys. Plasmas **8**, 2446 (2001)
Robey, H.F., Zhou, Y., Buckingham, A.C., Keiter, P., Remington, B.A., Drake, R.P.: Phys. Plasmas **10**, 614 (2003)
Ryutov, D.D., Drake, R.P., Kane, J., Liang, E., Remington, B.A., Wood-Vasey, M.: Astrophys. J. **518**, 821 (1999)
Taylor, S.G.: Proc. R. Soc. A **201**, 192 (1950)

Astrophys Space Sci (2009) 322: 57–63
DOI 10.1007/s10509-008-9953-x

ORIGINAL ARTICLE

Modeling of multi-interface, diverging, hydrodynamic experiments for the National Ignition Facility

M.J. Grosskopf · R.P. Drake · C.C. Kuranz · A.R. Miles · J.F. Hansen · T. Plewa · N. Hearn · D. Arnett · J.C. Wheeler

Received: 20 June 2008 / Accepted: 18 November 2008 / Published online: 9 December 2008

Abstract The National Ignition Facility (NIF) will soon provide experiments with far more than ten times the energy than has been previously available on laser facilities. In the context of supernova-relevant hydrodynamics, this will enable experiments in which hydrodynamic instabilities develop from multiple, coupled interfaces in a diverging explosion. This paper discusses the design of such blast-wave-driven explosions in which the relative masses of the layers are scaled to those within the star. It reports scaling simulations with CALE to model the global dynamics of such an experiment. CALE is a hybrid, Arbitrary Lagrangian-Eulerian code. The simulations probed the instability growth and multi-interface interactions in mass-scaled systems using different materials. The simulations assist in the target design process and in developing an experiment that can be diagnosed.

M.J. Grosskopf (✉) · R.P. Drake · C.C. Kuranz
University of Michigan, Ann Arbor, MI 48109, USA
e-mail: mikegros@umich.edu

A.R. Miles · J.F. Hansen
Lawrence Livermore National Laboratory, Livermore, CA 94550, USA

T. Plewa
Florida State University, Tallahassee, FL 32306, USA

N. Hearn
University of Chicago, Chicago, IL 60637, USA

D. Arnett
University of Arizona, Tucson, AZ 85721, USA

J.C. Wheeler
University of Texas, Austin, TX 78712, USA

Keywords Supernovae · Hydrodynamic instabilities · Laboratory astrophysics

1 Introduction

Hydrodynamic instabilities develop when blast waves drive stellar explosions (Bandiera 1984; Colgate and White 1966; Cowie 1975) and in many other astrophysical contexts. Specifically the Rayleigh-Taylor (RT) and Richtmyer-Meshkov (RM) instabilities (Meshkov 1969; Rayleigh 1900; Richtmyer 1960; Taylor 1950) enable the outward penetration of dense, high-Z elements and the inward penetration of low-Z elements. The Kelvin-Helmholtz instability (Chandrasekhar 1961) may also contribute, especially if the blast wave is non-spherical (Kifonidis et al. 2006). Efforts to understand the observations of Supernova (SN) 1987A have led to a sequence of laboratory-astrophysics experiments at high energy density to attempt to better understand the behavior of these instabilities in blast-wave-driven systems (Drake et al. 2004; Kane et al. 1997; Miles et al. 2004; Robey et al. 2001). With the National Ignition Facility (NIF) laser entering its last phases before experiments, we have begun to explore its potential to accomplish more complex and more relevant experiments than those which can be achieved on present lasers. Preparation is critical because of the expense and difficulty of doing experiments on a large-scale facility such as NIF. We have used CALE, an adaptive Lagrangian-Eulerian hydrodynamics code, to model the potential experiments, refine the target design, and assess possible diagnostics. By running a series of simulations, we have been developing initial target parameters for a spherically-diverging, multi-interface, supernova-relevant Rayleigh-Taylor experiment for the NIF laser facility.

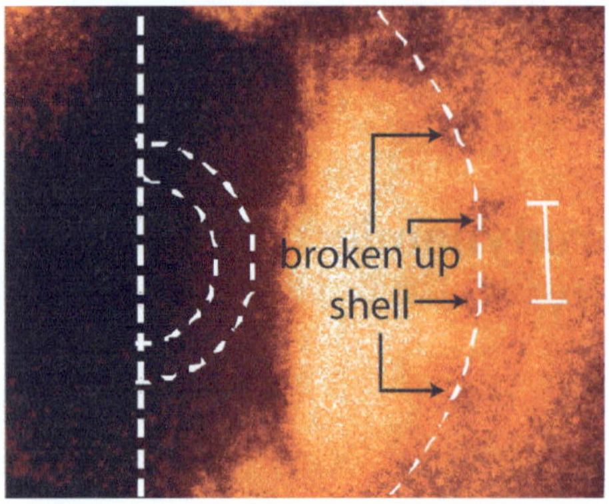

Fig. 1 Radiograph from spherically diverging experiment on Omega. Adapted from Drake et al. (2002)

Prior experiments have been focused primarily on a planar geometry and a single interface, equivalent in a scaled sense to a very small patch within the star, over timescales when the effects of divergence are not significant (Robey et al. 2001; Ryutov et al. 1999) In these cases the shock has been driven axially down a cylindrical tube. Among the advantages of this approach is the ability to use backlit radiography to acquire detailed information about the Rayleigh-Taylor structure. Using a planar geometry also allows the available drive energy to push the dynamical evolution further by limiting energy losses due to divergence. Recent planar experiments used 10 beams on the Omega laser (Boehly et al. 1995) to drive a blast wave across a patterned interface. Between 2004 and 2006, many successful experiments were carried out with 3D single-mode, "egg-crate" initial conditions and with combinations of this and other modes (Kuranz et al. 2007) (also see Kuranz et al., this volume). Most recently a 51-mode interface motivated by results obtained from a stellar-evolution code developed by Dave Arnett and Casey Meakin at the University of Arizona was shot on the Omega laser in May 2008.

One prior experiment (Drake et al. 2002) used diverging, spherical targets at the Omega facility. They contained a germanium-doped spherical plastic shell followed by a cylinder of low-density carbonized resorcinol formaldehyde (CRF) foam. The hemisphere had an inner diameter of 440 microns. The experiments were successful in creating a nearly spherically divergent system, but the shell broke up very quickly as it diverged. Figure 1 shows data from one laser shot. A thicker spherical shell would have been needed to keep the shell structure intact long enough to have sufficient development of the Rayleigh-Taylor instabilities, but unfortunately, the available energy was insufficient to drive the Rayleigh-Taylor evolution to later states. Another prior experiment observed the evolution of two structured interfaces coupled by a propagating blast wave (Kane et al. 2001).

2 Fundamental approach to the NIF experiment design

The NIF laser will provide the additional energy needed to do supernova-relevant experiments involving both divergence and multiple interfaces. This will enable us, in the presence of a diverging system, to look for interaction between the mixing layers. Past simulations of SN 1987A, with varying initial conditions and specific models (Fryxell et al. 1991; Kifonidis et al. 2003, 2006), have seen a variety of interactions, from large-scale penetration to very little penetration to some penetration. In order to design a multi-interface, diverging experiment, one needs to first ask in what way the experiment can be scaled to the SN. The same equations will apply to the SN and the laboratory experiment, as has been the case previously (Ryutov et al. 1999), but the spatial structure of the two systems cannot be identical. Specifically, it is not feasible to create the initial stellar density profile in which the density decreases by orders of magnitude as radius increases. However, the dynamics of blast waves and their instabilities is sensitive primarily to the amount of mass that has been swept up by the blast wave. This determines the net deceleration. Because of this, we have chosen to scale the relative masses of the material within the C/He boundary, the He layer, and the H layer. The thicknesses of the various layers are scaled to the 15 solar mass progenitor of Kifonidis et al. (2003). The mass interior to the second interface is $2.8\times$ the mass of the initial layer and the total mass is $10\times$ the mass of the initial layer. The mass of the initial layer was established as is described below.

The design simulations were accomplished with CALE, a 2-D Arbitrary Lagrangian-Eulerian radiation hydrodynamics code written in C at the Lawrence Livermore National Laboratory. It can be run with an Eulerian, Lagrangian, or adaptive mesh. The code allows for many different forms to handle equation of state (EOS) information, and in these runs uses a tabular equation-of-state from a Livermore EOS library. The laser absorption is modeled in a basic fashion, with the ray tracked until it reaches a zone above the "critical density" where the plasma resonant frequency equals the laser light frequency, then its entire energy is deposited. The images below are color-contour plots created with a built-in graphical interface in CALE.

The simulations were all run in 2-D CALE using an R–Z coordinate system with polar meshing. CALE uses a Van Leer advection scheme that is second-order accurate in space (Barton 1985). Although CALE is an ALE code, these runs were done with a purely Eulerian mesh. The $R = 0$ axis is an axis of symmetry; all other boundaries are free-outflow boundaries. There are 250 zones in the azimuthal direction and 496 zones in the radial direction. This equates to 0.36 degrees per zone in the azimuthal direction. The radial resolution varies; there is the same number of zones in each material, but the radial extent of the material is set to maintain the mass scaling relation described above. Lastly, for the higher irradiance case, the inner diameter of the sphere is 600 μm, and for the low irradiance cases, the inner diameter is 1456 μm.

The thicknesses of the inner layer is set by the requirement that the interfaces be driven by a blast wave. The first layer must be thick enough to allow the rarefaction that develops after the laser pulse ends to catch the shock before it reaches the first interface. It was found that the shock would form a blast wave if the thickness of the first layer was double the radial distance the shock traveled while the laser was on. In practice, a series of preliminary runs were done to find the distance the shock would travel through the initial material while the laser was on. The first layer was then given twice that thickness. The mass scaling requirements were then used to define the thicknesses of the other layers. The interfaces between the layers were each given a perturbation to seed a particular mode of Rayleigh-Taylor growth. The perturbation had a wavelength of $\pi/50$ radians azimuthally and had a $k * a$ of 1, where k is the wave number of the perturbation and a is the amplitude. This perturbation is purely illustrative at the present phase of the design; the actual perturbation to be used for an experiment will be determined later, informed by simulations of presupernova convective structure (Meakin and Arnett 2006). In addition, while the present simulations incorporated nearly spherical shocks, we would intend as part of an actual experiment to use a nonconcentric inner layer and/or nonuniform irradiation to launch aspherical shocks.

A silver wedge was placed along the back side of the target covering all but the surface area where the laser will be incident. The dense material was placed there to inhibit the flow of material in the negative axial direction. Containing the material allowed the shock to maintain sphericity longer in time. The shock still lost some sphericity because of the manner in which the laser was incident on the target. Additional zones with low density were placed in the area from 200 μm ID to the target surface. These zones were to approximate vacuum while allowing the ablating material from the inner layer of the target to interact with the laser rays while the laser is on. Figure 2 shows the basic layout of the target for the simulations.

We chose to run these simulations using a laser pulse that delivered 100 kJ of 0.35-μm wavelength laser light in 3 ns. Factors affecting this choice are as follows. First, one needs enough energy to accomplish the desired experiment. Second, experience shows that the actual laser energy needed to produce a given ablation pressure can be larger by some factor than a specific simulation indicates, presumably due to inaccuracies in laser and/or other energy transport. Early experiments can help to clarify this, but the 100 kJ we use here might in reality correspond to significantly more energy. Third, we desire to use less than one-quarter of NIF for the drive, and another one-quarter for the diagnostics, allowing one entire experiment to be done by firing only half of NIF. We ran simulations covering a range of different materials and two potential irradiances. The low irradiance was near that of the previous planar experiments, delivering 100 kJ in 3 ns to the 1.456 mm inner diameter of the target resulting in an irradiance of around 5×10^{14} W/cm^3. The high-irradiance case also delivered the same laser power to a 0.6 mm inner diameter, resulting in a laser irradiance of 3×10^{15} W/cm^3. This lead to a factor of 3.3 increase in simulated ablation pressure over the low-irradiance case.

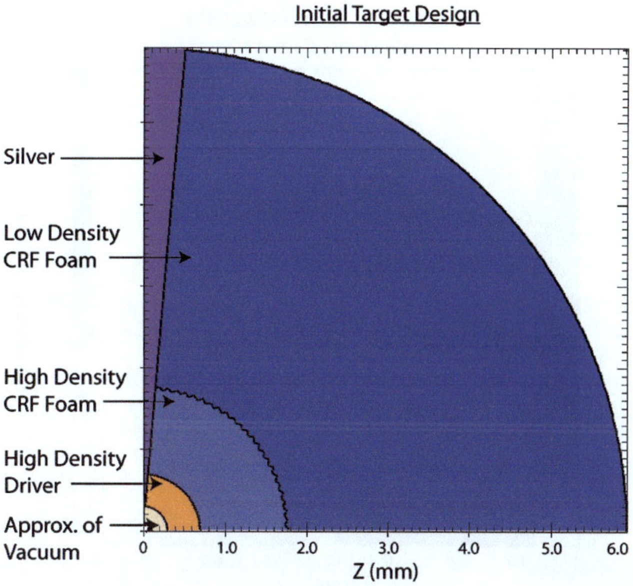

Fig. 2 Initial configuration of the laser target

3 Diagnostic possibilities

An important element of the NIF experiment design is to consider what diagnostics may prove viable. In previous experiments we used backlit radiography to image the Rayleigh-Taylor growth. This worked well in planar systems where the diagnostic looks down a row of Rayleigh-Taylor spikes and where a tracer strip of more-opaque material can be incorporated in the midplane of the target. Such tracer strips are unlikely to be feasible in the NIF target, because of a combination of materials constraints and cost. When the diagnostic looks through a diverging system, the image is dominated by the spike tips and the bubbles become difficult to discern. Figure 3 shows a simulated radiograph generated by CALE graphical tools, at 100 ns for a run using the low laser irradiance and involving layers of titanium, 500 mg/cc foam, and 50 mg/cc foam. The perturbations corresponding to the system as simulated in 2D are ring-shaped. While the Rayleigh-Taylor growth on the second interface is somewhat distinguishable, the streaks created by the spikes blur

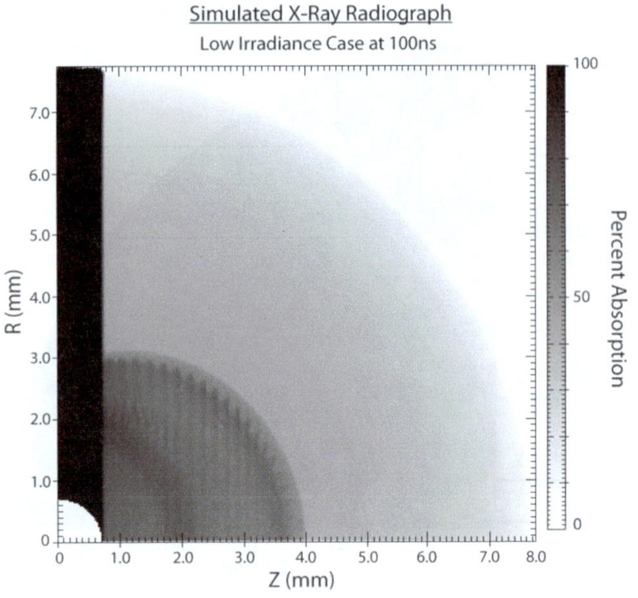

Fig. 3 Simulated radiograph

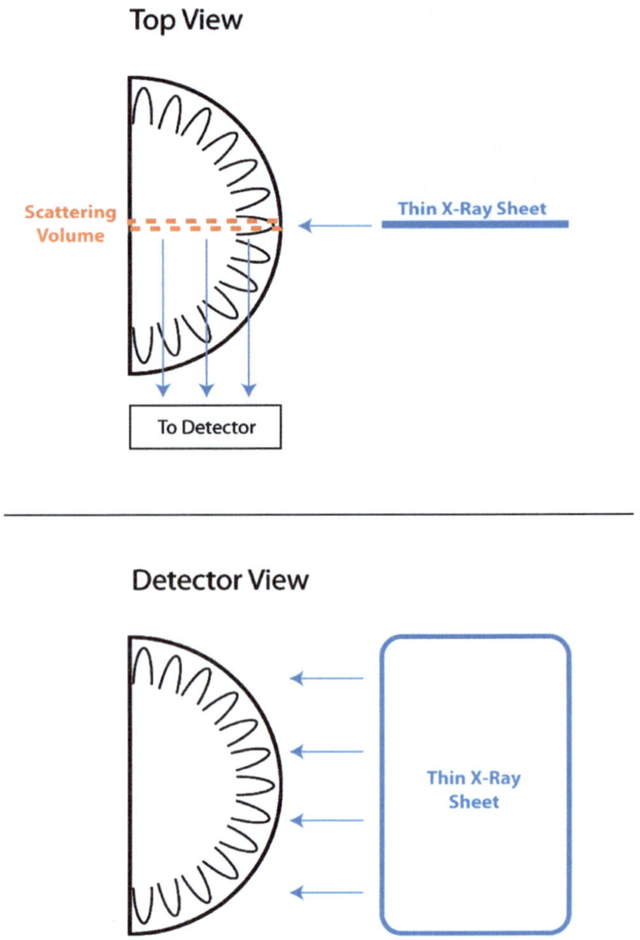

Fig. 4 Imaging x-ray Thomson scattering schematic

any view of the interior structure. By selection of the backlighter energy, one will be able to observe the maximum outward penetration of the spikes of either interior layer. This will accomplish the minimum goal of the experiment.

X-ray Thomson scattering (Glenzer and Redmer 2008) provides an alternative possible technique that might provide a more detailed and more local image of the structure of the mixing regions. A large number of additional NIF beams would be used to produce He-α x-rays from a backlighter target material, of an energy that readily penetrates the primary target. Shields would clip these x-rays to form a thin sheet that would travel through the target. These x-rays would be scattered in all directions by the target materials. The detector would image these x-rays (through a K-α filter) in the direction perpendicular to the x-ray sheet. Because the scattering is approximately proportional to the density of the target material, this diagnostic would obtain an image of the density structure. A schematic diagram of this is shown in Fig. 4. In a previous experiment, 4.75 keV x-rays were scattered off beryllium plasma with an electron density of 10^{23} e^-/cc, spectrally dispersed by a Bragg crystal and imaged with a gated framing camera (Glenzer and Redmer 2008). This successfully demonstrated x-ray Thomson scattering, and showed that 4.75 keV x-rays could be used to diagnose dense plasmas. An evaluation of the parameters for NIF shows that, for regions having total electron densities above 10^{22} cm^{-3}, it should be possible to a obtain spatial resolution of some tens of microns using this technique. Smaller resolutions are possible at higher densities.

4 Results from different cases

A series of different runs investigated a range of possible materials for the target, each with three distinct layers. Most cases were done looking for a density drop of approximately a factor of 10 between each layer. The foams in each case used the equation-of-state table for polystyrene, but the exact choice of foam would likely be low-density CRF foam, which can be machined. Initial simulations of each case were done at 18 kJ with Molybdenum-Aluminum-Foam (MAF), Molybdenum-Polystyrene-Foam (MPF), Nickel-Foam-Foam (NFF), Titanium-Foam-Foam (TFF), and Aluminum-Foam-Foam (AFF). The Titanium and Aluminum driver cases were expanded to the 100 kJ drive laser case because they showed substantial R-T development and they were promising for x-ray scattering diagnostics.

The high-irradiance TFF case has an initial radius of 0.3 mm, with the thickness of the 4.5 g/cc Ti layer set at 0.598 mm. The following foam layers are 1.395 mm of 500 mg/cc foam and 5.449 mm of 50 mg/cc foam. A 0.02 mm layer of polystyrene is added to the inner surface

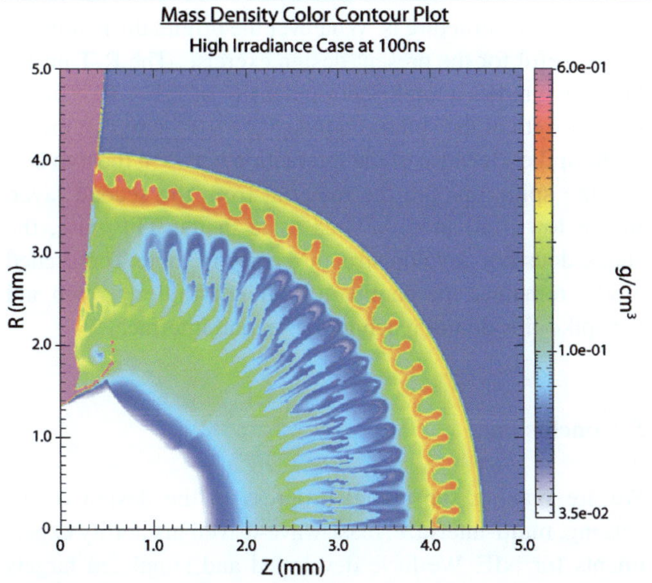

Fig. 5 High-irradiance case with Titanium-Foam-Foam having a CH ablator layer, at 100 ns

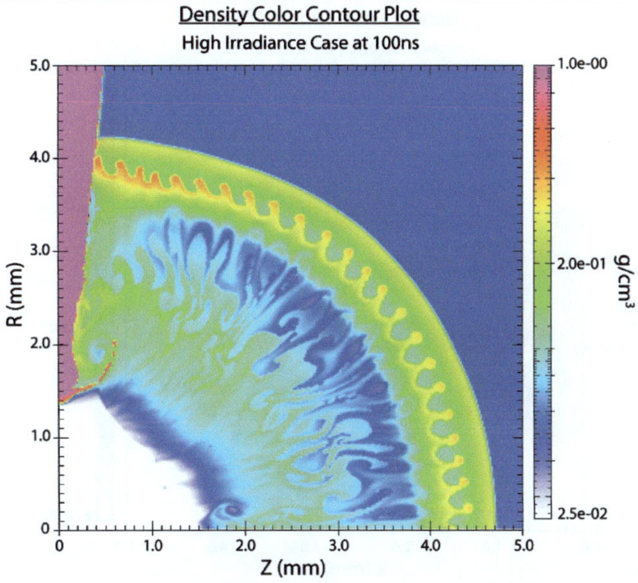

Fig. 6 Same as Fig. 5 without CH ablator layer

of the titanium layer to prevent direct irradiation of the high-Z metal surface. As the laser pulse ends, a uniform pseudo-spherical shock wave has formed nearly halfway through the initial layer, which forms a blast wave as it reaches the first interface. By 50 ns, the shock is just reaching the second interface and has already developed a great deal of well-defined Rayleigh-Taylor growth on the first interface behind the shock. At 100 ns (Fig. 5), the second interface has begun to develop unstable growth in the seeded mode, and there is significant growth on the titanium-foam interface as well. R-T spikes in the inner mixing region are 1 mm or greater in length and have well-formed cap structures. The titanium spikes reach near to the back of the second mixing layer, but there is minimal interaction between layers. Knowing the material in the spikes, the total electron density can be calculated from the mass density. Inside the structure in the inner interface the total electron density is found to be around 2.5×10^{22} cm^{-3}, and around 1×10^{23} cm^{-3} in the outer spike tips, therefore this case should be viable for x-ray Thomson scattering.

The high-irradiance TFF case was also run without the thin polystyrene layer, directly irradiating the inner surface of the titanium. Perturbations are observed on the incident shock and in the post-shock flow before the shock reaches the first interface, the origin of which is unclear. This seeds a sporadic set of modes onto the initial interface, producing an erratic structure. By 25 ns, substantial structure has formed behind the shock in the mixing region, though the shock has smoothed out and become uniform. At 50 ns, as the shock is just passing the second interface, large features in the R-T growth are nearly perturbing the second interface as it begins to evolve. At 100 ns (Fig. 6), the structure in the outer mixing region has begun to grow well-ordered Rayleigh-Taylor spikes, while the inner mixing region has become disorganized, with tips of material protruding to the back of the outer interface. We suspect that this disorganized structure is the result of some numerical effect, and are doing further simulations to explore its origins.

The low-irradiance TFF case had an initial radius of 0.728 mm, with the thickness of the Ti layer set to 0.312 mm and with no inner CH ablator. The following foam thicknesses are 1.321 mm and 5.382 mm respectively. As the laser pulse ends, a shock has again formed nearly halfway through the first layer; however, as opposed to the high-irradiance case, the shock is uniform and without the sporadic structure. By 25 ns, the shock has traveled 1.5 equal-mass-radii and ordered R-T spikes have begun to grow. At 100 ns (Fig. 7), the shock has already passed through the second interface, though it has not yet reached the radius in the third material where it has swept up a mass equal to the mass of the interior shells. The structure on the first interface is well organized and the relevant modes show a good deal of growth. The spikes have growth a bit of a "mushroom cap", which is somewhat suppressed as we move away from the center axis, possibly due to the large number of modes or to the greater shear across the mixing zone at higher angles. In this low-irradiance case, the spike-tips have not penetrated as far, and have not reached the back of the second layer. The electron density is $\sim 5 \times 10^{22}$ cm^{-3} in the spike tips in the inner mixing region, and is $\sim 1.5 \times 10^{23}$ cm^{-3} in the outer spike tips. Again, this is sufficient for x-ray scattering to be a viable diagnostic for this experiment.

The high-irradiance AFF case had an initial radius of 0.3 mm, with the thickness of the Al layer set at 0.764 mm. Following foam layers are 1.667 mm of 300 mg/cc foam and

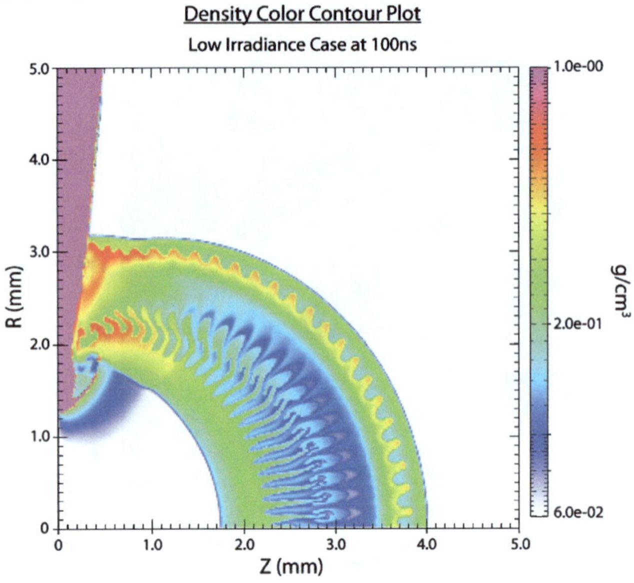

Fig. 7 TFF Case without CH ablator layer. Low irradiance at 100 ns

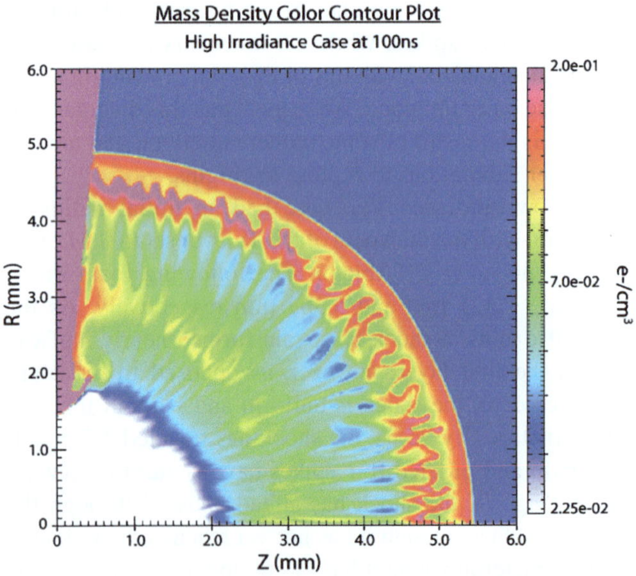

Fig. 8 AFF Case. High irradiance at 100 ns

6.488 mm of 30 mg/cc foam. This case also had the same abnormal structure in the shock just as the laser turned off that the high-irradiance TFF case with no ablator layer showed. By 25 ns, this has led to structure that actually reaches all the way to the shock and continues to perturb the shock. As the shock passes through the second interface, the structure from the first interface overtakes the second layer's perturbations, seeding its own modes as the dominant growth. However, the second interface slows the growth of the initial perturbations and allows the shock to begin to distance itself from the structures and smooth out (Fig. 8). As in the TFF case discussed above, it is as yet unclear what the origin of the abnormal structure is. Whatever the origin, the results remain useful for the present design exercise. The R-T unstable structure has a total electron density of 2.3×10^{22} cm^{-3} in the length of the spike and then 5.2×10^{22} cm^{-3} at the tip of the spike. Because of the interaction between the two layers, the spike tips are the material from the second layer. In the low-irradiance AFF case, as in the TFF case, the shock does not develop structure early in time. The seeded modes remained the dominant Rayleigh-Taylor growth and the spike-tips do not reach to the second interface.

5 Conclusion

We are making substantial progress in the design of diverging, multi-interface, blast-wave-driven instability experiments for NIF. We have developed and simulated targets for which the relative masses of the three layers are scaled to the stellar case. Both the TFF and AFF cases show significant Rayleigh-Taylor growth in a spherically diverging geometry. These targets can be shot and diagnosed using as few as half the NIF beams. Electron density data from the structure also shows that it should be feasible to use x-ray Thomson scattering as an imaging diagnostic in these experiments, in addition to radiography. Some of the simulations show structure that is probably not real; we are exploring its origin in continuing work. Further simulations will explore the possibility of replacing either or both of the foam layers in the target with more machinable materials having a better-understood equation-of-state.

Acknowledgements This research was sponsored by LLNL through contract LLNL B56128 and by the NNSA through DOE Research Grant DE-FG52-04NA00064.

References

Bandiera, R.: Astron. Astrophys. **139**, 368 (1984)
Barton, R.T.: Numer. Astrophys. **482** (1985)
Boehly, T.R., et al.: Rev. Sci. Instr. **66**, 508 (1995)
Chandrasekhar, S.: Hydrodynamic and Hydromagnetic Stability. Dover, New York (1961)
Colgate, S.A., White, R.H.: Astrophys. J. **143**, 626 (1966)
Cowie, L.L.: Mon. Not. R. Astron. Soc. **173**, 429 (1975)
Drake, R.P., et al.: Astrophys. J. **564**, 896 (2002)
Drake, R.P., et al.: Phys. Plasmas **11**, 2829 (2004)
Fryxell, B., Muller, E., Arnett, D.: Astrophys. J. **367**, 619 (1991)
Glenzer, S.H., Redmer, R.: Rev. Mod. Phys. (2008)
Kane, J., et al.: Astrophys. J. **478**, L75 (1997)
Kane, J.O., et al.: Phys. Rev. E. **63**, 055401R (2001)
Kifonidis, K., Plewa, T., Janka, H.-T., Muller, E.: Astron. Astrophys. **408**, 621 (2003)
Kifonidis, K., Plewa, T., Sheck, L., Janka, H.-T., Muller, E.: Astron. Astrophys. **457**, 963 (2006)
Kuranz, C.C., et al.: Astrophys. Space Sci. **307**, 115 (2007)
Meakin, C., Arnett, D.: Astrophys. J. **637**, L53 (2006)

Meshkov, E.E.: Sov. Fluid Dyn. **4**, 101 (1969)
Miles, A.R., Braun, D.G., Edwards, M.J., Robey, H.F., Drake, R.P., Leibrandt, D.R.: Phys. Plasmas **11**, 3631 (2004)
Rayleigh, L.: Scientific Papers II. Cambridge University Press, Cambridge (1900)
Richtmyer, R.D.: Commun. Pure. Appl. Math. **13**, 297 (1960)
Robey, H.F., et al.: Phys. Plasmas **8**, 2446 (2001)
Ryutov, D.D., Drake, R.P., Kane, J., Liang, E., Remington, B.A., Wood-Vasey, M.: Astrophys. J. **518**, 821 (1999)
Taylor, S.G.: Proc. R. Soc. A **201**, 192 (1950)

Astrophys Space Sci (2009) 322: 65–70
DOI 10.1007/s10509-008-9928-y

ORIGINAL ARTICLE

Shock waves in the large-scale structure of the Universe

Dongsu Ryu · Hyesung Kang

Received: 12 June 2008 / Accepted: 8 October 2008 / Published online: 31 October 2008

Abstract Cosmological shock waves are induced during hierarchical formation of large-scale structure in the universe. Like most astrophysical shocks, they are collisionless, since they form in the tenuous intergalactic medium through electromagnetic viscosities. The gravitational energy released during structure formation is transferred by these shocks to the intergalactic gas as heat, cosmic-rays, turbulence, and magnetic fields. Here we briefly described the properties and consequences of the shock waves in the context of the large-scale structure of the universe.

Keywords Cosmic-rays · Large-scale structure of universe · Magnetic fields · Shock waves · Turbulence

1 Introduction

Shock waves are ubiquitous in astrophysical environments; from solar winds to the largest scale of the universe (Miniati et al. 2000; Ryu et al. 2003; Furlanetto and Loeb 2004; Pavlidou and Fields 2006). In the current paradigm of the cold dark matter (CDM) cosmology, the large-scale structure of the universe forms through hierarchical clustering of matter. Deepening of gravitational potential wells causes gas to move supersonically. Cosmological shocks form when the gas accretes onto clusters, filaments, and sheets, or as a consequence of the chaotic flow motions of the gas inside the nonlinear structures. The gravitational energy released during the formation of large-scale structure in the universe is transferred by these shocks to the intergalactic medium (IGM).

Cosmological shocks are collisionless shocks which form in a tenuous plasma via collective electromagnetic interactions between baryonic particles and electromagnetic fields (Quest 1988). They play key roles in governing the nature of the IGM through the following processes: in addition to the entropy generation, cosmic-rays (CRs) are produced via diffusive shock acceleration (DSA) (Bell 1978; Blandford and Ostriker 1978), magnetic fields are generated via the Biermann battery mechanism (Kulsrud et al. 1997) and Weibel instability (Medvedev et al. 2006) and also amplified by streaming CRs (Bell 2004), and vorticity is generated at curved shocks (Binney 1974).

Cosmological shocks in the intergalactic space have been studied in details using various hydrodynamic simulations for the cold dark matter cosmology with cosmological constant (ΛCDM) (Ryu et al. 2003; Pfrommer et al. 2006; Kang et al. 2007; Skillman et al. 2008; Vazza et al. 2008). In this contribution, we describe the properties of cosmological shocks and their implications for the intergalactic plasma from a simulation using a PM/Eulerian hydrodynamic cosmology code (Ryu et al. 1993) with the following parameters: $\Omega_{BM} = 0.043$, $\Omega_{DM} = 0.227$, and $\Omega_\Lambda = 0.73$, $h \equiv H_0/(100$ km/s/Mpc$) = 0.7$, and $\sigma_8 = 0.8$. A cubic region of comoving size 100 h^{-1} Mpc was simulated with 1024^3 grid zones for gas and gravity and 512^3 particles for dark matter, allowing a uniform spatial resolution of $\Delta l = 97.7\,h^{-1}$ kpc. The simulation is adiabatic in the sense that it does not include radiative cooling, galaxy/star formation, feedbacks from galaxies/stars, and reionization of the

D. Ryu (✉)
Department of Astronomy and Space Science,
Chungnam National University, Daejeon 305-764, Korea
e-mail: ryu@canopus.cnu.ac.kr

H. Kang
Department of Earth Sciences, Pusan National University,
Pusan 609-735, Korea
e-mail: kang@uju.es.pusan.ac.kr

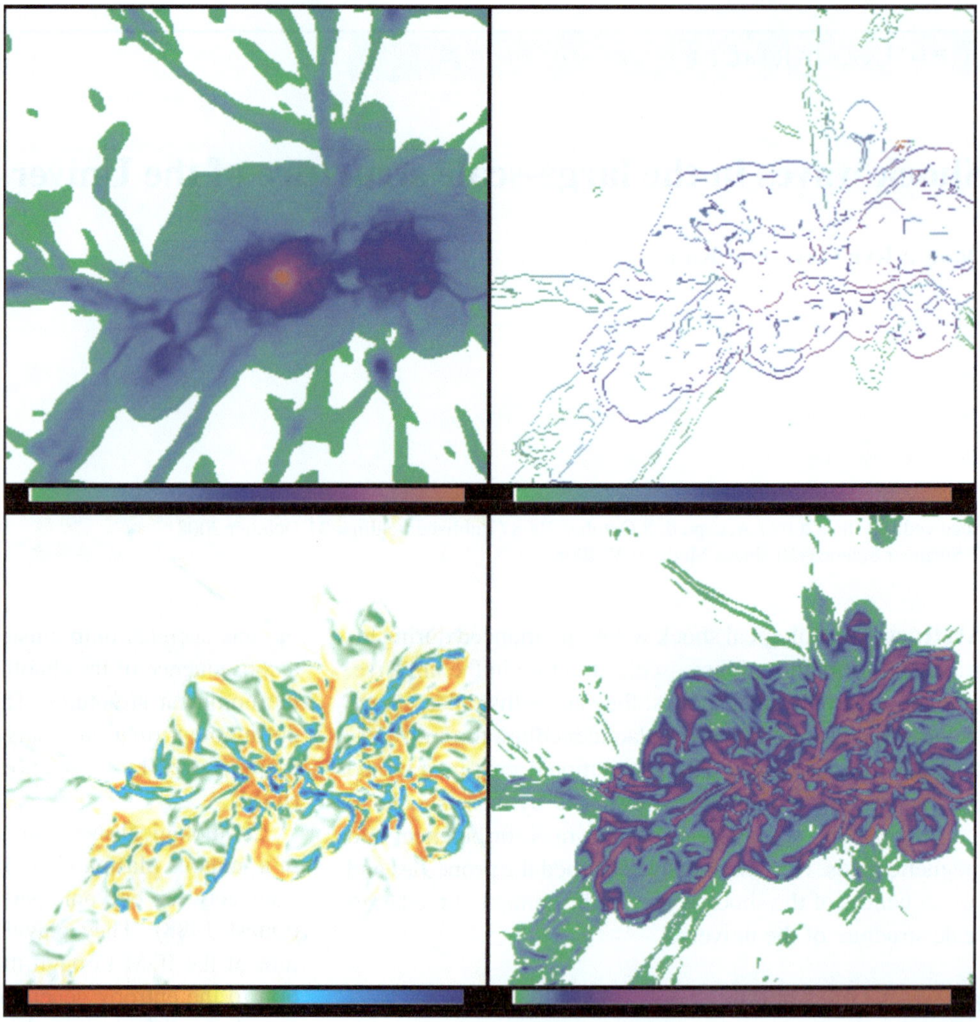

Fig. 1 Two-dimensional images showing X-ray emissivity (*top left*), locations of shocks with color-coded shock speed V_s (*top right*), perpendicular component of vorticity (*bottom left*), and magnitude of vorticity (*bottom right*) in the region of $(25\,h^{-1}\,\mathrm{Mpc})^2$ around a galaxy cluster at present ($z = 0$). Color codes V_s from 15 (*green*) to 1,800 km s^{-1} (*red*)

IGM. A temperature floor was set to be the temperature of cosmic background radiation.

2 Properties of cosmological shocks

As a post-processing step, shocks in the simulated volume were identified. The procedure to identify shocks was described in details in Ryu et al. (2003). A zone was tagged as a shock zone whenever the following criteria are met: 1) the gradients of gas temperature and entropy have the same sign, 2) the local flow is converging with $\vec{\nabla} \cdot \vec{v} < 0$, and 3) $|\Delta \log T| \geq 0.11$ corresponding to the temperature jump of a shock with $M \geq 1.3$. Typically, a shock is represented by a jump spread over 2–3 tagged zones. Hence, a shock center was identified within the tagged zones, where $\vec{\nabla} \cdot \vec{v}$ is minimum, and this center was labeled as part of a shock surface. The Mach number of the shock center, M, was calculated from the temperature jump across the entire shock zones, and the shock speed, V_s, was calculated with the Mach number and the pre-shock sound speed. To avoid confusion, only those portions of shock surfaces with $M \geq 1.5$ were kept and used for the analysis of shocks properties.

In the top panels of Fig. 1, we compare the locations of cosmological shocks with the X-ray emissivity in the region around a cluster of galaxies, both of which are calculated from the simulation data at redshift $z = 0$. External shocks encompass this complex nonlinear structure and define the outermost boundaries up to $\sim 10\ h^{-1}$ Mpc from the cluster core, far beyond the region observable with X-ray of size $\sim 1\ h^{-1}$ Mpc. Internal shocks are found within the region bounded by external shocks. External shocks have high Mach numbers of up to $M \sim 10^3$ due to the low temperature of the accreting gas in the void region. Internal shocks, on the other hand, have mainly low Mach numbers of $M \lesssim 3$, because the gas inside nonlinear structures has been previously heated by shocks and so has high temperature.

The frequency of cosmological shocks in the simulated volume is represented by the quantity S, the area of shock surfaces per unit comoving volume, in other words, the

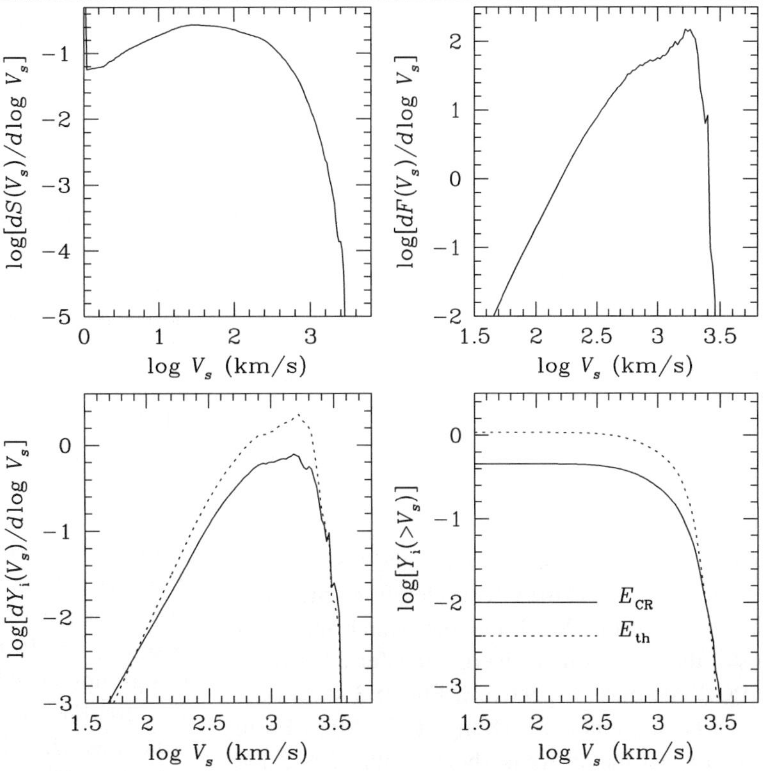

Fig. 2 (*Top left*) Reciprocal of the mean comoving distance between shock surfaces at $z = 0$ in units of $1/(h^{-1}\,\mathrm{Mpc})$. (*Top right*) Kinetic energy flux passing through shock surfaces per unit comoving volume at $z = 0$ in units of 10^{40} erg s^{-1} $(h^{-1}\,\mathrm{Mpc})^{-3}$. (*Bottom left*) Thermal energy dissipated (*dotted line*) and CR energy generated (*solid line*) at shock surfaces, integrated from $z = 5$ to 0. (*Bottom right*) Cumulative energy distributions. The energies in the *bottom panels* are normalized to the total thermal energy at $z = 0$. All quantities are plotted as a function of shock speed V_s

reciprocal of the mean comoving distance between shock surfaces. In the top left panel of Fig. 2, we show $S(V_s)$ per unit logarithmic shock speed interval at $z = 0$. We note that the frequency of low speed shocks with $V_s < 15\,\mathrm{km\,s^{-1}}$ is overestimated here, since the temperature of the intergalactic medium is unrealistically low in the adiabatic simulation without the cosmological reionization process. Although shocks with $V_s \sim$ a few $\times$ $10\,\mathrm{km\,s^{-1}}$ are most common, those with speed up to several $\times 10^3\,\mathrm{km\,s^{-1}}$ are present at $z = 0$. The mean comoving distance between shock surfaces is $1/S \sim 3\,h^{-1}$ Mpc when averaged over the entire universe, while it is $\sim 1\,h^{-1}$ Mpc inside the nonlinear structures of clusters, filaments, and sheets.

In order to evaluate the energetics of cosmological shocks, the incident shock kinetic energy flux, $f_{\mathrm{kin}} = (1/2)\rho_1 V_s^3$, is calculated. Here ρ_1 is the preshock gas density. Then the average kinetic energy flux through shock surfaces per unit comoving volume, F, is calculated. The top right panel of Fig. 2 shows $F(V_s)$ per unit logarithmic shock speed interval. Energetically the shocks with $V_s > 10^3\,\mathrm{km\,s^{-1}}$, which form in the deepest gravitation potential wells in and around clusters of galaxies, are most important. Those responsible for most of shock energetics are the internal shocks with relatively low Mach number of $M \sim$ 2–4 in the hot IGM, because they form in the high-density gas inside nonlinear structures. On the other hand, external shocks typically form in accretion flows with the low-density gas in voids, so the amount of the kinetic energy passed through the external shocks is rather small.

3 Energy dissipation at cosmological shocks

In addition to the gas entropy generation, the acceleration of CRs is an integral part of collisionless shocks, in which electromagnetic interactions between plasma and magnetic fields provide the necessary viscosities. Supra-thermal particles are extracted from the shock-heated thermal particle distribution (Malkov and Drury 2001). With 10^{-4}–10^{-3} of the particle flux passing through the shocks injected into the CR population, up to ~60% of the kinetic energy of strong quasi-parallel shocks can be converted into CR ions and the nonlinear feedback to the underlying flow can be substantial (Kang and Jones 2005). At perpendicular shocks, however, the CR injection and acceleration are expected to be much less efficient, compared to parallel shocks, since the transport of low energy particles normal to the average field direction is suppressed. So the CR acceleration depends sensitively on the mean magnetic field orientation.

Time-dependent simulations of DSA at quasi-parallel shocks with a thermal leakage injection model and a Bohm-type diffusion coefficient have shown that the evolution of CR modified shocks becomes self-similar, after the particles are accelerated to relativistic energies and the precursor compression reaches a time-asymptotic state (Kang and Jones 2005; Kang and Jones 2007). The self-similar evolution of CR modified shocks depends somewhat weakly on the details of various particle–wave interactions, but it is mainly determined by the shock Mach number. Based on

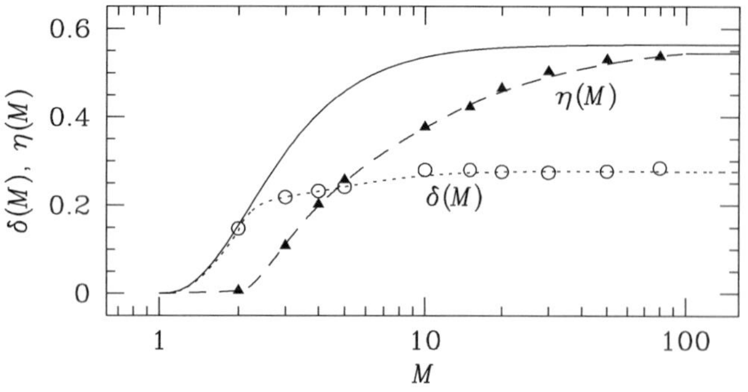

Fig. 3 Gas thermalization efficiency, $\delta(M)$, and CR acceleration efficiency, $\eta(M)$, as a function of Mach number. *Symbols* are the values estimated from numerical simulations based on a DSA model and *dotted* and *dashed lines* are the fits. *Solid line* is for the gas thermalization efficiency for shocks without CRs

this self-similar evolution, we can estimate the gas thermalization efficiency, $\delta(M)$, and the CR acceleration efficiency, $\eta(M)$, as a function shock Mach number, which represent the fractions of the shock kinetic energy transferred into the thermal and CR energies, respectively. Figure 3 shows the results of such DSA simulations (Kang et al. 2007). From the figure, we expect that at weak shocks with $M \lesssim 3$ the energy transfer to CRs should be $\lesssim 10\%$ of the shock kinetic energy at each shock passage, whereas at strong shocks with $M \gtrsim 3$ the transfer is very efficient and the flow should be significantly modified by the CR pressure.

Adopting the efficiencies shown in Fig. 3, the thermal and CR energy fluxes dissipated at each cosmological shock can be estimated as $\delta(M) f_{\rm kin}$ and $\eta(M) f_{\rm kin}$, respectively. Then the total energies dissipated through the surfaces of cosmological shocks during the large-scale structure formation of the universe can be calculated (Ryu et al. 2003; Kang et al. 2007). The bottom panels of Fig. 2 show the thermal and CR energies, integrated from $z = 5$ to 0, normalized to the total gas thermal energy at $z = 0$, as a function of shock speed. Here we assume CRs are freshly injected at shocks without a pre-existing population. The shocks with $V_s > 10^3$ km s^{-1} are most responsible for the shock dissipation into heat and CRs. The figure shows that the shock dissipation can count most of the gas thermal energy in the IGM (Kang et al. 2005). The ratio of the total CR energy, $Y_{CR}(> V_{s,\rm min})$, to the total gas thermal energy, $Y_{\rm th}(> V_{s,\rm min})$, dissipated at cosmological shocks throughout the history of the universe is about 0.4 (where $V_{s,\rm min}$ is the minimum shock speed), giving a rough estimate for the energy density of CR protons relative to that of thermal gas in the IGM as $\varepsilon_{\rm CRp} \sim 0.4 \varepsilon_{\rm therm}$.

Because of uncertainties in the DSA model such as the field obliquity and the injection efficiency, however, it is not meant to be an accurate estimate of the CR energy in the IGM. Yet the results imply that the IGM could contain a dynamically significant CR population.

4 Turbulence induced by cosmological shocks

Vorticity can be generated in the IGM either directly at curved cosmological shocks or by the baroclinity of flow. The baroclinity is resulted mostly from the entropy variation induced at cosmological shocks. Therefore, the baroclinic vorticity generation also can be attributed to the presence of cosmological shocks. A quantitative estimation of the vorticity in the IGM was made using the data of the simulation described in Sect. 1 (Ryu et al. 2008). The bottom panels of Fig. 1 show the distribution of the vorticity around a cluster complex. The distribution closely matches that of shocks, as expected.

The top panels of Fig. 4 show the magnitude of the vorticity in the simulated volume as a function of gas temperature and density. Here the vorticity magnitude is given in units of the reciprocal of the age of the universe, $1/t_{\rm age}$. There is a clear trend that the vorticity is larger in hotter and denser regions. At the present epoch, the rms vorticity is $\omega_{\rm rms} t_{\rm age} \sim 10$ to 30 in the regions associated with clusters/groups of galaxies ($T > 10^7$ K) and filaments ($10^5 < T < 10^7$ K), whereas it is on the order of unity in sheetlike structures with $10^4 < T < 10^5$ K and even smaller in voids with $T < 10^4$ K.

With $t_{\rm eddy} = 1/\omega$ interpreted as the local eddy turnover time, $\omega \times t_{\rm age}$ represents the number of eddy turnovers of vorticity in the age of the universe. It takes a few turnovers for vorticity to decay into smaller eddies and develop into turbulence. So with $\omega_{\rm rms} t_{\rm age} \sim 10$–30, the flows in clusters/groups and filaments is likely to be in the state of turbulence. On the other hand, with $\omega_{\rm rms} t_{\rm age} \lesssim 1$ the flow in sheetlike structures and voids is expected to be mostly nonturbulent.

In order to estimate the energy associated with the turbulence in the IGM, the curl component of flow motions, $\vec{v}_{\rm curl}$, which satisfies the relation $\vec{\nabla} \times \vec{v}_{\rm curl} \equiv \vec{\nabla} \times \vec{v}$, is extracted from the velocity field. As vorticity cascades and develops into turbulence, the energy $(1/2)\rho v_{\rm curl}^2$ is transferred to turbulent motions, so it can be regarded as the turbulence energy, $\varepsilon_{\rm turb}$. As shown in Ryu et al. (2008), $\varepsilon_{\rm turb} < \varepsilon_{\rm therm}$ in clusters/groups. In particular, the mass-averaged value is $\langle \varepsilon_{\rm turb}/\varepsilon_{\rm therm} \rangle_{\rm mass} = 0.1 - 0.3$ in the intracluster medium (ICM), which is in good agreement with the observationally inferred values in cluster core regions (Schuecker et al. 2004). In filaments, this ratio is estimated to be $0.5 \lesssim \langle \varepsilon_{\rm turb}/\varepsilon_{\rm therm} \rangle \lesssim 1.5$ and it increases with decreasing temperature.

5 Intergalactic magnetic field

How have the intergalactic magnetic fields (IGMFs) arisen? The general consensus is that there was no viable mechanism to produce strong, coherent magnetic fields in the IGM

Fig. 4 Volume fraction with given temperature and vorticity magnitude (*top left*), with given gas density and vorticity magnitude (*top right*), with given temperature and magnetic field strength (*bottom left*), and with given gas density and magnetic field strength (*bottom right*) at present ($z = 0$). Here, $t_{\rm age}$ is the present age of the universe

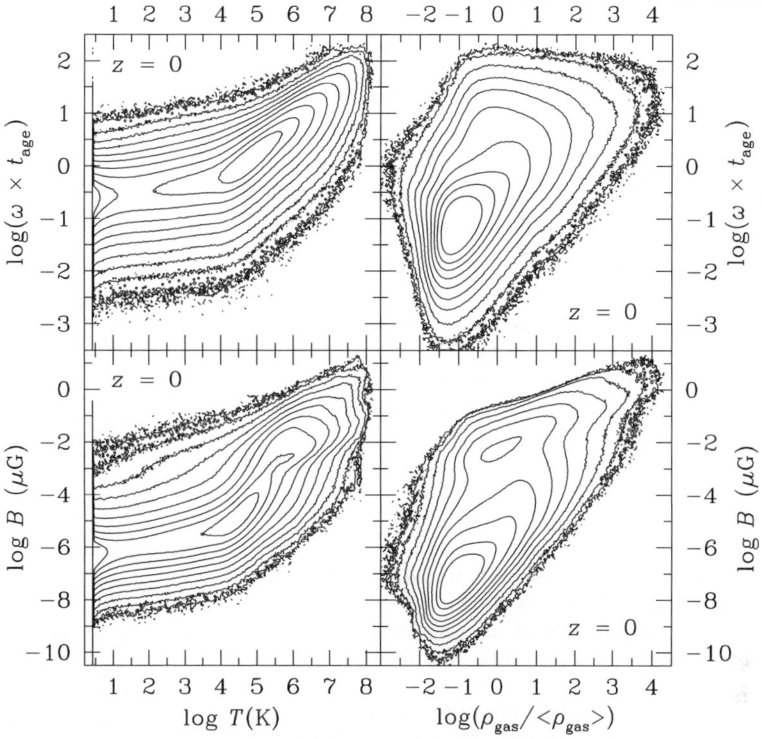

prior to the formation of large-scale structure and galaxies (Kulsrud and Zweibel 2008). However, it is reasonable to assume that weak seed fields were created in the early universe. A number of mechanisms, including the Biermann battery mechanism (Kulsrud et al. 1997) and Weibel instability (Medvedev et al. 2006) working at early cosmological shocks, have been suggested (Kulsrud and Zweibel 2008). The turbulence described in Sect. 4 then can amplify the seed fields in the IGM through the stretching of field lines, a process known as the turbulence dynamo. In this scenario the evolution of the IGMFs should go through three stages: (i) the initial exponential growth stage, when the back-reaction of magnetic fields is negligible; (ii) the linear growth stage, when the back-reaction starts to operate; and (iii) the final saturation stage (Cho and Vishniac 2000; Cho et al. 2008).

In order to estimate the strength of the IGMFs resulted from the dynamo action of turbulence in the IGM, we model the growth and saturation of magnetic energy as:

$$\frac{\varepsilon_B}{\varepsilon_{\rm turb}} = \begin{cases} 0.04 \times \exp\left[(t'-4)/0.36\right] & \text{for } t' < 4, \\ (0.36/41) \times (t'-4) + 0.04 & \text{for } 4 < t' < 45, \\ 0.4 & \text{for } t' > 45, \end{cases}$$

based on a simulation of incompressible magnetohydrodynamic turbulence (Ryu et al. 2008; Cho et al. 2008). Here $t' = t/t_{\rm eddy}$ is the number of eddy turnovers. This provides a functional fit for the fraction of the turbulence energy, $\varepsilon_{\rm turb}$, transfered to the magnetic energy, ε_B, as a result of the turbulence dynamo.

The above formula is convoluted to the data of the simulation described in Sect. 1, setting $t' \equiv \omega \times t_{\rm age}$, and the strength of the IGMFs is calculated as $B = (8\pi \varepsilon_{\rm B})^{1/2}$. The resulting magnetic field strength is presented in the bottom panels of Fig. 4. On average the IGMFs are stronger in hotter and denser regions. The strength of the IGMFs is $B \gtrsim 1$ μG inside clusters/groups (the mass-averaged value for $T > 10^7$ K), ~ 0.1 μG around clusters/groups (the volume-averaged value for $T > 10^7$ K), and ~ 10 nG in filaments (with $10^5 < T < 10^7$ K) at present. The IGMFs should be much weaker in sheetlike structures and voids. But as noted above, turbulence has not developed fully in such low density regions, so our model is not adequate to predict the field strength there.

We note that in addition to the turbulence dynamo, other processes such as galactic winds driven by supernova explosions and jets from active galactic nuclei can further strengthen the magnetic fields to the IGM (for references, see Ryu et al. 2008).

6 Conclusion

Shocks are inevitable consequences of the formation of large-scale structure in the universe. They heat gas, accelerate cosmic-ray particles, produce vorticity and turbulence, and generate and amplify magnetic fields in the IGM. By applying detailed models of the DSA and the turbulence dynamo to the data of a cosmological hydrodynamic simulation of a concordance ΛCDM universe, we have made the following quantitative estimates: $\varepsilon_{\mathrm{CRp}}/\varepsilon_{\mathrm{therm}} \sim 0.4$ in the whole IGM, $\langle\varepsilon_{\mathrm{turb}}/\varepsilon_{\mathrm{therm}}\rangle_{\mathrm{mass}} = 0.1$–$0.3$ in the ICM, $\langle\varepsilon_{\mathrm{turb}}/\varepsilon_{\mathrm{therm}}\rangle \sim 0.5$–$1.5$ in filaments, $B \gtrsim 1$ μG inside clusters/groups, $B \sim 0.1$ μG around clusters/groups, and $B \sim 10$ nG in filaments at present. Our results suggest that the nonthermal components can be energetically significant in the intergalactic plasma of large-scale structure, as in the interstellar plasma inside Our Galaxy.

Acknowledgements The work of DR was supported by the Korea Research Foundation Grant funded by the Korean Government (MOEHRD) (KRF-2007-341-C00020). The work of HK was supported for two years by Pusan National University Research Grant.

References

Bell, A.R.: The acceleration of cosmic rays in shock fronts. I. Mon. Not. R. Astron. Soc. **182**, 147–156 (1978)

Bell, A.R.: Turbulent amplification of magnetic field and diffusive shock acceleration of cosmic rays. Mon. Not. R. Astron. Soc. **353**, 550–558 (2004)

Binney, J.: Galaxy formation without primordial turbulence: mechanisms for generating cosmic vorticity. Mon. Not. R. Astron. Soc. **168**, 73–92 (1974)

Blandford, R.D., Ostriker, J.P.: Particle acceleration by astrophysical shocks. Astrophys. J. **221**, L29–L32 (1978)

Cho, J., Vishniac, E.T.: The generation of magnetic fields through driven turbulence. Astrophys. J. **538**, 217–225 (2000)

Cho, J., Vishniac, E.T., Beresnyak, A., Lazarian, A., Ryu, D.: Growth of magnetic fields induced by turbulent motions. Astrophys. J. (2008, submitted)

Furlanetto, S.R., Loeb, A.: Large-scale structure shocks at low and high redshifts. Astrophys. J. **611**, 642–654 (2004)

Kang, H., Jones, T.W.: Efficiency of nonlinear particle acceleration at cosmic structure shocks. Astrophys. J. **620**, 44–58 (2005)

Kang, H., Jones, T.W.: Self-similar evolution of cosmic-ray-modified quasi-parallel plane shocks. Astropart. Phys. **28**, 232–246 (2007)

Kang, H., Ryu, D., Cen, R., Song, D.: Shock-heated gas in the large-scale structure of the universe. Astrophys. J. **620**, 21–30 (2005)

Kang, H., Ryu, D., Cen, R., Ostriker, J.P.: Cosmological shock waves in the large scale structure of the Universe: non-gravitational effects. Astrophys. J. **669**, 729–740 (2007)

Kulsrud, R.M., Cen, R., Ostriker, J.P., Ryu, D.: The Protogalactic origin for cosmic magnetic fields. Astrophys. J. **480**, 481–491 (1997)

Kulsrud, R.M., Zweibel, E.G.: On the origin of cosmic magnetic fields. Rep. Prog. Phys. **71**, 1–33 (2008). 046901

Malkov, M.A., Drury, L.O'C.: Nonlinear theory of diffusive acceleration of particles by shock waves. Rep. Prog. Phys. **64**, 429–481 (2001).

Medvedev, M.V., Silva, L.O., Kamionkowski, M.: Cluster magnetic fields from large-scale structure and galaxy cluster shocks. Astrophys. J. **642**, L1–L4 (2006)

Miniati, F., Ryu, D., Kang, H., Jones, T.W., Cen, R., Ostriker, J.P.: Properties of cosmic shock waves in large scale structure formation. Astrophys. J. **542**, 608–621 (2000)

Pavlidou, V., Fields, B.D.: Analytical models for the energetics of cosmic accretion shocks, their cosmological evolution, and the effect of environment. Astrophys. J. **642**, 734–745 (2006)

Pfrommer, C., Springel, V., Enßlin, T.A., Jubelgas, M.: Detecting shock waves in cosmological smoothed particle hydrodynamics simulations. Mon. Not. R. Astron. Soc. **367**, 113–131 (2006)

Quest, K.B.: Theory and simulation of collisionless parallel shocks. J. Geophys. Res. **93**, 9649–9680 (1988)

Ryu, D., Ostriker, J.P., Kang, H., Cen, R.: A cosmological hydrodynamic code based on the total variation diminishing scheme. Astrophys. J. **414**, 1–19 (1993)

Ryu, D., Kang, H., Hallman, E., Jones, T.W.: Cosmological shock waves and their role in the large-scale structure of the universe. Astrophys. J. **593**, 599–610 (2003)

Ryu, D., Kang, H., Cho, J., Das, S.: Turbulence and magnetic fields in the large-scale structure of the universe. Science **320**, 909–912 (2008)

Schuecker, P., Finoguenov, A., Miniati, F., Böhringer, H., Briel, U.G.: Probing turbulence in the Coma galaxy cluster. Astron. Astrophys. **426**, 387–397 (2004)

Skillman, S.W., O'Shea, B.W., Hallman, E.J., Burns, J.O., Norman, M.L.: Cosmological shocks in adaptive mesh refinement simulations and the acceleration of cosmic rays. Astrophys. J. (2008, in press). arXiv:0806.1522

Vazza, F., Brunetti, G., Gheller, C.: Shock waves in Eulerian cosmological simulations: main properties and acceleration of cosmic rays. Mon. Not. R. Astron. Soc. (2008, submitted). arXiv:0808.0609

Astrophys Space Sci (2009) 322: 71–75
DOI 10.1007/s10509-009-9996-7

ORIGINAL ARTICLE

Analytical structure of steady radiative shocks in magnetic cataclysmic variables

E. Falize · C. Michaut · C. Cavet · S. Bouquet · M. Koenig · B. Loupias · A. Ravasio · C.G. Gregory

Received: 17 June 2008 / Accepted: 27 January 2009 / Published online: 5 March 2009

Abstract In this paper we study specific classes of radiating shocks which are widely spread in astrophysical environments. We present new analytical solutions available for any exponents of typical power-law cooling function which generalize the five special cases (corresponding to specific values of these exponents) of radiating shocks structure and proceed to the analytical determination of physical quantities. Then an application of the results for the accretion shock in polar is realized. A discussion of the reproducibility of cooling layer in laboratory is proposed using scaling laws.

Keywords Radiative shock · Scaling laws · Polars · Magnetic cataclysmic variables · Laboratory astrophysics

1 Introduction

From galaxy collisions to the magnetospherical accretion of T-Tauri (Calvet and Gulbring 1998), radiative shocks are fundamental components of various astronomical phenomena and environments with a wide spread of spatial scales. The interstellar medium is endlessly perturbed by violent phenomena such as ejected matter by stellar objects or supernovae explosions which lead to the formation of a shock. When the shock is strong enough the radiation can play most important role in the dynamics and stability of the post-shock matter and lead to the formation of cooling layer (Chevalier and Imamura 1982; Kylafis and Lamb 1982; Michaut et al. 2008). Quantitatively such layer is formed when the cooling time is approximately equal to the dynamical time or when the cooling time is comparable to or less than the sound-crossing time (Toth and Draine 1993). This type of shock has became very common in laboratory (Michaut et al. 2007; Drake 2007) with the spectacular advent of high-energy-density experimental facilities (Remington et al. 2006). They allow to have new spatial scales which permit to explore experimentally the properties of radiative shocks. This paper focus on the study of a specific situation of an accreting shock in magnetic cataclysmic variables (MCVs) (Mignone 2005; Chevalier and Imamura 1982). Cataclysmic variables (CVs) stars are closed binaries containing an accreting white dwarf (primary star that we note WD) from a late type Roche-lobe filling secondary star (Warner 1995; Wu 2000; Cropper 1990). These objects constitute an important class of low-luminosity (10^{29}–10^{32} erg s^{-1}), compact galactic X-ray sources (Imamura and Wolff 1990; Falanga et al. 2005) and provide an excellent probe of the accretion process under extreme astrophysical conditions. A sub-class of CVs exhibits an optical emission which is strongly polarized and constitutes a proof that an important magnetic field exists in these objects. This field can be sufficiently strong to impose the WD rotating synchronously

E. Falize (✉) · S. Bouquet
Département de Physique Théorique et Appliquée (DPTA), CEA DIF/DPTA, 91297 Arpajon cedex, France
e-mail: emeric.falize@obspm.fr

E. Falize · C. Michaut · C. Cavet · S. Bouquet
Laboratoire Univers et Théories (LUTH) CNRS, UMR8102, INSU Observatoire de Paris Université Denis Diderot—Paris VII, 5 Place Jules Janssen, 92190 Meudon, France

M. Koenig · B. Loupias · C.G. Gregory
LULI, Ecole Polytechnique, CNRS, Université Paris 6, 91128 Palaiseau, France

A. Ravasio
CEA-Saclay DSM/DRECAM/SPAM, 91191 Gif-sur-Yvette, France

and not only to guide the flows but could also dictate the radiative loss processes. Thus, we can divide the CVs in three sub-classes: the non magnetic CVs ($B < 1$ MG), the intermediate polars or *DQHer* (1 MG $< B <$ 10 MG) and the polars ($B >$ 10 MG) or *AMHer*. As in any magnetic accreting system the matter forms an accretion disk until it reaches the magnetospheric radius R_{mag}. In polars the magnetic field is so strong that $R_{mag} > R_{L_1}$ where R_{L_1} is the position of the first Lagrange point. Consequently, the accreted matter is directly attached to magnetic field lines and it follows almost radially with free-fall velocity towards the magnetic poles of WD. The in-falling matter is shocked near the surface of the WD, and radiates over most of the electromagnetic spectrum from X-ray to infrared. In this paper we study the structure of the cooling layer which is formed when the shock is reaching the steady regime. In the first part, we start by reminding the hypothesis of physical models. In the second part we propose a complete analytical solution which generalizes the five known solutions and we determine several physical quantities. Before concluding, we examine in the last part the scaling laws problem for laboratory astrophysics.

2 Radiative shock and cooling layer

When the in-falling matter hits the photosphere of the WD a reverse shock is created (see Fig. 1). The kinetic energy of the material in the accretion stream, acquired from the Roche lobe, is converted into radiation. As the gas below the front is generally not cool, it expands. The equilibrium height x_s of the shock is determined by the requirement that the post-shock flow must have sufficient time to cool and to decelerate. The radiative cooling is essentially produced by electrons, firstly by bremsstrahlung cooling and secondly by cyclotron cooling (i.e. the electrons spiral around magnetic field lines). In order to model this specific zone of in-falling matter, we consider a plane-parallel,[1] collisional shock[2] with a post-shock medium which can be described by a single temperature.[3] Moreover, we suppose that the local gravitational field does not modify the shock structure.[4] Finally, we assume that the shock is in a steady-state.[5] In this model we neglect the albedo of the white dwarf, the

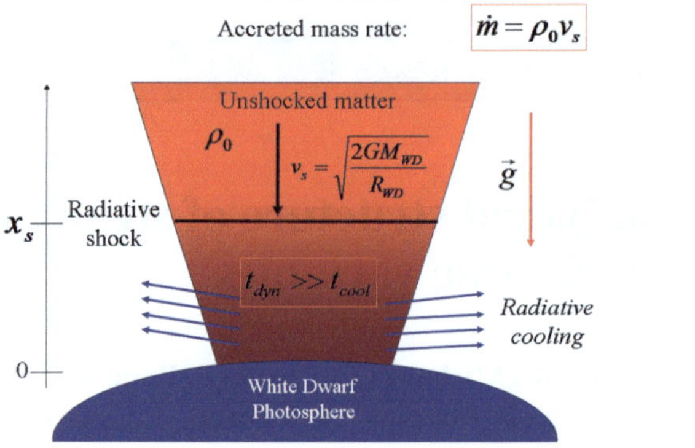

Fig. 1 Convention and radiative shock near white dwarf surface. The steady radiative shock is located at x_s and the post shock region corresponds to $0 \le x \le x_s$

Compton cooling, thermal conductivity effects and nuclear reactions which can be produced on WD surface and give birth to the phenomena of novae. Thus, the cooling layer can be modeled by hydrodynamic equations (Kylafis and Lamb 1982):

$$\frac{d}{dx}[\rho v] = 0, \quad \frac{d}{dx}[\rho v^2 + P] = 0, \tag{1}$$

$$v\left[\frac{dP}{dx} - \gamma \frac{P}{\rho}\frac{d\rho}{dx}\right] = -(\gamma - 1)\Lambda(\rho, P, x), \tag{2}$$

where x, ρ, v, P, γ, Λ are respectively the spatial coordinate, the density, the velocity, the pressure, the polytropic index and the cooling function. Although these equations are theoretically consistent, we must specify an equation of state in order to connect the microscopic phenomena to the cooling function. We can consider an equation of state given by: $P = \varepsilon_0[Z]\rho^{\mu}T^{\nu}$ where T, $\varepsilon_0[Z]$, μ and ν are respectively the temperature, a function of ionization Z and two free exponents. We must impose $\gamma(1-\nu) = (\mu - \nu)$ in order to assure that the entropy is a perfect differential in the post-shock region. We suppose that the cooling function is:

$$\Lambda(\rho, P, x) = \Lambda_0 \rho^{\epsilon} P^{\zeta}(x + X_0)^{\theta}, \tag{3}$$

where Λ_0, ϵ, ζ, θ and X_0 are five constants (Drake 2006; Michaut et al. 2007). This form generalizes the optically thin case where $\theta = 0$. The arbitrarity of ϵ and ζ is motivated by the fact that $\Lambda \propto \kappa_P \sigma T^4$ where κ_P is the Planck opacity which can be modeled by a power law of ρ and T at high temperature and by the fact that several continuous processes can be modeled exactly or approximatively by power laws too. This form includes the bremsstrahlung process which is given by $\Lambda^{brem} \propto \rho^2 T^{1/2}$ (Rybicki and Lightman 1979), the optically thick cyclotron cooling $\Lambda_1^{cyc} \propto \rho^{0.15}T^{2.5}$ (Saxton et al. 1998) or $\Lambda_2^{cyc} \propto \rho^{0.15}T^{2.5}x^{-0.85}$ (Chanmugam et al. 1985). Moreover, we suppose that $L_{cool} \gg \lambda_s$, where L_{cool} and λ_s are respectively the cooling length and the shock thickness, in such a

[1] i.e. $x_s \ll R_{WD}$ where x_s is the thickness of the cooling layer and R_{WD} is the white dwarf radius implying $\partial/\partial y = \partial/\partial z = 0$.

[2] i.e., $t_{ii} \ll t_{dyn}$ where t_{ii} is the characteristic time of collision between ions and t_{dyn} is the dynamical time.

[3] i.e., $t_{ei} \ll t_{cool}$ where t_{ei} is the characteristic time of energy exchanged between electrons and ions and t_{cool} is the radiative cooling time.

[4] i.e., $g_* x_s \ll v_s^2$ where g_* and v_s are respectively the gravitational field and the shock velocity.

[5] i.e., $t_{cool} \ll t_{dyn}$ which means that cooling effects are faster than dynamical ones.

way that the shock can be considered as a pure discontinuity and the Rankine-Hugoniot relation are satisfied just behind the shock front:

$$\frac{\rho[x_s^-]}{\rho_0} = \frac{(\gamma+1)M^2}{(\gamma-1)M^2+2}, \tag{4}$$

$$\left|\frac{v[x_s^-]}{v_s}\right| = \frac{(\gamma-1)M^2+2}{(\gamma+1)M^2}, \tag{5}$$

$$\frac{P[x_s^-]}{\rho_0 v_s^2} = \frac{2\gamma M^2-(\gamma-1)}{\gamma(\gamma+1)M^2}, \tag{6}$$

where M and x_s^- are respectively the Mach number and the position just behind the shock front (see Fig. 1).

3 Analytical solutions and physical quantities

In order to solve (1–3) , we begin by introducing a normalized spatial coordinate $\xi = x/x_s$ where x_s is the thickness of cooling zone and we must elaborate a transformation which includes the boundary conditions. It is easy to check that the following transformation:

$$\rho[\xi] = \frac{\rho_0}{\eta[\xi]}, \quad v[\xi] = -v_s \eta[\xi], \tag{7}$$

where ρ_0 and v_s are respectively the ambiant medium density and shock velocity (with $\eta[1] = [2+(\gamma-1)M^2]/[(\gamma+1)M^2]$), obeys the Rankine-Hugoniot as well as (1). Injecting (7) into the second equation of system (1) gives the transformed pressure:

$$P[\xi] = \rho_0 v_s^2\left(1+\frac{1}{\gamma M^2}-\eta[\xi]\right). \tag{8}$$

In the limit $M \to \infty$, (7)–(8) are equivalent to the Bertschinger transformation (Bertschinger 1986). In order to determine the intermediate function η we introduce (7–8) into (2) leading to:

$$\left[(\gamma+1)\eta - \gamma\left(1+\frac{1}{\gamma M^2}\right)\right]\frac{d\eta}{d\xi} = -(\gamma-1)\frac{x_s}{\rho_0 v_s^3}\Lambda. \tag{9}$$

From the form (3) of the cooling function, we can write (9) in form of quadratures for which the solution is given by (13). This solution generalizes the five particular solutions that we can find in Laming (2004) and Chevalier and Imamura (1982) and gives an analytical expression of Wu et al. integral form (Wu et al. 1994). The first physical quantity that we can evaluate is the thickness of cooling layer x_s. We obtain in the general case (14) that we can rewrite in the standard form:

$$x_s = \Delta[\gamma, M, \epsilon, \zeta] \times v_s t_{cool}, \tag{10}$$

Table 1 Accretion density column Ξ-parameter in function of cooling processes and polytropic index. Notice that the parameter θ is zero

$(\epsilon,\zeta)/\gamma$	5/3	7/5	4/3	1.1	1.06
(−2.35, 2.50)	1.991	2.894	3.406	11.387	19.022
(1.50, 0.50)	6.967	10.296	11.961	35.288	57.509
(2.50, −0.50)	5.869	8.651	10.047	29.629	48.292
(2.00, 0.00)	6.286	9.273	10.769	31.756	51.752
(1.00, 1.00)	8.302	12.313	14.316	42.327	68.994
(3.00, −1.00)	5.585	8.232	9.560	28.208	45.983

where t_{cool} is the cooling time defined in the post-shock medium and Δ is a function of different parameters. In the situation of polar the accretion rate is given by $\dot{m} = \rho_0 v_s$ and the shock velocity by the free fall velocity $v_s = \sqrt{2GM_{WD}/R_{WD}}$ where M_{WD} is the WD mass. Consequently we can write x_s in terms of M_{WD}, $\dot{m}$, and the properties of white dwarf. An application of this relation to the case of polars with only bremsstrahlung cooling gives:

$$x_s = 7.6\times10^6\ \text{cm}\left[\frac{\dot{m}}{4\ \text{g}\,\text{cm}^{-2}\,\text{s}^{-1}}\right]^{-1}\left[\frac{M_{WD}}{0.5M_\odot}\right]^{3/2} \times\left[\frac{R_{WD}}{10^9\ \text{cm}}\right]^{-3/2}, \tag{11}$$

which is compatible with the value given by Wu et al. (1994). Moreover using the Nauenberg relation which gives the radius of white dwarf in terms of its mass, we can evaluate x_s with $\dot{m}$ and M_{WD} only. Thus, for a mean white dwarf $0.8M_\odot$ which gives a radius around $R_{WD} = 7\times10^8$ cm and $\dot{m} = 4\ \text{g}\,\text{cm}^{-2}\,\text{s}^{-1}$, we obtain $x_s = 2.6\times10^7$ cm. The second important physical quantity is the accreting column density which is defined by:

$$\Xi = \left(\int_0^{x_s}\rho(x)\cdot dx\right)\Big/\left(\int_0^{x_s}\rho_0\cdot dx\right). \tag{12}$$

The analytical expression of Ξ is given by (15). This relation is an extension form of Laming (2004) results. We report in the Table 1 different asymptotic ($M\to\infty$) value of Ξ in function of cooling process and polytropic index.

4 Scaling laws and cooling layers

Since the extension of the cooling zone is small (few hundred kilometers) it is difficult to observe it. Consequently the experimental exploration of this zone is very interesting in order to study its stability and to validate the model which searches to explain the QPOs phenomena (Chevalier and Imamura 1982) in terms of development of global thermal instability. We know that the Rankine-Hugoniot relations are invariant by homothetic group (Falize et al. 2008a).

Table 2 Scaling laws for cooling layer in magnetic cataclysmic variables

Physical ratio	BC	BC + CC1
$r/\tilde{r}$	$a^{\delta_6-2\delta_5}$	$a^{-3\delta_5/40}$
$\rho/\tilde{\rho}$	a^{δ_5}	a^{δ_5}
$P/\tilde{P}$	a^{δ_6}	$a^{77\delta_5/40}$
$t/\tilde{t}$	$a^{(\delta_6-3\delta_5)/2}$	$a^{-43\delta_5/80}$
$v/\tilde{v}$	$a^{(\delta_6-\delta_5)/2}$	$a^{37\delta_5/80}$
$T/\tilde{T}$	$a^{(\delta_6-\delta_5)}$	$a^{37\delta_5/40}$
$M/\tilde{M}$	$a^{\delta_6-\delta_5}$	$a^{37\delta_5/40}$
$\alpha/\tilde{\alpha}$	$a^{\delta_6-\gamma\delta_5}$	$a^{(77/40-\gamma)\delta_5}$
$\mathcal{Q}_{0,1}/\tilde{\mathcal{Q}}_{0,1}$	1	1
$\mathcal{Q}_{0,2}/\tilde{\mathcal{Q}}_{0,2}$	0	1

Using the result presented by Falize et al. (2008b), in the general case of optically thin and thick plasma, we demonstrate that we have three effective parameters in global invariance and two or one free parameters in absolute invariance respectively in cases of one and two cooling functions. These results are presented in Table 2. We use the conventions written in Falize et al. (2008b).

According to this result, we see that it is possible to reproduce potentially this particular system in laboratory. Moreover laboratory plasma have a dynamical time which is around nanosecond, thus we must have a cooling time in the same magnitude order. Since Chevalier-Imamura instability developed in the cooling time we should be able to observe experimentally its development in laboratory.

5 Conclusion

In this paper we propose new analytical solutions for the problem of steady radiative shocks which constitutes a generalization of five specific solutions known in literature. We speak about new solutions because we solved the full system taking into account three parameters which are the exponents used in the general cooling function of power-law type. Since the five specific solutions have been found directly introducing respectively in equations the five couples of values for the two first exponents (ϵ, ζ), while $\theta = 0$ is implicit. Of course with our analytical solutions these specific solutions can be retrieved, but also any solutions can be calculated for any values of $(\epsilon, \zeta, \theta)$. Moreover we proceed to the determination of physical quantities which is consistent with numerical evaluation. This simple model allows to interpret the main physical properties of the accretion column steady structure which are usually used in astronomy to determine the main quantities of magnetic cataclysmic variables. Regardless the complex dynamics and stability properties of optically thin plasmas can affect the establishment of stationary regime. Our discussion on the scaling laws of cooling layer allows to affirm that we can potentially reproduce the cooling layer in laboratory. Consequently, the experimental study will permit to explore the unsteady phase, the physics of this unobservable specific zone and test astrophysical model.

Acknowledgements The authors would like to thank Martine Mouchet for several fruitful discussions.

Appendix: Analytic expression

$$\begin{aligned}
&cste + (\gamma+1)\eta^{\epsilon+2}\left[1+\frac{1}{\gamma M^2}\right]^{-\zeta}\frac{\Gamma[\epsilon+2]}{\Gamma[\epsilon+3]}\\
&\quad\times F_{21}\left[\zeta, \epsilon+2; \epsilon+3; \frac{\eta}{1+1/[\gamma M^2]}\right]\\
&\quad-\gamma\eta^{\epsilon+1}\left[1+\frac{1}{\gamma M^2}\right]^{1-\zeta}\frac{\Gamma[\epsilon+1]}{\Gamma[\epsilon+2]}\\
&\quad\times F_{21}\left[\zeta, \epsilon+1; \epsilon+2; \frac{\eta}{1+1/[\gamma M^2]}\right]\\
&\quad=\begin{cases}\kappa_0 \ln[\xi + X_0/x_s]; & \theta=-1\\ \frac{\kappa_0}{\theta+1}[\xi+X_0/x_s]^{\theta+1}; & \theta\neq-1,\end{cases}
\end{aligned}\tag{13}$$

where $\kappa_0 = -(\gamma-1)x_s^{\theta+1}\Lambda_0\rho_0^{\epsilon+\zeta-1}v_s^{2\zeta-3}$ and F_{21} is the Gauss hypergeometric function and Γ is the Euler function.

$$\begin{aligned}
x_s &= \frac{\rho_0 v_s^3}{\Lambda_0(\gamma-1)\rho_0^{\epsilon}[\rho_0 v_s^2]^{\zeta}}\left\{\gamma\left(\frac{2+(\gamma-1)M^2}{(\gamma+1)M^2}\right)^{\epsilon+1}\right.\\
&\quad\times\left[1+\frac{1}{\gamma M^2}\right]^{1-\zeta}\frac{\Gamma[\epsilon+1]}{\Gamma[\epsilon+2]}\\
&\quad\times F_{21}\left[\zeta, \epsilon+1; \epsilon+2; \frac{2+(\gamma-1)M^2}{(\gamma+1)M^2(1+1/[\gamma M^2])}\right]\\
&\quad-(\gamma+1)\left(\frac{2+(\gamma-1)M^2}{(\gamma+1)M^2}\right)^{\epsilon+2}\\
&\quad\times\left[1+\frac{1}{\gamma M^2}\right]^{-\zeta}\frac{\Gamma[\epsilon+2]}{\Gamma[\epsilon+3]}\\
&\quad\left.\times F_{21}\left[\zeta, \epsilon+2; \epsilon+3; \frac{2+(\gamma-1)M^2}{(\gamma+1)M^2(1+1/[\gamma M^2])}\right]\right\},
\end{aligned}\tag{14}$$

$$\begin{aligned}
\Xi &= \left\{\gamma\left[1+\frac{1}{\gamma M^2}\right]\frac{\Gamma(\epsilon)}{\Gamma(\epsilon+1)}\right.\\
&\quad\times F_{21}\left[\zeta, \epsilon; \epsilon+1; \frac{2+(\gamma-1)M^2}{(\gamma+1)M^2(1+1/[\gamma M^2])}\right]\\
&\quad-\left[\frac{2+(\gamma-1)M^2}{M^2}\right]\frac{\Gamma(\epsilon+1)}{\Gamma(\epsilon+2)}
\end{aligned}$$

$$
\times F_{21}\left[\zeta,\epsilon+1;\epsilon+2;\frac{2+(\gamma-1)M^2}{(\gamma+1)M^2(1+1/[\gamma M^2])}\right]\Bigg\}
$$

$$
\Bigg/\Bigg\{\gamma\left[1+\frac{1}{\gamma M^2}\right]\left[\frac{2+(\gamma-1)M^2}{(\gamma+1)M^2}\right]\frac{\Gamma(\epsilon+1)}{\Gamma(\epsilon+2)}
$$

$$
\times F_{21}\left[\zeta,\epsilon+1;\epsilon+2;\frac{2+(\gamma-1)M^2}{(\gamma+1)M^2(1+1/[\gamma M^2])}\right]
$$

$$
-(\gamma+1)\left[\frac{2+(\gamma-1)M^2}{(\gamma+1)M^2}\right]^2\frac{\Gamma(\epsilon+2)}{\Gamma(\epsilon+3)}
$$

$$
\times F_{21}\left[\zeta,\epsilon+2;\epsilon+3;\frac{2+(\gamma-1)M^2}{(\gamma+1)M^2(1+1/[\gamma M^2])}\right]\Bigg\}. \quad (15)
$$

References

Bertschinger, E.: Astrophys. J. **304**, 154 (1986)

Calvet, N., Gulbring, E.: Astron. Astrophys. **509**, 802 (1998)

Chanmugam, G., Langer, S.H., Shaviv, G.: Astrophys. J. **299**, L87 (1985)

Chevalier, R.A., Imamura, J.N.: Astrophys. J. **261**, 543 (1982)

Cropper, M.: Space Sci. Rev. **54**, 195 (1990)

Drake, R.P.: High Energy Density Physics: Fundamentals, Inertial Fusion and Experimental Astrophysics. Springer, Berlin (2006)

Drake, R.P.: Phys. Plasmas **14**, 043301 (2007)

Falanga, M., Bonnet-Bidaud, J.M., Suleimanov, V.: Astron. Astrophys. **444**, 561 (2005)

Falize, E., Bouquet, S., Michaut, C.: J. Phys. Conf. Series **112**, 042016 (2008a)

Falize, E., Bouquet, S., Michaut, C.: Astrophys. Space Sci., Proc. of HEDLA'08 held in St. Louis, MO, USA, 12–15 April 2008, in this issue (2008b). doi:10.1007/s10509-009-9983-z

Imamura, J.N., Wolff, M.T.: Astrophys. J. **355**, 216 (1990)

Kylafis, N.D., Lamb, D.Q.: Astrophys. J. Suppl. Ser. **48**, 239 (1982)

Laming, J.M.: Phys. Rev. E **70**, 057402 (2004)

Michaut, C., Falize, E., Cavet, C., et al.: Astrophys. Space Sci., Proc. of HEDLA'08 held in St. Louis, MO, USA, 12–15 April 2008, in this issue (2008)

Michaut, C., Vinci, T., Boireau, L., et al.: Astrophys. Space Sci. **307**, 159 (2007)

Mignone, A.: Astrophys. J. **626**, 373 (2005)

Remington, B.A., Drake, R.P., Ryutov, D. D.: Rev. Modern Phys. **78**, 755 (2006)

Rybicki, G.B., Lightman, A.P.: In: Radiative Processes in Astrophysics, Chap. 5 (1979)

Saxton, C.J., Wu, K., Pongracic, H., Shaviv, G.: Mon. Not. R. Astron. Soc. **299**, 862 (1998)

Toth, G., Draine, B.T.: Astrophys. J. **413**, 176 (1993)

Warner, B.: Cataclysmic Variables. Cambridge Univ. Press, Cambridge (1995)

Wu, K.: Space Sci. Rev. **93**, 611 (2000)

Wu, K., Chanmugam, G., Shaviv, G.: Astrophys. J. **426**, 664 (1994)

Astrophys Space Sci (2009) 322: 77–84
DOI 10.1007/s10509-008-9966-5

ORIGINAL ARTICLE

Classification of and recent research involving radiative shocks

C. Michaut · E. Falize · C. Cavet · S. Bouquet · M. Koenig · T. Vinci · A. Reighard · R.P. Drake

Received: 17 June 2008 / Accepted: 9 December 2008 / Published online: 23 December 2008

Abstract Radiative shocks (RS) occur in astrophysical systems and in high-energy density laboratory experiments. Aided by three dimensionless parameters, we propose a classification of RS into four types, integrating previous work that has focused independently on optical depth and on Mach number. Specific terms, such as a cooling function, a radiative flux, or radiative energy and pressure must be added to the Euler equations in order to model these various kinds of shocks. We examine how these terms correspond to the radiative classification regimes. In astrophysics, observed RS arise generally in optically thin material. Thus, radiation escapes without interaction with the surrounding gas, except perhaps to ionize it, and the energy loss in such shocks can be modeled by a cooling function Λ. In this case only the post-shock region is structured by the radiation cooling. We found the analytical solution for hydrodynamic equations including $\Lambda \propto \rho^{\varepsilon} P^{\zeta} x^{\theta}$ for arbitrary values of ε, ζ, θ. This is a completely new result. An application of this calculation for the accretion shock in cataclysmic variables of polar type is given in astrophysical terms. We also draw a parallel between RS experiments performed using the LULI2000 laser facility, in France and the Omega laser Facility, in USA. RS developed in these laboratories are more or less optically thick. These high-Mach number RS present a radiative precursor.

Keywords Radiative shocks · High-power lasers · Cooling function

C. Michaut (✉) · E. Falize · C. Cavet · S. Bouquet
LUTH, Observatoire de Paris, CNRS, Université Paris-Diderot, 92190 Meudon, France
e-mail: Claire.Michaut@obspm.fr

E. Falize · S. Bouquet · T. Vinci
Département de Physique Théorique et Appliquée, CEA-DIF, 91297 Arpajon, France

M. Koenig · T. Vinci
LULI, Ecole Polytechnique, CNRS, Université Paris 6, 91128 Palaiseau, France

A. Reighard · R.P. Drake
Department of Atmospheric Oceanic and Space Sciences, University of Michigan, Ann Arbor, MI 48109, USA

A. Reighard
Lawrence Livermore Laboratory, Livermore, CA 94550, USA

1 Introduction

Shock waves are ubiquitous in astrophysics, arising in most circumstances where an energy release or a collision creates a supersonic interaction. Such shock waves are often difficult to model theoretically or computationally. Important complications include radiative effects (Zeldovich and Raizer 1966), collisionless dynamics (Sagdeev 1966), and either axial (Chevalier and Imamura 1982) or lateral (Vishniac 1983) instabilities. These complications warrant firstly an effort to classify such shock waves in order to identify their fundamental physical regimes, and secondly the development of experiments that can directly probe the shock physics and can provide validation of the complex computer models used to interpret astrophysical observations. In the specific area of radiative shocks (RS), of interest here, both classification and experiments have begun to develop only recently. The present paper discusses our recent progress in these areas, and specifically presents new progress on the general classification of RS.

In the first part of this paper, a new classification of different radiative regimes is presented in order to understand how to use radiative terms in the radiative-hydrodynamic equations for specific applications. In fact, astrophysical RS occur a very wide range of optical depth, radiation energy, energy density, and collisionality. Therefore, RS can usefully be divided in four large classes: RS in optically thin media, a hybrid radiative regime in media that are in some regions optically thick, and an optically thick regime that may or may not be radiation-dominated. But most observed astrophysical RS are optically thin with low radiation energy densities. In addition, the fact that whether a shock is collisional or collisionless has almost no relevance to its radiative hydrodynamic behavior and, as a consequence, in this paper we treat the shock transition as abrupt by comparison to the radiative cooling distance. Our results are relevant to all collisional RS and to collisionless shocks unless the spatial scale of effects such as heating by reflected particles becomes comparable to the radiative cooling distance.

The second part discusses experiments to study radiative shocks, including primarily a short review of some RS experiments which point out how the two regions in front of and behind the shock, which are respectively pre-shock (upstream) and post-shock (downstream) regions, are altered by radiation and are experimentally produced and probed. We have performed two types of experiments use high-power lasers, but due to experimental limitations they can not diagnose entire RS showing structure both in pre- and post-shock regions. Taking advantage of these two types of experiments, we are able to propose, for the future, complete experiments on RS using new experimental facilities such as LIL in France.

In addition, we discuss recent theoretical work which gives recently new analytic solutions specifically for RS in optically thin media. In this case, radiation is modeled by a cooling function Λ representing an energy loss. Therefore, this theoretical work is motivated by potential experiments to measure the downstream flow characteristics. Moreover, using the same formalism, we attempt to predict the precursor length in the upstream flow: in that case, we will try to find an appropriate expression of Λ to model a radiation flux, propagating towards the upstream flow in the precursor, instead of the standard cooling function.

The connection between experimental situations and astrophysical ones requires scaling laws. Previous work has given scaling laws in pure hydrodynamic phenomena, sometimes with a magnetic field and sometimes with a cooling function (Ryutov et al. 1999, 2000, 2001; Murakami and Iida 2002). We have extended these scaling laws in the case of the radiative regime.

In the final part, we show how to apply these analytical results to an astrophysical situation. The chosen example is the accretion shock in cataclysmic variables of the polar type.

2 Theoretical considerations

We discuss the structures of RS to clarify the various types that arise in astrophysics. RS are in all cases strong enough to emit a significant fraction of the mechanical energy driving the shock as radiation, but the interaction between radiation and matter depends on the state of matter. We can distinguish optically thin RS, an intermediate class that we label as hybrid RS (HRS), and optically thick RS, that split into two families according to the relative importance of radiation flux and radiation energy density. We describe shocks in which radiation energy density matters as radiation-dominated shocks. On the other hand, in literature, one can often find that the optical depth τ is a key dimensionless parameter used to categorize the nature of RS. Recall that τ is the integral of the absorption per unit length along a path through the physical system, approximately equal to the ratio between the characteristic length of the system and a characteristic photon mean free path. According to us (see below) this parameter is not sufficient to completely characterize RS. It is important to classify RS in order to know precisely which physical ingredients have to be included in equations in order to design experiments and also to understand the shock physics. Moreover, from our current knowledge and progresses, codes coupling radiative transfer and hydrodynamics are so difficult to develop from both physical and numerical viewpoints that approximations are required. In other words, it is not yet possible to treat simultaneously several options (physical situations) with a single formalism (or model). For that reason, laboratory experiments are of great interest and help to go further in a unique (global) direction and in the elaboration of a general model and testing it in various conditions (i.e. from optically thin to optically thick regimes with intermediate situations).

2.1 New classification of RS

In this work, we are interested in generalizing the classes of RS beyond the previous specific cases discussed by Bouquet et al. (2000) and Drake (2005). In order to better delimit these regimes, we propose to use simultaneously three dimensionless parameters: the cooling parameter χ, a modified Boltzmann number Bo and a ratio R (see below for definitions). This approach has not been taken previously. After the introduction of these parameters, we discuss specific cases.

2.1.1 Key parameters

We introduce three key parameters to help in classifying RS. First of all we use a cooling parameter, χ, then a modified Boltzmann number, Bo, and finally the ratio R of material to radiation energy densities. These parameters can help determine which relevant quantities, such as a cooling function

Λ, a radiative flux F_{rad}, radiation energy density E_{rad}, and radiation pressure P_{rad} must be added to the Euler equations in a given case and thus provide a quantitative basis for classification.

The cooling parameter, χ, compares the time t_{cool} from which the cooling begins to have an effect on the material with the characteristic time L/v of the hydrodynamics (L is a characteristic length, v is a characteristic velocity). The parameter χ can be expressed in terms of the post-shock parameters density ρ, velocity v, and temperature T as

$$\chi \equiv \frac{t_{cool}}{L/v} = \frac{1}{\gamma - 1}\frac{k_B}{\mu m_H}\frac{\rho v T}{\Lambda(\rho, T) L}, \tag{1}$$

where γ is the polytropic index, k_B the Boltzmann constant and μm_H the reduced mass. We note that the cooling function Λ and the opacity or specific opacity can be related to one another (Drake 2006). This definition is only valid for optically thin plasmas, because the use of a cooling function in the energy equation is a very simple approximation to the effects of radiative transfer that becomes invalid in systems that are not optically thin. For optically thick plasmas, $\Lambda \times L$ should be replaced by F_{rad} and gives:

$$\chi \equiv \frac{t_{cool}}{L/v} = \frac{1}{\gamma - 1}\frac{k_B}{\mu m_H}\frac{\rho v T}{F_{rad}}. \tag{2}$$

If χ is very large ($\chi \gg 1$), the shock is not radiative. But if χ is very low ($\chi \ll 1$), radiation cooling dominates, therefore cooling must be taken into account.

A second tool is the modified Boltzmann number Bo which compares the material energy flux to the radiative flux. Using the same variables, we compare Bo to one to determine if the radiative flux plays a role in the modeling:

$$Bo \equiv \frac{(\rho e + P)\|\overrightarrow{v}\|}{\|\overrightarrow{F_{rad}}\| f(\tau)} = \frac{\gamma}{\gamma - 1}\frac{k_B}{\sigma \mu m_H}\frac{\rho v}{T^3 f(\tau)}, \tag{3}$$

where ρe and P are the internal gas energy and the thermal pressure, respectively, where σ is the Stefan-Boltzmann constant, and where τ is the optical depth of the region heated by the shock. The function $f(\tau)$ is the ratio of actual radiative flux to σT^4, accounting for optical depth and geometry. For very thin layers, $f(\tau)$ equals τ while for very thick layers it approaches 1. For planar layers the general formula (Drake 2006) is $f(\tau) = 1 + e^{-\tau}(\tau - 1) - \tau^2 E_1(\tau)$, where $E_1(\tau)$ is the exponential integral of order 1. If $Bo \ll 1$ then energy transport by radiative fluxes is significant and must be included in any theory.

A third useful parameter is R, which is the ratio of the internal gas energy density ρe to the radiative energy density E_{rad}, modified from that discussed by Mihalas and Weibel-Mihalas (1999) to allow for effects of optical depth. As E_{rad} and P_{rad} have the same order of magnitude, this ratio allows one to determine if they play a role in the shock structure and, consequently if the shock is radiation-dominated ($R \ll 1$) or not ($R \gg 1$):

$$R \equiv \frac{\rho e}{E_{rad} g(\tau)} = \frac{1}{\gamma - 1}\frac{1}{a_R}\frac{k_B}{\mu m_H}\frac{\rho}{T^3 g(\tau)}, \tag{4}$$

where a_R is the first radiation constant and $g(\tau)$ is the ratio of actual radiation energy density to $a_R T^4$, accounting for optical depth and geometry. For very thin layers, $g(\tau)$ equals τ while for very thick layers it approaches 1. For planar layers the general formula is $g(\tau) = 1 - e^{-\tau} + \tau E_1(\tau)$.

In the following, we are going to use these three generalized radiation dimensionless numbers to classify the various types of RS.

2.2 Specific cases

2.2.1 Optically thin RS

Our first family, which we will label "optically thin RS", corresponds to RS in media that are optically thin both upstream and downstream, which is the usual situation in observed astrophysical shocks. Such shocks have $\chi \ll 1$. In this regime the shocked matter is negligibly affected by emission from any part of the system, shocked or unshocked. (Even so, photons from the shocked matter may ionize and heat the upstream or downstream material, if it is or becomes neutral. This however does not affect the shock dynamics.) Theoretically such systems can be treated as having no coupling between matter and radiation. The effects of radiation can be modeled by a cooling function $\Lambda(\rho, T)$ which depends both on the density ρ and the temperature T. This function represents the energy loss due to escaping radiation. In these RS, the shock front is governed by classical Rankine-Hugoniot relations. In addition, the analysis of the RS discussed here is not invalidated under circumstances that may require modification of the Rankine-Hugoniot relations to incorporate effects such as magnetic fields or non-isotropic particle distributions. Behind the shock, the density increases as shown in Fig. 1. In the optically thin limit, the

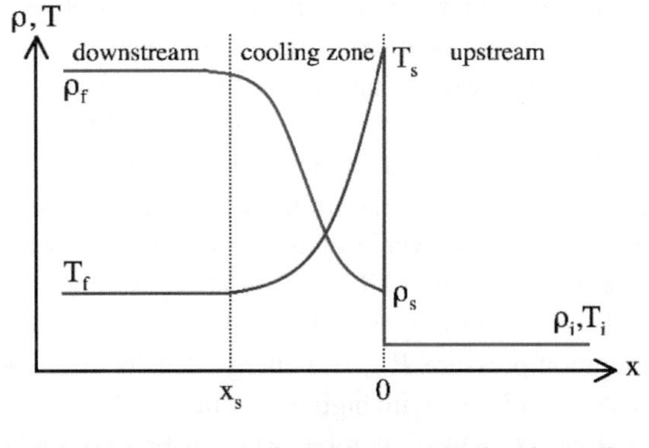

Fig. 1 Schematic structure of a RS in an optically thin medium with radiative losses; the subscripts i, s and f stands respectively of "initial" (upstream), "shock" and "final" (downstream)

temperature T_f will equal T_i and both will be set by the balance of radiative cooling with heating from all surrounding radiation sources. A complete analytical study of the structure of these shocks is described in Falize et al. (2008c). Optically thin RS might in principle have $Bo \gg 1$ or even $R \gg 1$ but because of the lack of radiation-matter coupling it is sufficient to treat them using the Euler equations with the addition of a cooling function.

2.2.2 *Hybrid radiative shocks*

A second category of RS, namely hybrid radiative shocks (HRS, see beginning of Sect. 2) develops when the medium becomes optically thick in one but not both sides of the shock discontinuity. In the optically thick regions, we have $\chi \ll 1$ and $Bo \ll 1$. The typical case, and the case in current experiments, is optically thick downstream and optically thin upstream. Such shocks have a similar structure as shown in Fig. 1, except that T_f is much higher than T_i. Present-day experiments with $Bo \ll 1$ that produce sufficient radiation to ionize all the upstream matter are in this regime. Strong enough HRS could have $R \ll 1$ and thus be radiation-dominated. Shocks breaking out of supernovae would initially be in this regime. Present-day laboratory shocks are not.

Our third and fourth classes exist when the medium becomes optically thick both upstream and downstream of the density jump. In this case the emitted radiation interacts with the gas in all directions. When $\chi \ll 1$ (see (2)) one will find $Bo \ll 1$ and one must take into account the radiative flux F_{rad} in the equations. Depending on the Mach number, these RS are of two types (Bouquet et al. 2000).

2.2.3 *Optically-thick, flux-dominated RS*

For the first type, with sufficiently low Mach number ($M < M_{crit}$), the radiative energy E_{rad} and pressure P_{rad} do not need to be included because $R \gg 1$. We call this class of shock "optically-thick, flux-dominated RS". In these shocks a precursor forms due to the upstream absorption of the radiation emitted from the downstream flow. Some present-day experiments have optically thick upstream precursors (Koenig et al. 2006; Michaut et al. 2007) because their shocks produce radiation that does not propagate through the upstream material, including those produced to date using the high-power laser at the LULI laboratory. Until now, the energy of the LULI facility allows the production of shocks in which only F_{rad} has to be taken into account (the radiation pressure P_{rad} and energy density E_{rad} can be neglected). For lasers with higher energies, higher shock velocities (and also Mach number M) can be expected and, therefore, P_{rad} and E_{rad} cannot be anymore neglected as their magnitudes increase with M (see next subsection).

2.2.4 *Optically-thick, flux-dominated RS*

If the Mach number is greater ($M > M_{crit}$), then $R \ll 1$ and one cannot neglect E_{rad} and P_{rad} in the equations. As radiative energy density and pressure are of the same order of magnitude, these terms appear always together. We classify these types of RS as "optically-thick, radiation-dominated shocks." In such shocks, E_{rad} and P_{rad} have a major role in determining the shock structure. Note that the optical depth τ is not always sufficient to define the shock structure because this parameter does not separate these two classes of optically thick RS. The relevant parameter in this case is the Mach number. When optical depth is very large $\tau \gg 1$, the diffusion approximation is correct to model the radiative transfer except near the density jump (Bouquet et al. 2000; Drake 2007). In this case, terms proportional to $a_R \times T^4$ appear in equations instead of E_{rad}, P_{rad}.

In conclusion regarding classification, the knowledge of the three parameter χ, Bo and R (or ρ, v and T), both in front of and behind the shock front are required correctly classify the regimes of RS.

3 Review of some experiments

We now turn from the theoretical classification of RS toward some experimental results. It might seem a simple task to devise experiments that addressed the various regimes described in the previous section, or at least to know what might be possible in experiments. In fact neither is simple, for several reasons. The primary one is that the laboratory study of RS is relatively new and difficult, and the invention of new approaches and techniques is very much to be expected. Beyond this, the realm of the possible changes frequently as the capabilities of facilities and diagnostics evolve.

We summarize below results from two extended sequences of experiments. In these experiments, it has proven feasible to produce shock waves that become many radiation mean free paths thick downstream of the shock and are few radiation mean free paths thick upstream of the shock. The length of the upstream precursor then depends on the radiation flux. Such shocks are in the dimensionless regime of shock waves emerging from thick, dense environments such as supernovae. Such astrophysical shocks have long been calculated (Klein and Chevalier 1978; Ensman and Burrows 1992) and are now beginning to be observed (Soderberg et al. 2008). The calculation of these shocks involves numerical, radiative-hydrodynamics codes that are quite complex essentially due to the required coupling between radiative transfer and hydrodynamics. Such codes are also required to understand stellar evolution where RS are always involved. The experiments will at minimum be useful for validating such codes, but also have the potential to reveal new physics about these shocks, as the numerical models do not fully reflect reality and as within the scope of modeling large-scale astrophysical systems, such shocks are often very poorly resolved.

These recent experiments specifically illustrate that one can diagnose different zones in RS: specifically the pre-shock and post-shock regions. The first RS experiment was achieved in 1986 (Bozier et al. 1986) using the Octal laser at CEA in France. It revealed the existence of a radiative precursor. Then, similar experiments were performed on the LULI 6 beams facility (Fleury et al. 2002; Bouquet et al. 2004) and LULI2000 facility (Koenig et al. 2006; Michaut et al. 2007) at LULI. These latest experiments contain more simultaneous diagnostics allowing to measure several physical quantities (Koenig et al. 2005; Vinci et al. 2005). Another type of RS experiments have been performed using the Omega laser facility (Reighard et al. 2006a, 2007).

3.1 Observation of the precursor region

Figure 2 is an image of a RS with its radiative precursor obtained using the LULI2000 laser. The 2D behavior and the limited extent of the precursor are evident. A rise of the electronic density was measured ahead of the shock i.e. in the upstream region. The corresponding ionization is due to the absorption of the radiation emitted by the shock front.

The precursor formation is also found by 2D simulations (DUED, code elaborated by Atzeni et al. 2005 and modified by Vinci et al. 2006). In Fig. 3, the precursor reaches the steady-state 2 ns after the shock formation in xenon and its length is $\sim 200\ \mu$m.

3.2 Observation of the downstream region

Figure 4 shows a radiographic image of the downstream shock structure in xenon obtained using the Omega laser. The dark features correspond to the dense xenon gas, which is moving to the right. The shock front is at the right edge of the observed dark area and the beryllium/xenon interface is at the left edge. The shocked material exhibits spatial modulation. In this radiographic image, the beryllium driver is transparent. The post-shock zone is the part between the beryllium/xenon interface and the shock front at the right edge of this area. Its length is ~ 100 µm. In another experiment, in argon, the leading edge of the shocked region was probed using UV Thomson scattering (Reighard et al. 2006b). The measured, instantaneous shock velocity was 110 ± 7 km/s. This system was simulated in 1D using the HYADES code (Larsen and Lane 1994). Figure 5 shows the results. This calculation shows the profiles behind the shock front. We see, in blue line (labelled "mass density"), how the density increases behind the shock as cooling removes energy from the shocked plasma. In the same region, the temperature is decreasing away from the shock front in the post-shock cooled region. In these two types of experi-

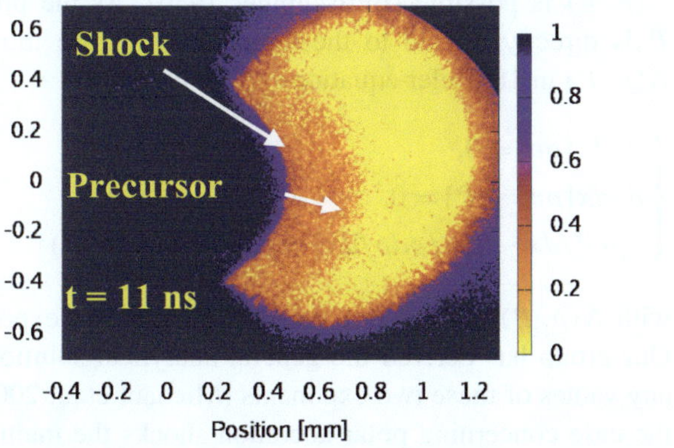

Fig. 2 Optical emission from a xenon shock and its precursor obtained using LULI2000 ($E \sim 300$ J, $v_{shock} \sim 60$ km/s, $\rho_i = 10^{-3}$ g/cm^3), observed using a gated optical imager

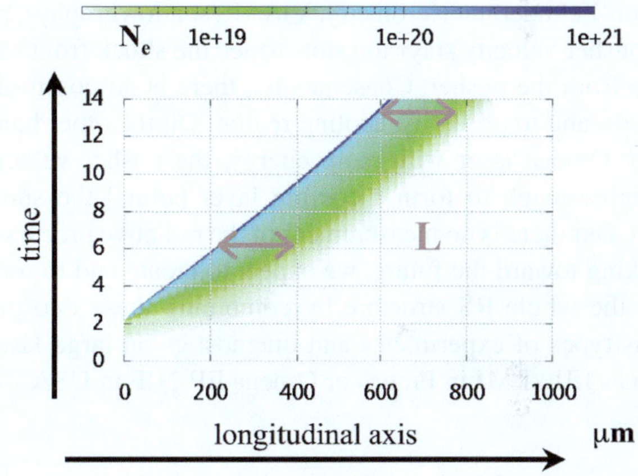

Fig. 3 2D-simulation of RS and its precursor in same conditions as in Fig. 2; the symbol N_e stands for the electronic density; the ionization Z^* varies from 0 to 8 as one moves from the head of the precursor (*right border* of the precursor) to the shock jump (*left border*, almost a straight line)

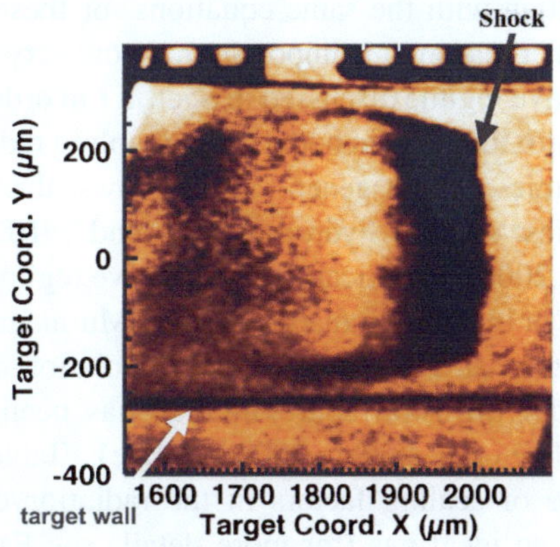

Fig. 4 Radiographic image of a xenon RS ($v_{shock} \sim 140$ km/s) driven by the Omega laser

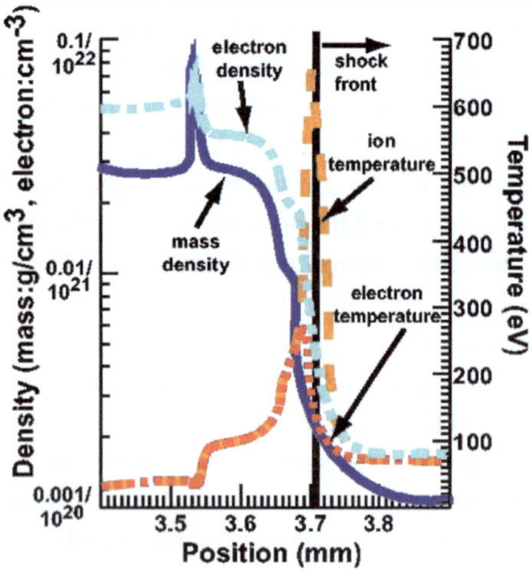

Fig. 5 Simulated RS structure in argon (1D, HYADES), showing profiles in the cooling zone

ments on laboratory RS, note that using LULI2000, one can generate a precursor and measure it with visible diagnostics (Mach-Zehnder interferometry, VISAR, shadowgraphy), but the pusher velocity stays too slow to see the shock front separate from the pusher. Consequently, there is no possibility to form and to observe a cooling region. On the other hand, using Omega laser with more energy, the pusher velocity is high enough to form a cooling layer behind the shock front. But there is no measurement of the radiative precursor. Looking toward the future, we expect to create and to measure the whole RS structure by combining these two previous types of experiments and diagnostics on large lasers such as LIL, LMJ in France or Omega EP, NIF in USA.

4 Connection between laboratory and astronomy

Scaling laws are tools to connect an experimental situation with an astrophysical one. But to pass from a system to the other, very strong requirements are imposed: the equations have to be invariant! Explicitly, models of RS must be strictly written with the same equations for these two systems. For that reason, it is important to know very well what is the radiative regime discussed in Sect. 2.1 in order to make correct approximations in both cases, coupling radiation and hydrodynamics. About general scaling laws, there are previous studies: in pure hydrodynamics and MHD (Ryutov et al. 1999, 2000), in optically thin radiative regime (Ryutov et al. 2001) or in hybrid radiative regime (Murakami and Iida 2002; Falize et al. 2008d). Recently, a radiation-dominated regime under diffusion approximation has been also discussed (Falize et al. 2008b, 2008d, 2008e). Table 1 shows an example of scaling factors in the radiation-dominated regime for an ideal gas (for more details see Falize et al. 2008d). In these relations, parameters m and n represent exponents used by Falize et al. (2008e) in the description of

Table 1 Radiative hydrodynamic scaling laws in optically thick ideal gas (where $\alpha = P_{th}\rho^{-\gamma}$); laboratory and astrophysical quantities are respectively written with and without symbol "tilde"

Physical ratio	Scaling factor
$r/\tilde{r}$	$a^{([m+1/2]+[n-5]/3)\delta_5}$
$t/\tilde{t}$	$a^{(m+1+[n-7]/3)\delta_5}$
$v/\tilde{v}$	$a^{\delta_5/6}$
$\rho/\tilde{\rho}$	a^{δ_5}
$P_{th}/\tilde{P}_{th}$	$a^{(4/3)\delta_5}$
$T/\tilde{T}$	$a^{\delta_5/3}$
$E_{rad}/\tilde{E}_{rad}$	$a^{(4/3)\delta_5}$
$F_{rad}/\tilde{F}_{rad}$	$a^{(3/2)\delta_5}$
$P_{rad}/\tilde{P}_{rad}$	$a^{(4/3)\delta_5}$
$\alpha/\tilde{\alpha}$	$a^{(4/3-\gamma)\delta_5}$

the radiative conductivity ($\kappa_{rad} \propto \rho^m T^n$) (see also Drake 2006). Therefore m and n values are imposed by physical processes. The parameter δ_5 is free in order to make the experiment representative of an astrophysical configuration. Its value is calculated, for example, taking into account the density ratio $\rho/\tilde{\rho}$ between the astrophysical system and the experimental one. Once it is determined, the scaling factor for all the other astrophysical and experimental quantities can be derived. These scaling laws provide a rigorous and correct procedure to prove the relevance of a specific laboratory experiment to a specific astrophysical system.

5 Accretion shock in cataclysmic variable of polar type

Now, we apply the RS theory to well known astrophysical problem and for the first time, we derive the general analytical solution. In astrophysics, RS occur in all radiative regimes, but they are easier to observe in optically thin media. In this case, RS modeling with a cooling function $\Lambda(\rho, T)$ is possible (Bertschinger 1986). As the pressure P is directly related to the temperature T, we introduce $\Lambda(\rho, P)$ in 1D-Euler equations:

$$\begin{cases} d/dx[\rho v] = 0, \\ d/dx[\rho v^2 + P] = 0, \\ v[dP/dx - \gamma P/\rho d\rho/dx] = -(\gamma - 1)\Lambda(\rho, P) \end{cases} \tag{5}$$

with $\Lambda(\rho, P) = \Lambda_0 \rho^{\epsilon} P^{\zeta}$, ϵ and ζ are arbitrary exponents. Our group has derived the general analytical solutions for any values of these two exponents (Michaut et al. 2008). In the case concerning polar accretion shocks the main cooling process is bremsstrahlung, which gives following value for exponents: $\epsilon = 3/2$, $\zeta = 1/2$. A polar is a binary stellar system in interaction, in which a white dwarf presenting a very strong magnetic field ($B > 1000$ T) accretes matter from its companion star. The matter falls along magnetic field lines with a free-fall velocity. When the falling matter collides the white dwarf surface a reflected shock appears, which moves back along the accretion column evolving toward quasi-steady state. The gas is cooled by radiation in the region located between the white dwarf surface and the shock front. The expression of the cooling length x_s can be obtained rigourously in the general case by solving (5) analytically for any exponents ϵ and ζ in the cooling function Λ (see calculation details in Falize et al. 2008a). Once this general solution is obtained we can restrict to the case of bremsstrahlung ($\epsilon = 3/2$ and $\zeta = 1/2$) since it is relevant for polar type stars. On the other hand, astronomical observations give an estimation of the accretion luminosity L_{acc}, the accretion rate $\dot{m}$, the white dwarf mass M_{wd}, the white dwarf radius R_{wd} and, therefore, they are suitable variable to express x_s. We find:

$$x_s = 7.60 \times 10^4 \left[\frac{\dot{m}}{40}\right]^{-1} \left[\frac{M_{wd}}{0.5 M_\odot}\right]^{3/2} \left[\frac{R_{wd}}{10^7}\right]^{-3/2} \quad (6)$$

where x_s, $\dot{m}$ and R_{wd} are expressed in MKSA. We recover quite well the result of Wu et al. (1994) since from our general expression we obtain the numerical coefficient 7.60 instead of 7.59.

Furthermore, in previous work (Chanmugam et al. 1985; Imamura 1985; Drake 2006), it has been shown that Λ can be written with a x-axis dependance:

$$\Lambda(\rho, P, x) = \Lambda_0 \rho^\epsilon P^\zeta (x + x_0)^\theta \quad (7)$$

where θ is a third exponent describing the spatial dependence of the cooling. With this form of Λ, we found analytical solutions for (5) and for any exponent values (Falize et al. 2008c; Michaut et al. 2008). The interest of choosing a spatial dependance lies in the fact that we can also take into account more types of cooling, as cyclotron cooling for example (Chanmugam et al. 1985). Of course, we can show that (6) is a particular case of the general solution with $\theta = 0$. Thus we have exhibited that our general solution for any exponents $(\epsilon, \zeta, \theta)$ permits to recover the particular solution given by Wu et al. (1994) in the case of an accretion shock in a polar. Furthermore the general solution allows to study many astrophysical cases with different kinds of cooling processes.

Strengthened by this theoretical approach, we are able to propose a new experiment (scheduled in 2009 on the LULI2000), in which we will try to discriminate the main cooling process and its major effect on the stability of the accretion column.

6 Conclusion

In order to study RS, we are pursuing a complete approach based on experiments, theory and numerical simulations. The connection from laboratory to astrophysics is developed by scaling laws. We developed adapted ones for the case of the radiation-dominated regime. Modelling RS occurring in laboratory or in the Universe is possible under the condition that the radiative regime is very well known so that correct approximations are made. It is for this reason that we have described a radiative classification aided by three parameters (χ, Bo and R). We have shown that we can create and diagnose the radiative precursor ahead the shock front and the cooling zone behind it. We are now equipped to diagnose the entire shock structure, probing simultaneously post- and pre-shocks regions using new generation of high-power lasers.

References

Atzeni, S., Schiavi, A., Califano, F., Cattani, F., Cornolti, F., Del Sarto, A., Lisseikina, T., et al.: Comput. Phys. Commun. **169**, 153 (2005)

Bertschinger, E.: Astrophys. J. **304**, 154 (1986)

Bouquet, S., Teyssier, R., Chièze, J.-P.: Astrophys. J. Suppl. Ser. **127**, 245 (2000)

Bouquet, S., Stehlé, C., Koenig, M., Chièze, J.P., Benuzzi-Mounaix, A., Batani, D., et al.: Phys. Rev. Lett. **92**, 225001 (2004)

Bozier, J.C., Thiell, G., Le Breton, J.P., Azra, S., Decroisette, M., Schirmann, D.: Phys. Rev. Lett. **57**, 1304 (1986)

Chanmugam, G., Langer, S.H., Shaviv, G.: Astrophys. J. **299**, L87 (1985)

Chevalier, R.A., Imamura, J.N.: Astrophys. J. **261**, 543 (1982)

Drake, R.P.: Astrophys. Space Sci. **298**, 49 (2005)

Drake, R.P.: High Energy Density Physics: Fundamentals, Inertial Fusion and Experimental Astrophysics. Springer, Berlin (2006)

Drake, R.P.: Phys. Plasmas **14**, 043301 (2007)

Ensman, L., Burrows, A.: Astrophys. J. **393**, 742 (1992)

Falize, E., Michaut, C., Bouquet, S., Cavet, C.: In: Bouvier, J., Chalabaev, A., Charbonnel, C. (eds.) SF2A-2007: Proc. of the Annual meeting of the French Society of Astronomy and Astrophysics, Grenoble, France, 2–6 July 2007, p. 469. EdP-Sciences (2008a). arXiv:0804.3238F

Falize, E., Bouquet, S., Michaut, C.: In: Bouvier, J., Chalabaev, A., Charbonnel, C. (eds.) SF2A-2007: Proc. of the Annual meeting of the French Society of Astronomy and Astrophysics, Grenoble, France, 2–6 July 2007, p. 473. EdP-Sciences (2008b)

Falize, E., Michaut, C., Cavet, C., Koenig, M., Loupias, B., Ravasio, A., Gregory, C.D., Bouquet, S.: Analytical structure of steady radiative shocks in magnetized cataclysmic variables. In: Proc. of HEDLA'08 held in St Louis, MO, USA, 12–15 April 2008. Accepted in Astrophys. Space Sci. (2008c)

Falize, E., Bouquet, S., Michaut, C.: Scaling laws for radiative and magnetic fluids. In: Proc. of HEDLA'08 held in St Louis, MO, USA, 12–15 April 2008. Accepted in Astrophys. Space Sci. (2008d)

Falize, E., Bouquet, S., Michaut, C.: J. Phys. Conf. Ser. **112**, 042016 (2008e)

Fleury, X., Bouquet, S., Stehlé, C., Koenig, M., Batani, D., Benuzzi-Mounaix, A., Chièze, J.P., et al.: Laser. Part. Beams **20**, 263 (2002)

Imamura, J.N.: Astrophys. J. **296**, 128 (1985)
Klein, R.I., Chevalier, R.A.: Astrophys. J. **223**, L109 (1978)
Koenig, M., Vinci, T., Benuzzi-Mounaix, A., Lepape, S., Ozaki, N., Bouquet, S., Boireau, L., et al.: Astrophys. Space Sci. **298**, 69 (2005)
Koenig, M., Vinci, T., Benuzzi-Mounaix, A., Ozaki, N., Ravasio, A., Boireau, L., Michaut, C., et al.: Phys. Plasmas **13**, 056504 (2006)
Larsen, J.T., Lane, S.M.: J. Quant. Spectrosc. Radiat. Transfer **51**, 179 (1994)
Michaut, C., Vinci, T., Boireau, L., Koenig, M., Bouquet, S., Benuzzi-Mounaix, A., Ozaki, N., et al.: Astrophys. Space Sci. **307**, 159 (2007)
Michaut, C., Falize, E., Cavet, C., Bouquet, S., Koenig, M., Vinci, T., Loupias, B.: J. Phys. Conf. Ser. **112**, 042013 (2008)
Mihalas, D., Weibel-Mihalas, B.: Foundations of Radiation Hydrodynamics. Dover, New York (1999)
Murakami, M., Iida, S.: Phys. Plasmas **9**, 2745 (2002)
Reighard, A.B., Drake, R.P., Dannenberg, K.K., Kremer, D.J., Grosskopf, M., Harding, E.C., Leibrandt, D., et al.: Phys. Plasmas **13**, 08290 (2006a)
Reighard, A.B., Drake, R.P., Donjakowski, T., Grosskopf, M., Dannenberg, K.K., Froula, D., Glenzer, S., et al.: Rev. Sci. Instrum. **77**, 105E04 (2006b)
Reighard, A.B., Drake, R.P., Mucino, J.E., Knauer, J.P., Busquet, M.: Phys. Plasmas **14**, 056504 (2007)
Ryutov, D.D., Drake, R.P., Kane, J., Liang, E., Remington, B.A., Wood-Vasey, W.M.: Astrophys. J. **518**, 821 (1999)
Ryutov, D.D., Drake, R.P., Remington, B.A.: Astrophys. J. Suppl. Ser. **127**, 465 (2000)
Ryutov, D.D., Remington, B.A., Robey, H.F., Drake, R.P.: Phys. Plasmas **8**, 1804 (2001)
Sagdeev, R.Z.: Rev. Plasma Phys. **4**, 23 (1966)
Soderberg, A.M., et al.: Nature **453**, 469 (2008)
Vinci, T., Koenig, M., Benuzzi-Mounaix, A., Boireau, L., Bouquet, S., Leygnac, S., Michaut, C., et al.: Astrophys. Space Sci. **298**, 333 (2005)
Vinci, T., Koenig, M., Benuzzi-Mounaix, A., Ozaki, N., Ravasio, A., Boireau, L., Michaut, C., et al.: J. Phys. IV **133**, 1039 (2006)
Vishniac, E.T.: Astrophys. J. **274**, 152 (1983)
Wu, K., Chanmugam, G., Shaviv, G.: Astrophys. J. **426**, 664 (1994)
Zeldovich, Y.B., Raizer, Y.P.: Physics of Shock Waves and High-Temperature Hydrodynamic Phenomena. Dover, New York (2002). Academic Press (1966)

Astrophys Space Sci (2009) 322: 85–90
DOI 10.1007/s10509-008-9936-y

ORIGINAL ARTICLE

Simulations of the supersonic radiative jet propagation in plasmas

X. Ribeyre · P. Nicolaï · S. Galera · V.T. Tikhonchuk

Received: 4 June 2008 / Accepted: 24 October 2008 / Published online: 6 November 2008

Abstract Supersonic plasma jets are ubiquitous in astrophysics. Our study focus on the jets emanated from Herbig-Haro (HH) objects. They have velocities of a few hundred km/s and are extending over the distances more than a parsec. Interaction of the jets with surrounding matter produces two specific structures in the jet head: the bow shock and the Mach disk. The radiative cooling of these shocks affects strongly the jet dynamics.

A tool to understand the physics of these jets is the laboratory experiment. A supersonic jet interaction with surrounding plasma was studied on the PALS laser facility. A collimated high-Z plasma jet with a velocity exceeding 400 km/s was generated and propagated over a few millimeters length. Here we report on study the effect of radiative cooling on the head jet structure with a 2D radiative hydrodynamic code. The simulation results demonstrated the scalability of the experimental observations to the HH jets.

Keywords Hydrodynamics · Plasmas jets · Herbig-Haro object

1 Introduction

Numerical simulations of high Mach number jet propagation in ambient medium have been reported in Blondin et al. (1990), Stone and Norman (1993). It was shown that the radiative cooling modifies the jet structure and its evolution. In particular, it shows a tendency to enhance the jet collimation and density. In the case of a strong cooling the material is accumulated in the jet head and forms a plow of a cold gas. These simulations are relevant for stellar jets and HH objects (Reipurth and Bally 2001). However, the validity of physical assumptions and numerical simulations has to be verified. Laboratory experiments, offer such a possibility if the scaling conditions are fulfilled.

Coker et al. (2007) report on experiments that were conducted on the high-energy laser facility to study the supersonic jet propagation in plasmas. The jet deflection by a high density ambient medium cloud (Coker et al. 2007), allowed by using the appropriate scaling to model a larger astrophysical jet such as HH110 where the outflow is colliding with a molecular cloud. However, in this experiment the radiative cooling was not important. Following the Blondin's work (Blondin et al. 1990), we investigated experimentally the radiative cooling effect on a high-Z supersonic jet propagation in uniform plasmas (Nicolaï et al. 2008).

In this paper, we present the results of numerical simulations of the supersonic jet propagation in plasmas using the 2D radiative code CHIC for the experimental conditions of Nicolaï et al. (2008). Simulations of the adiabatic and radiative jets for various density ratios between the jet and the ambient medium are performed. The bow shock and Mach disk evolution depend strongly on the radiation cooling effect that can be controlled by choice of jet material. Finally, radiative hydrodynamic scaling to the astrophysical objects is demonstrated.

2 Simulation code description and initial conditions

The numerical simulations were performed with the CHIC code (Maire et al. 2007). This is a 2-D radiative ALE (Arbi-

X. Ribeyre (✉) · P. Nicolaï · S. Galera · V.T. Tikhonchuk
Centre Lasers Intenses et Applications, UMR 5107,
CNRS—Université Bordeaux 1—CEA, Université Bordeaux 1,
351, Cours de la Libération, 33405 Talence, France
e-mail: ribeyre@celia.u-bordeaux1.fr

Table 1 Comparison of Herbig-Haro and laboratory jets

Quantity	Herbig-Haro jet	Simulation laboratory
Jet length (cm)	3×10^{17}	0.5
Jet radius (cm)	7.5×10^{15}	0.015
Jet velocity (km/s)	100	500
Atomic number Z	2	15
Mass number A	4	40
Particle density (cm^{-3})	20	1.5×10^{18}
Density ($g\ cm^{-3}$)	1.3×10^{-22}	10^{-4}
Temperature T (eV)	1	100
Pressure ($dyne\ cm^{-2}$)	10^{-10}	4×10^{9}
Jet sound speed (km/s)	10	80
Mach number	9	6

trary Lagrangian Eulerian) code based on the Godunov discretization scheme with a two-temperature hydrodynamics. The CHIC includes several physical models. For our simulations we retain the following options: (i) the Spitzer-Härm model for electron and ion heat conduction; (ii) a perfect gas equation of state; (iii) the Thomas-Fermi model for ionization and (iv) the multi-group radiative diffusion with LTE opacities.

The jet simulation parameters are chosen to be close to the experimental values (Nicolaï et al. 2008). The simulations are performed with a non-structured mesh with 170×60 grid cells in the axisymmetric geometry and the mesh is refined around the jet head. Table 1 presents the main simulation parameters.

The initial conditions for all simulations are: the jet and ambient material are the same: argon, the jet velocity and radius are $v_j = 500$ km/s and $r_j = 150$ µm with a local Mach number about 6, jet density is $\rho_j = 10^{-4}$ g/cc. The initial jet length is 2 mm. The density contrast is defined by the ratio $\eta = \rho_j/\rho_a$ where ρ_a is the ambient density. Three configurations are studied, a jet propagation in a low-density medium, $\eta = 10$, equal densities, $\eta = 1$, and a high density medium, $\eta = 0.1$. The variation of η is realized by changing ρ_a and the ambient temperature is fixed to 10 eV.

3 Radiative cooling effect on working surface evolution

We focus our study on the jet head evolution. Figure 1a shows the jet classical working surface (WS) structure: the Mach-disk, the shocked jet—the contact discontinuity between the jet and ambient medium—the shocked ambient medium and the bow shock (Hartigan 1989). Figures 1 and 2 show the density and the electron temperature distributions after 2 and 7 ns of jet propagation for different values of η. For each case, two simulations were conducted: one

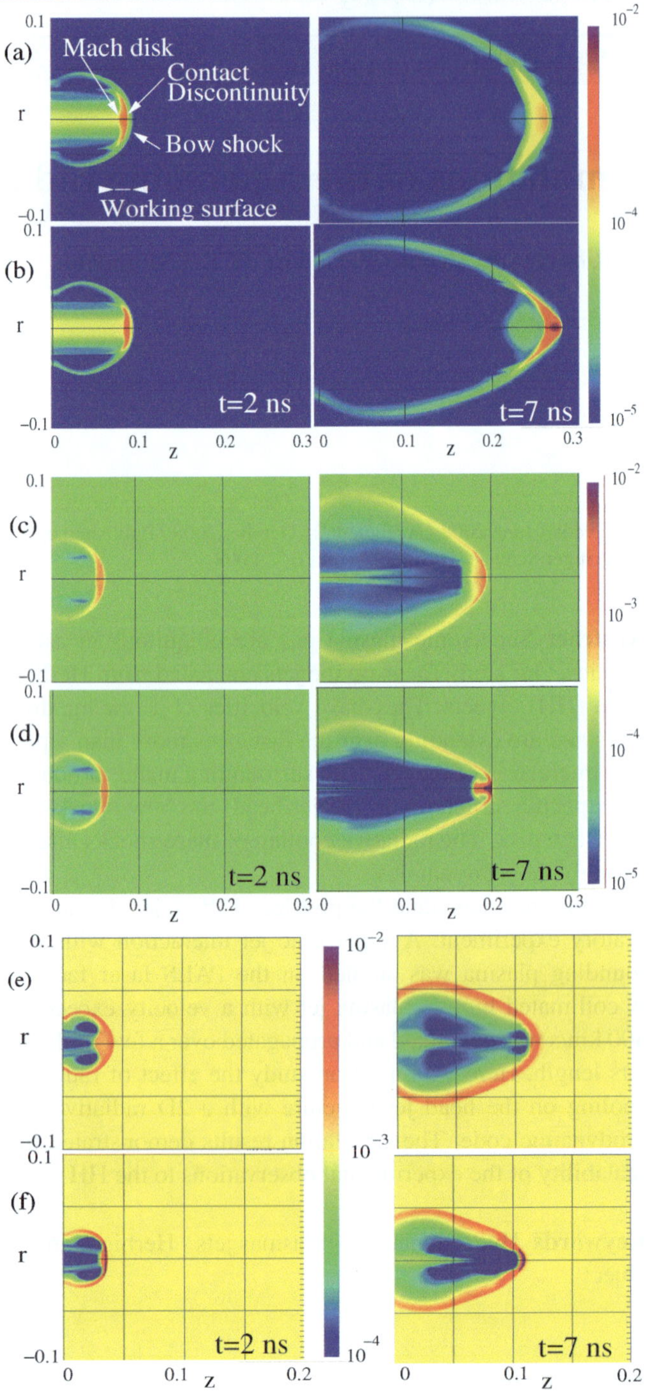

Fig. 1 Density in the logarithmic scale, for 2 and 7 ns of jet propagation. This figure shows the effect of density variation and the radiative effect on the jet evolution for different density ratio on the temporal evolution of the jet head. *Upper row*: for $\eta = 10$, (**a**) in adiabatic, (**b**) in radiative case. *Middle row*: $\eta = 1$, (**c**) in adiabatic, (**d**) in radiative case. *Bottom row*: for $\eta = 0.1$, (**e**) in adiabatic, (**f**) in radiative case

for the adiabatic case (where the radiative transport switched off) and another one for the radiative case. As η increases, the WS propagates over a longer distance on the axis and also the jet extends radially and the WS extension in the

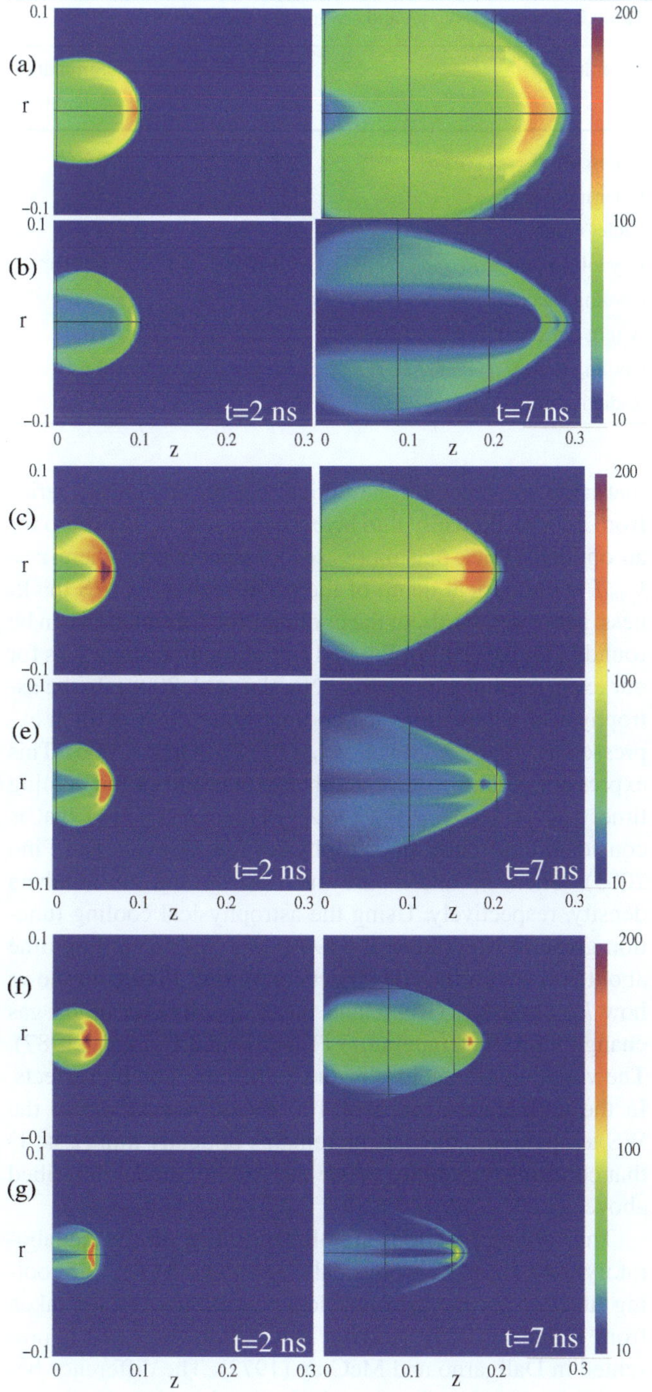

Fig. 2 Distributions of the electron temperature for the same parameters as shown in Fig. 1

axial direction increases. Whatever η, the radiative cooling decreases the WS temperature, collapses the jet head and reduces the jet radial extension (see Fig. 2). The substructure that appears at the WS in the adiabatic case at 7 ns in Fig. 1a is due to a mesh effect. For these three values of η we observe significant differences in the jet evolution. For $\eta = 10$ the cooling enhances the bullet-like (Hartigan 1989) shape of the WS; i.e., the Mach disk becomes denser and propagates like a bullet. Conversely, with $\eta = 0.1$, the jet shape resembles a bubble and the bow shock is denser forms a thin dense shell. Moreover, in the adiabatic case the oblique shock modulates the axial jet density. For $\eta = 1$, we observe an intermediate regime, the bow shock and the Mach disk remain denser. In all radiative simulations, the mean WS density increases by a factor about five comparing to the adiabatic case, while the mean WS electron temperature decreases by a factor about two. As the pressure is approximately the same, this implies that the ionization level decreases in the radiative case. The radiative cooling time is the same order of the hydrodynamic characteristic time. The pertinence of this observation to HH jet conditions will be confirmed later by the radiative scaling analysis.

Theses simulation were performed with a multi-group radiative diffusion model, this implies that a photon emitted in the bow shock can be absorbed in the Mach disk and vice versa. This treatment is more detailed that a simple energy radiative losses considered in Blondin et al. (1990). It tends to equilibrate the temperature in the bow shock and to decrease to some extent the radiation losses.

4 Temporal evolution of jet head structure

A supersonic jet propagating into an uniform ambient medium develops a double shock pattern as it described in previous section. The velocity of the bow shock, v_{bs}, can be estimated by assuming a stationary propagation and by equalizing the momentum flux of the jet material at the jet head with the momentum flux generated by the ambient medium: $(\rho_j(v_j - v_{bs})^2 + p_j)r_j^2 \simeq (\rho_a v_{bs}^2 + p_a)r_a^2$, where, v_j is the jet velocity, r_a is the radius at the head of the jet, p_j and p_a are the thermal pressure of the jet and the ambient gas, respectively (see for example Lind 1989). For high Mach number, the thermal gas pressure can be neglected and if we assume, according to the simulation results shown in Figs. 1 and 2 that $r_j \simeq r_a$, we obtain a simple relation between the velocities of the bow shock and the jet:

$$v_{bs} \simeq v_j\,(1 + \eta^{-1/2})^{-1}. \qquad (1)$$

Figure 3 represents the time evolution of the Mach disk and the bow shock for different density ratios η. One can see that indeed the Mach disk and bow shock are close to the stationary state, the velocity difference between them is very small. The WS is approximately in a ballistic motion, i.e., with a constant velocity. Moreover, as we see previously in Figs. 1 and 2, denser is the ambient medium (η increases) more the bow shock velocity decreases. The bow shock position is in good agreement with the model described by (1) (the black circles in Fig. 3). In particular, given the jet velocity $v_j = 500$ km/s, (1) yields for $\eta = 10$, $v_{bs} = 380$ km/s, $\eta = 1$,

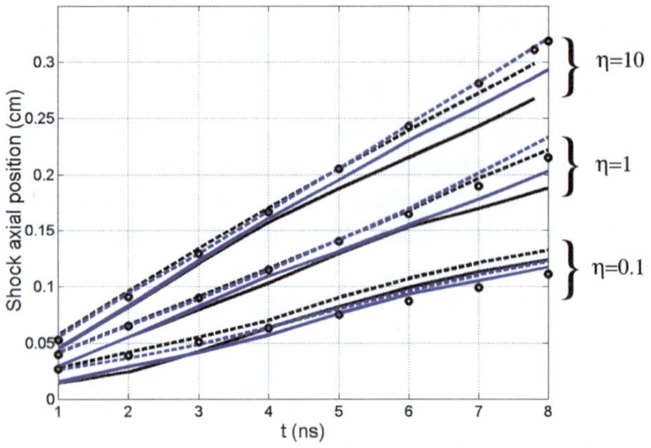

Fig. 3 Comparison of working surface position found in simulations with the estimate given by (1) for the Mach disk (*solid line*) and the bow shock (*dashed line*), for different density ratios η in adiabatic (*black lines*) and radiative case (*blue lines*). *Black dots* show the bow shock position

$v_{bs} = 250$ km/s and $\eta = 0.1$, $v_{bs} = 120$ km/s. This simple model is relevant both in the laboratory that in HH jet system. The calculated positions of the radiative bow shock are in good agreement with observations (Nicolaï et al. 2008).

5 Jet scaling with the effect of radiation losses

The laboratory experiment can be re-scaled to the astrophysical HH jet if the polytropic Euler equations are relevant for both systems and if these equations are invariant under transformations that preserve the Euler number, Eu (Ryutov et al. 1999). Such a scaling is valid if the dissipative mechanisms: magnetic field, thermal diffusion and viscosity are negligible, i.e., if $Pe \gg 1$ and $Re \gg 1$ (see Table 2). Then the Euler equations contain all the relevant physics and the experiment can be scaled to arbitrary dimension (Ryutov et al. 1999). However, in our case, the radiative dissipation is not negligible and must be taken into account. The comparison of the cooling parameter: t_s/t_{cool} between the two systems ensure that the radiative dissipation have the same dynamics. This parameter is the ratio between the hydrodynamic timescale t_s and the cooling time τ_{cool} (see Table 2). In laboratory system, to calculate the cooling time, we assume that the WS can be considered as a thin shell plasma layer, i.e., the photon mean free path is greater than the WS thickness, this assumption will be confirmed later. The cooling time is given by: $\tau_{cool} = (Z+1)k_B/2(\gamma-1)Am_p\sigma k_{pm}(T)T^3 f(\tau)$, where, k_B, m_p, σ are the Boltzmann constant, the proton mass and the Stephan-Boltzmann constant, respectively. The parameters γ, A, Z are the polytropic coefficient ($\gamma = 5/3$ in our case), the average atomic mass number and average level ionization, respectively. The mean Planck opacity is k_{pm}, and T is the temperature. The correction coefficient $f(\tau)$ that takes into account the optical plasma thickness, varies from 2 for an optically thin layer (our case $\tau \ll 1$) to zero for an optically thick plasma ($\tau \gg 1$), respectively, where $\tau = k_{pm}d$ is the optical depth of the plasma, d is the WS thickness. More details about the cooling time formulation can be found in Drake (2006). We find a cooling time about 1 ns for our experimental parameters (Nicolaï et al. 2008). In the astrophysical publications the cooling function Λ is often expressed as: $\Lambda = 2\sigma A^2 m_p^2 T^4 k_{pm}(T)/Z\rho$ (Drake 2006). This expression allows to give another expression of the cooling time: $\tau_{cool} = (n_e+n_i)k_BT/(\gamma-1)n_en_i\Lambda(T)f(\tau)$ that is considered currently in astrophysics (de Gouveia Dal Pino 2005), where n_e and n_i are the electron and ion numbers density, respectively. Using the astrophysical cooling function given in Blondin et al. (1990) one finds a cooling time about 300 yrs for the HH object conditions. Figure 4a shows how k_{pm} and k_{rm} (Rosseland mean opacity) for argon gas change with the temperature (Tsakiris and Eidmann 1987). The variation of opacities is due to the emission line effects. In the temperature range of 100 eV corresponding to the WS temperature, the plasma remains optically thin ($\tau \ll 1$) that confirms possibility to use the cooling model described above.

Table 2 Characteristic jet and scaling law parameters

Quantity	Herbig-Haro jet	Simulation laboratory
Timescale $t_s = r_j\sqrt{\rho/p}$	300 yrs	2 ns
Euler number $E_u = v_j/\sqrt{p/\rho}$	11	8
Péclet number Pe	10^5	9
Reynolds number Re	5×10^7	6×10^6
Cooling function		
Λ (erg cm^3 s^{-1})	10^{-23}	10^{-19}
Cooling time τ_{cool}	300 yrs	1 ns
Radiative effect t_s/τ_{cool}	1	2

The cooling functions are plotted in Fig. 4b for the laboratory, $\Lambda(T)$, and astrophysical conditions, $\tilde{\Lambda}(\tilde{T})$. The cooling function for the astrophysical conditions $\tilde{\Lambda}(\tilde{T})$ is taken from Kafatos (1973). Another calculation for $\tilde{\Lambda}(\tilde{T})$ is presented in Dallgarno and McCray (1972). The difference between the two function is only in the domain of temperature below 1 eV, which is not of the interest for our analysis here. The emission intensity increases strongly in the range 1–10 eV (1 eV $\simeq 10^4$ K), saturate at 20 eV and decreases slowly for higher temperatures. This cooling function was calculated for optically thin medium where the cooling is mainly due to molecular vibration, rotation, fine line structure, resonance line, bremsstrahlung, recombination, dust emission. Its temperature dependence is strongly different from a black body emission (Dallgarno and McCray 1972). The temperature dependence of the cooling function has a very strong effect of the WS morphology.

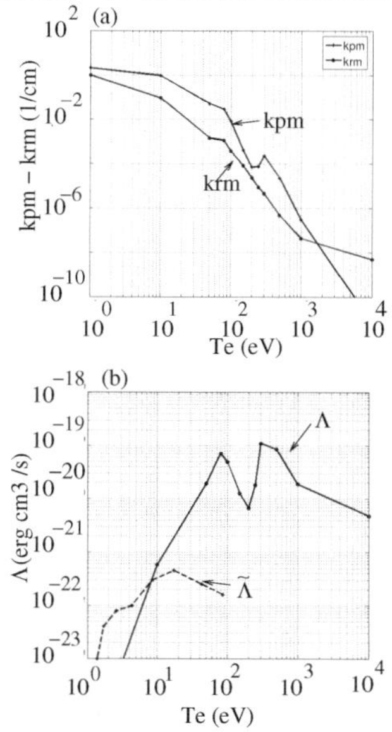

Fig. 4 (**a**) Plot of the Planck and Rosseland mean opacities, respectively k_{pm}, k_{rm} versus the electronic temperature for an argon density of 10^{-4} g/cc. (**b**) Plot of the cooling function Λ versus the temperature in laboratory Λ for 10^{-4} g/cc and for HH-jet $\tilde{\Lambda}$ is given in Blondin et al. (1990)

The shape of the cooling function presented in Fig. 4b shows a possibility of thermal instability of the WS at high temperatures. According Sutherland et al. (2003), Chevalier and Imamura (1982) the criterion for the thermal instability is $d\ln(\Lambda)/d\ln T \leq 1-2$. For the astrophysical conditions this criterion is satisfied for the temperatures about 10 eV. For the laboratory plasma the instability may occur for the temperature above 300 eV. For lower temperatures, the temperature dependence of the cooling function is complicated and the stability analysis needs to be performed separately. In our simulations the temperature in the WS does not exceeds 200 eV and we did not observe the instability. It may append that in our simulation the instability is suppressed by the electron thermal conduction (Clark and Carswell 2007) which is much stronger than in corresponding astrophysical conditions.

The shapes $\Lambda(T)$ and $\tilde{\Lambda}(\tilde{T})$ are similar, although the positions of the maxima and their magnitudes are different. Let us consider the scaling of energy equation from the laboratory to astrophysical system including the radiative cooling. If we assume that the plasma is optically thin, the energy conservation equation reads:

$$\partial_t p + \mathbf{v}\cdot\nabla p + \gamma p \nabla\cdot\mathbf{v} = -(\gamma-1) n_e n_i \Lambda(T), \qquad (2)$$

where, $\mathbf{v}$ is the velocity vector, ∂_t is the time derivative and ∇ the gradient vector. This relation can be used to connect two systems: the laboratory system is defined by the following set of variable $(r, t, v, p, \rho, Z, A, T, \Lambda)$ and the astrophysical system is defined by: $(\tilde{r}, \tilde{t}, \tilde{v}, \tilde{p}, \tilde{\rho}, \tilde{Z}, \tilde{A}, \tilde{T}, \tilde{\Lambda})$. For each variable, a scaling factor (noticed with a star) is defined as the ratio between the laboratory and astrophysical variables, e.g.: $r^* = r/\tilde{r}$. The scaling for the velocity is deduced from the momentum conservation and the time scaling from the mass conservation. Two systems demonstrate the similar hydrodynamic behavior if the following conditions are satisfied: $v^* = \sqrt{p^*/\rho^*}$, $t^* = r^*/v^*$. Equation (2) provides the relations between the two systems in terms of the temperature and the cooling rate. These two scaling factors read:

$$T^* = \frac{A^*}{Z^*}\frac{p^*}{\rho^*}, \qquad \Lambda^* = \frac{A^{*2}}{Z^* r^* \rho^*}\left(\frac{p*}{\rho^*}\right)^{3/2}. \qquad (3)$$

It is important to notice that these two parameters depend on the scaling factors Z^* and A^* differently. This implies that the radiative scaling can be adjusted by choosing the jet and target materials in the laboratory experiment. The parameters listed in Table 2 can be used to calculate the two scaling factors given by (3). We find $T^* \simeq 70$ and $\Lambda^* \simeq 2.5 \times 10^3$. These two factors correspond approximately to the temperature ratio and the cooling function ratio shown in Fig. 4b. Not only the ratio between the cooling functions is in agreement but also the shape of these two functions are similar. The scaling is still valid over a large range of temperature. Therefore, the argon gas provides a possibility to reproduce a macroscopic radiative behavior of a HH jet.

6 Conclusion

We have performed 2D numerical simulations for the parameters of jets produced at the PALS laser facility. These simulations demonstrate the radiative effect in the WS of the jet. The radiative cooling of the jet head decreases temperature by a factor two, while the density increases by a factor of five. The variation of the ratio between the ambient and jet density η is studied, the 2D simulations show the transition from a bullet to a bubble WS structure as the parameter η decreases. Moreover, for low density ratio $\eta \ll 1$, a complex axial jet internal structure is observed due to formation of oblique reverse shocks. From the momentum flux conservation, we estimated the dependence of the bow shock velocity on the density ratio η and jet velocity for a High Mach flow. This model equally applies for the laboratory and astrophysical conditions. The momentum model shows that the working surface evolves in a ballistic phase. Finally, the scaling of the radiation cooling term in the energy conservation equation allows to show that the similarity between laboratory and HH jet can be obtained by choosing an appropriate plasma composition. By choosing materials with specific radiation properties one can adjust the temperature dependence of the cooling function in a large temperature range. This allow to model the jet evolution in more details.

Acknowledgements This work is partly supported by the Aquitaine Region Council. X. Ribeyre thank G. Schurtz and O. Peyrusse for the fruitful discussions.

References

Blondin, J.M., et al.: Astrophys. J. **360**, 370 (1990)
Chevalier, R.A., Imamura, J.N.: Astrophys. J. **261**, 543 (1982)
Clark, C.J., Carswell, R.F.: Principles of Astrophysical Fluid Dynamics. Cambridge University Press, Cambridge (2007)
Coker, R.F., et al.: Astrophys. Space Sci. **307**, 57 (2007)
Dallgarno, K.R., McCray, R.A.: Annu. Rev. Astron. Astrophys. **10**, 375 (1972)
de Gouveia Dal Pino, E.M.: Adv. Space Res. **35**, 908 (2005)
Drake, R.P.: High-Energy-density Physics. Springer, Berlin, Heidelberg (2006)
Hartigan, P.: Astrophys. J. **339**, 987 (1989)
Kafatos, M.: Astrophys. J. **182**, 433 (1973)
Lind, K.R.: Astrophys. J. **344**, 89 (1989)
Maire, P.H., et al.: SIAM J. Sci. Comput. **29**, 1781 (2007)
Nicolaï, P., et al.: Supersonic plasma jet interaction with gases and plasmas. Astrophys. Space Sci. (2008) (ASTR1332, submitted in HEDLA volume)
Reipurth, B., Bally, J.: Annu. Rev. Astron. Astrophys. **39**, 403 (2001)
Ryutov, D., et al.: Astrophys. J. **518**, 821 (1999)
Stone, J.M., Norman, M.L.: Astrophys. Phys. J. **413**, 198 (1993)
Sutherland, R.S., et al.: Astrophys. J. **591**, 238 (2003)
Tsakiris, G.D., Eidmann, K.: J. Quant. Spectrosc. Rad. Transfert **38**, 353 (1987)

Astrophys Space Sci (2009) 322: 91–95
DOI 10.1007/s10509-008-9926-0

ORIGINAL ARTICLE

Theoretical and numerical studies of the Vishniac instability in supernova remnants

Cécile Cavet · Hung Chinh Nguyen · Claire Michaut · Emeric Falize · Serge Bouquet · Laurent Di Menza

Received: 17 June 2008 / Accepted: 3 October 2008 / Published online: 31 October 2008

Abstract Vishniac instability has been theoretically studied in supernova remnants where it is supposed to explain the fragmentation of the interstellar medium. However its role is not fully demonstrated in these astrophysical objects. Conditions and assumptions required for the instability growth are explained in detail in the present paper. In addition the HYDRO-MUSCL hydrodynamic code has been used to simulate this instability in order to compare the numerical growth rate with the Vishniac analytical solution.

Keywords Hydrodynamics · Vishniac instability · Shock waves · Supernova remnants

1 Introduction

Among the instabilities arising in astrophysical systems, and in particular in supernova remnants (SNRs), the Vishniac instability is not very well known. In their original analysis of global perturbations, Vishniac (1983) and Ryu and Vishniac (1987, 1989) identified the criteria that allows to compute the growth of a perturbation in a thin shell of shocked matter. The instability depends on the direction of two opposite forces: the thermal pressure p_{th} due to hot SNR gas that pushes outwards a dense shell and the ram pressure p_{ram} due to the accretion of the surrounding interstellar matter (ISM) on the shock front, that compresses inwards the expanding shell. If the ISM is uniform, the two pressures keep the same directions and the shock front is stable (Erpenbeck 1962); but in the case of non-uniform ISM, the shock front is distorted, the two forces do not counterbalance and oscillations can develop and grow. In a linear stability analysis, Vishniac (1983) used the (infinitely) thin shell approximation to study the perturbation equations and numerically solved the corresponding system. In a recent study (Cavet et al. 2008) an analytical solution for the growth rate has been determined in this case, but this approximation does not allow to access the general instability criteria. Aiming to investigate more physical cases such as thin shell with finite thickness, the original model of Ryu and Vishniac (1987) is presented and rewritten using the self similar formalism. The growth rate is also calculated through numerical simulations of perturbed radiative blast waves, and compared with the analytical one.

C. Cavet (✉) · H.C. Nguyen · C. Michaut · E. Falize · S. Bouquet
LUTH, Observatoire de Paris, CNRS, Université Paris-Diderot, Meudon, France
e-mail: cecile.cavet@obspm.fr

E. Falize · S. Bouquet
Département de Physique Théorique et Appliquée (DPTA), CEA Bruyères le Châtel, 91297 Arpajon Cedex, France

H.C. Nguyen · L. Di Menza
Laboratoire de Mathématiques, Université Paris-Sud, Orsay, France

2 Analytical model

2.1 Blast wave theory

The propagation of an SNR in the ISM is classically approached in four self similar stages. The initial ballistic stage is followed by the adiabatic Sedov phase, well-modeled by the strong explosion theory (an high amount of energy E_{in} is deposited in a small volume). With this model, Petruk (2000) has shown that the shock radius R_{sed} can be expressed versus t the age of the SNR which is in cylindrical coordinates (needed for the 2D-simulations):

$$R_{sed}(t) = \beta[E_{in}/\rho_i]^{1/4}\, t^{1/2}, \qquad (1)$$

where $\beta(\gamma) = [2(\gamma - 1)\gamma^2/\pi(\gamma + 1)]^{1/4}$ depends on the polytropic index γ and ρ_i is the mean ISM density. The third stage is the snowplow radiative phase: energy losses from the thin layer of accreted matter, behind the shock front, are no more negligible. This effect is taken into account via the decrease γ. For this phase, we have found in cylindrical coordinates an self-similar solution (based on spherical case (Liang and Keilty 2000)):

$$R_{sp}(t) = [\beta E_{in} R^2_{0,sed}/\rho_i]^{3/16} t^{3/8}, \tag{2}$$

where $R_{0,sed}$ is the radius at the end of the Sedov phase (10^4 years in (1)). Another manner to model this third phase is to take into account the radiative losses of the thin shell by introducing a cooling function which changes R_{sp} (Cioffi et al. 1988; Blondin et al. 1998). However, in this paper we intend to constrain only the parameter γ. In the fourth phase the turbulent shock matter is mixed with ISM. The Vishniac instability can appear and modify the shock front at an ordinary moment of the snowplow phase (Ryu and Vishniac 1989) and during the transition between the Sedov and snowplow phases Vishniac (1983), Ryu and Vishniac (1987).

2.2 Unperturbed solutions

Here we analyze self-similar solutions of hydrodynamic equations in order to identify the initial conditions of the Vishniac instability. Following Landau and Lifshitz (1959) (Sect. 106) and Kushnir et al. (2005), we introduce three dimensionless functions $G(\xi)$, $U(\xi)$, $P(\xi)$ (where $\xi = r/R_{sed}(t)$, r being the radial coordinate) respectively in terms of the density $\rho(r,t)$, the fluid velocity $u(r,t)$ and the pressure $p(r,t)$ as:

$$\rho = \rho_i G(\xi), \quad u = \dot{R}_{sed}\xi U(\xi), \quad p = \rho_i \dot{R}^2_{sed} P(\xi). \tag{3}$$

For $\xi = 1$, the Rankine-Hugoniot (RH) jump relations give the dimensionless functions:

$$G(1) = \frac{\gamma+1}{\gamma-1}, \quad U(1) = P(1) = \frac{2}{\gamma+1}. \tag{4}$$

Based on Guderley (1942) and taking the limit $\xi \to 0$, Kushnir et al. (2005) showed that:

$$G(\xi) = G(0)\xi^{a-2} + G(1)\xi^{2a-2}, \tag{5}$$

$$U(\xi) = P(\xi) = U(0) + U(1)\xi^a, \tag{6}$$

where $G(0) \simeq 0$, $U(0) = 1/\gamma$, $P(0) = p_{th}/p_{ram}$ and $a = (1+2\gamma)/(\gamma-1)$.

2.3 Global perturbation analysis

Vishniac instability can develop only as a consequence of an initial deformation induced by the misalignment of p_{th} and p_{ram} in the shock front shape. Initial deformations are the dimensionless quantities $\delta\widetilde{\rho_i}(\xi)$, $\delta\widetilde{u_i}(\xi)$ (with two components: radial $u_{i,r}$ and transverse $u_{i,T}$), $\delta\widetilde{p_i}(\xi)$. Using the separation of variables method, Ryu and Vishniac (1987) have written the evolution of perturbation dimensionless quantities $\delta\widetilde{\rho}$, $\delta\widetilde{\vec{u}}$, $\delta\widetilde{p}$ and $\delta\widetilde{R}$:

$$\delta\widetilde{\rho} = \delta\widetilde{\rho}_i(\xi) Y_{lm}(\theta, \Phi)^s, \tag{7}$$

$$\delta\widetilde{\vec{u}} = \left[\delta\tilde{u}_{i,r}(\xi)\vec{e_r} + \delta\tilde{u}_{i,T}(\xi)\xi\widetilde{\vec{\nabla}}_T\right] Y_{lm}(\theta, \Phi)t^s, \tag{8}$$

$$\delta\widetilde{p} = \delta\tilde{p}_i(\xi)\, Y_{lm}(\theta, \Phi)t^s, \tag{9}$$

$$\delta\widetilde{R} = Y_{lm}(\theta, \Phi)t^s, \tag{10}$$

where $\widetilde{\vec{\nabla}}_T$ is the tangential gradient operator and the spherical harmonic Y_{lm} represents the transverse oscillation on the shock front. The growth rate of the perturbation is s (with amplitude t^s). Ryu and Vishniac integrated numerically the perturbation hydrodynamic equations and thus determined the unstable perturbation criteria for a spherical thin shell: for an ionized gas the polytropic index is $\gamma < 1.2$, the perturbative mode number is $l \approx [8–80]$ ($Re(s) > 0$) and the growth rate of the perturbation reaches a maximal value $s_{theo} = 0.49 + 3.20i$ at $l_{theo} = 40$.

To complete our analysis we have overviewed other works studying the Vishniac instability. In the analytical work of Kushnir et al. (2005), a boundary condition has been added to the initial pattern. They found the same $l_{min} \simeq 10$ but different results when $l \to 0$. In a numerical study, Mac Low and Norman (1993) simulated only the case $l = 40$ and obtained results which are in good agreement with the theoretical s. In the field of laboratory astrophysics, Grun et al. (1991) have experimented the stability of spherical shock and they found $s \simeq 2.28 s_{theo}$ for $l \simeq 1/5 l_{theo}$ for a uniform xenon gas with $\gamma = 1.06$. This discrepancy is explained by Edens et al. (2004): the driving laser modifies the ambient gas conditions. With a different experimental setup, Edens et al. (2007) found two new experimental points with $\gamma = [1.1–1.2]$: $l \sim 16$, $s \sim 0.08$ and $l \sim 20$, $s \sim 0.3$. From this review it appears that the Vishniac instability is neither completely understood nor fully solved. In the next section we present numerical results aiming to analyze the physical process that generates the instability growth; furthermore in order to study a more astrophysical case the thin shell feedback is studied at low perturbative mode numbers.

3 Numerical simulations

Numerical simulations presented here are performed with the HYDRO-MUSCL code developed by our team. This

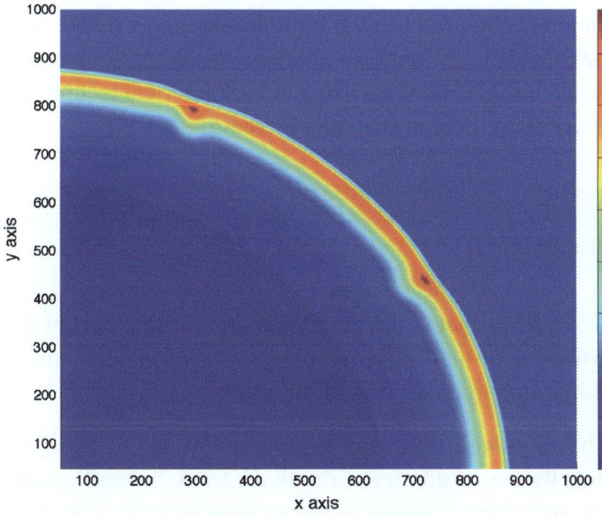

Fig. 1 Top view of the normalized density at time $2 \times t_0$

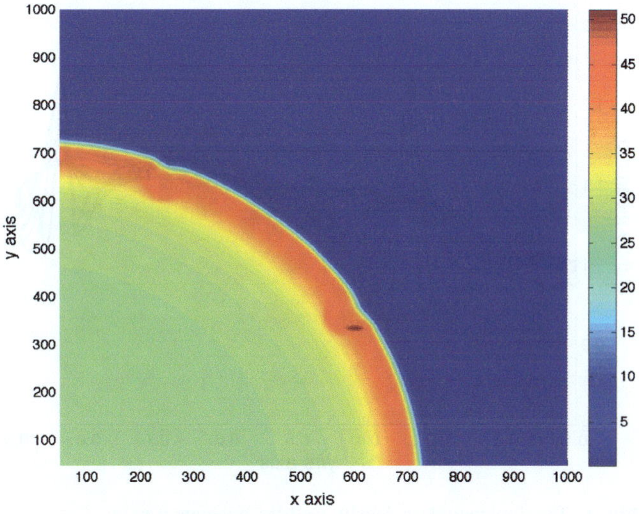

Fig. 2 Top view of the normalized pressure at time $2 \times t_0$

Eulerian code, solving hydrodynamic equations, uses a regular Cartesian grid and an adaptative time step calculated with the Courant-Friedrichs-Lewy (CFL) condition. The underlying numerical method is a MUSCL-Hancock finite volume scheme and a HLLC Riemann solver (Toro 1999). In order to perform 2D cylindrical hydrodynamic simulations for the propagation of a shock wave in the ISM, we induce an initial explosion by depositing a strong amount of energy in the form of p_{th} at the center of a box with a resolution 400×400 zones.

In a first simulation, we study the initial phase of the shock evolution when the shock radius can be approximated with the Sedov law given below. The gas is chosen adiabatic with $\gamma = 5/3$. The simulation is stopped when the radius evolves according to the Sedov self-similar solution (1) and when the density on the shock front reaches the strong shock limit (4). We note this state (t_0, r_0) and we use it as input data in the snowplow phase simulation.

In the second calculation, axial velocities are directly modified in data in order to obtain the snowplow radius evolution (2) and $\gamma = 1.1$. We introduce high-density spots ($\rho_{spot} \propto \rho_{shock}$) ahead of the shock wave to perturb the shock front and let the system evolve. To determine the profile of these perturbative spots, we have to evaluate $\delta\widetilde{\rho}_i(\xi)$ in (7). In Ryu and Vishniac (1987) this function is not explicitly estimated, but only determined by numerical means. We use an approximated value on the shock front (given by linearized RH equations) $\delta\widetilde{\rho}_i(\xi) \approx \delta\widetilde{\rho}_i(1)$, combined with (5) and we find $\delta\widetilde{\rho}_i(\xi) \approx -(2a-2)\xi^{2a-3}$. Thus in the numerical simulations we introduce a finite number of spots to take into account the perturbative mode number l and we add a bi-dimensional spot shape in order to reproduce the deformation profile according to r, θ, ϕ.

The numerical results are shown in the following snapshots (Figs. 1–3) displayed at time $2 \times t_0$, when the shock

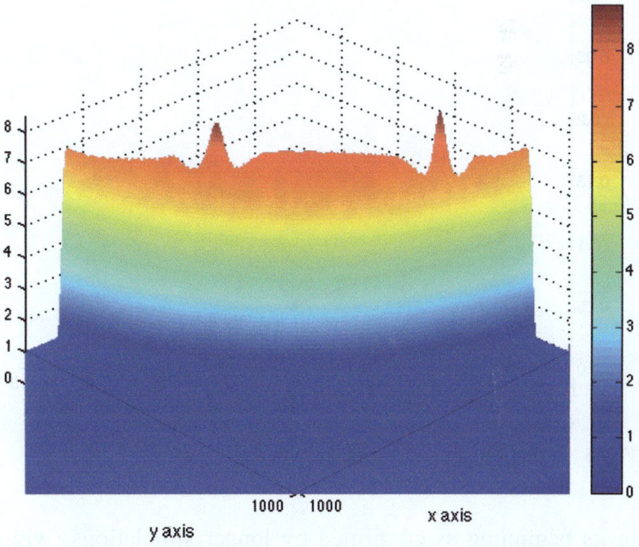

Fig. 3 Normalized density profile (given by *colorbar*) on the shock front in the snowplow phase at time $2 \times t_0$

front collides the high-density spots. We can see that the two introduced spots create radial oscillations both on density (Fig. 1) and pressure (Fig. 2), and density perturbation on the shock front (Fig. 3). In Fig. 2 delimited by the thin shell, the central p_{th} pushes material counterbalancing the high p_{ram} coming from the external accreted matter. Despite the theoretical analysis predicting that $\delta = p_{th}/(P(1)\, p_{ram}) = 0.27$, here we have $\delta = 0.60$ which describes an early stage with large magnitude for p_{th}. Therefore, the system has to evolve in order to decrease δ. Furthermore, in Fig. 2 the two local deformations of the shell shape where p_{th} and p_{ram} are not aligned (see Sect. 1). Thus in this numerical configuration all Vishniac instability criteria are fulfilled (γ and $\delta\widetilde{\rho}_i(\xi)\, Y_{lm}$), but the physical process of the instability growth is still

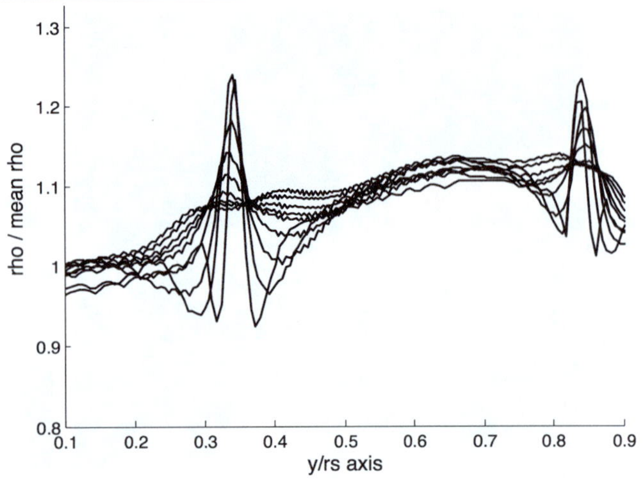

Fig. 4 Ten dimensionless density profiles $\tilde{\rho}$ versus y-axis

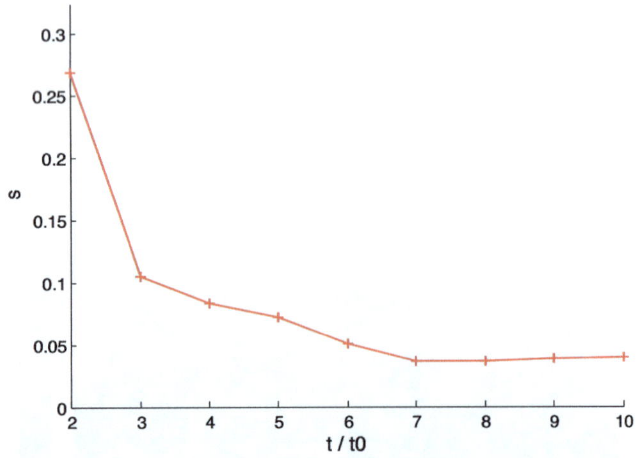

Fig. 5 Evolution of perturbation growth rate s versus t

in its beginning as confirmed by longer simulations (with $t > 7 \times t_0$).

In order to observe the evolution of the growth rate s of this small perturbations and to compare s with the theoretical results, we let run the simulation during $t = 10 \times t_0$. In Fig. 4 we superpose ten snapshots of the density profile projected on the y-axis and normalized for the shock front value which enables to measure the normalized peak oscillations $\delta\tilde{\rho}$ during this period. We estimate the growth rate by inversing (7): $s = [\ln(\delta\tilde{\rho}) - \ln(\delta\tilde{\rho}(1)\ Y_{lm})]/\ln t$. In the present case, $\delta\tilde{\rho}(1) = 1.8$. Figure 5 shows the evolution of the growth rate versus time and a stabilization to a limit value $s = 3.7 \times 10^{-2}$ is clearly observed. Compared to Vishniac prediction, this value seems too low. However, we have chosen a small mode number $l \sim 8$ due to the presence of only two high density spots. If we increase the number of initial spots, the mode number l is also increasing and can reach the optimal mode number $l = 40$ of (Ryu and Vishniac 1987). However due to numerical constraints (see Sect. 4) it is not easy to multiply initial high-density spots on the way of the thin shell. In future simulations, we will improve perturbative initial conditions using other numerical techniques for deforming the thin shell and increasing l.

4 Discussion

We have investigated the numerical criteria that permit the growth of an instability in the snowplow stage, and we stress out some important aspects of this study.

1. The *magnitude* of the initial perturbation has to be higher than the initial density of the shock front, but sufficiently low to prevent acoustic wave propagation in the thin shell which tends to decrease the perturbations.
2. The *number* of high density spots is chosen by the analytical study. Indeed in a SNR the mode number for a maximum growth is $l = 40$ ($l = 10$ for a quarter of circle). However, if perturbations are very close to each other, the acoustic waves interfere and produce a stable shock shape.
3. The *size* of perturbation has to be equal to the shell thickness in order to deform radially the thin shell.
4. As seen in Sect. 1, we have to fulfill p_{ram} not exactly aligned with p_{th} to allow the growth of the perturbation. Therefore we impose initial perturbations breaking the *angular symmetry* of the shell.
5. To ensure that the shock front evolve to the *self-similar phase*, we have to increase the simulation time ($100 \times t_0$). Indeed the instability cannot grow if initial perturbations are applied during the transient stage between Sedov and snowplow phases.

In conclusion, we have focused our work on the understanding of the theoretical analysis of Vishniac instability. We pointed out the unstable criteria and shape parameters of an initial perturbation in order to prepare numerical conditions. Then, taking into account all these constraints we have performed instability simulations using our 2D HYDRO-MUSCL code. For the case of cylindrical thin shell we calculated the growth rate s of the perturbation. In a future work, it is expected to get improvements of the stability study with use of a multi-processor version of HYDRO-MUSCL code, combined with profiles of perturbed shell designed by analytical tools that will enable a better control of spatial shape. Furthermore an experiment of the Vishniac instability realized by our team is planned on the LIL high-power facility (Bordeaux, France) in 2009.

Acknowledgements One of us is grateful for travel support provided by the APS for the HEDP/HEDLA 2008 conference.

References

Blondin, J.M., Wright, E.B., Borkowski, K.J., Reynolds, S.P.: Astrophys. J. **500**, 342 (1998)

Cavet, C., Michaut, C., Falize, E.: In: J. Bouvier, A. Chalabaev, C. Charbonnel (eds.) Vishniac instability: experimental preparation, SF2A-2007: Semaine de l'Astrophysique Française, meeting held in Grenoble, France, July 2–6, 2007. EDP-Sciences, Conference Series (2008, in press)

Cioffi, D.F., McKee, C.F., Bertschinger, E.: Astrophys. J. **334**, 252 (1988)

Edens, A.D., Ditmire, T., Hansen, J.F., Edwards, M.J., Adams, R.G., Rambo, P.K., Ruggles, L., et al.: Phys. Plasmas **11**, 4968 (2004)

Edens, A.D., Adams, R.G., Rambo, P.K., Smith, I.C., Porter, J.L., Ditmire, T.: Astrophys. Space Sci. **307**, 127 (2007)

Erpenbeck, J.J.: Phys. Fluids **5**, 1181 (1962)

Guderley, G.: Luftfahrtfarschung **19**, 302 (1942)

Grun, J., Stamper, J., Manka, C., Resnick, J., Burris, R., Crawford, J., Ripin, B.H.: Phys. Rev. Lett. **66**, 2738 (1991)

Kushnir, D., Waxman, E., Shvarts, D.: Astrophys. J. Suppl. Ser. **635**, 407 (2005)

Mac Low, M.-M., Norman, M.L.: Astrophys. J. **407**, 207 (1993)

Landau, L.D., Lifshitz, E.M.: Fluid Mechanics. Mir, Moscow (1959)

Liang, E., Keilty, K.: Astrophys. J. **533**, 890 (2000)

Petruk, O.: Astron. Astrophys. **313**, 820 (2000)

Ryu, D., Vishniac, E.T.: Astrophys. J. **313**, 820 (1987)

Ryu, D., Vishniac, E.T.: Astrophys. J. **337**, 917 (1989)

Toro, E.F.: Riemann Solvers and Numerical Methods for Fluid Dynamics. Springer, Berlin (1999)

Vishniac, E.T.: Astrophys. J. **274**, 152 (1983)

Astrophys Space Sci (2009) 322: 97–100
DOI 10.1007/s10509-008-9938-9

ORIGINAL ARTICLE

Design of jet-driven, radiative-blast-wave experiments for 10 kJ class lasers

R.P. Drake · J.P. Knauer

Received: 7 June 2008 / Accepted: 28 October 2008 / Published online: 25 November 2008

Abstract We discuss the design of jet-driven, radiative-blast-wave experiments for a 10 kJ class pulsed laser facility. The astrophysical motivation is the fact that jets from Young Stellar Objects are typically radiative and that the resulting radiative bow shocks produce complex structure that is difficult to predict. To drive a radiative bow shock, the jet velocity must exceed the threshold for strong radiative effects. Using a 10 kJ class laser, it is possible to produce such a jet that can drive a radiative bow shock in gas that is dense enough to permit diagnosis by x-ray radiography. We describe the design and simulations of such experiments. The basic approach is to shock the jet material and then accelerate it through a collimating hole and into a Xe ambient medium. We identify issues that must be addressed through experimentation or further simulations in order to field successful experiments.

Keywords Jets · Shock waves: radiative

1 Introduction

Very young stars typically produce supersonic jets, which drive bow shocks into the surrounding, ambient medium. It became clear in the 1980s that such jets are associated with bright knots known historically, for spectroscopic reasons, as Herbig-Haro objects. These bright knots are typically distributed along the jet. Subsequent observations, reviewed by Reipurth (Reipurth and Bally 2001) showed that the jets and the regions they affect are highly structured. Hubble Space Telescope observations have since revealed (Bally et al. 2002) "knots, filaments, small-scale bow shocks, and a complex velocity field." The driven shocks are necessarily radiative, as is discussed in the recent review by Dal Pino (2005), whose references include the large body of simulation work on these systems. Simulations indicate that this strong radiative cooling has a dramatic impact on the structure of jets. The present paper presents and discusses a design for an experiment to produce a jet that drives a radiative bow shock, intended for specific application to the Omega EP laser facility (Boehly et al. 1995) but also applicable in general to 10 kJ class, pulsed laser facilities.

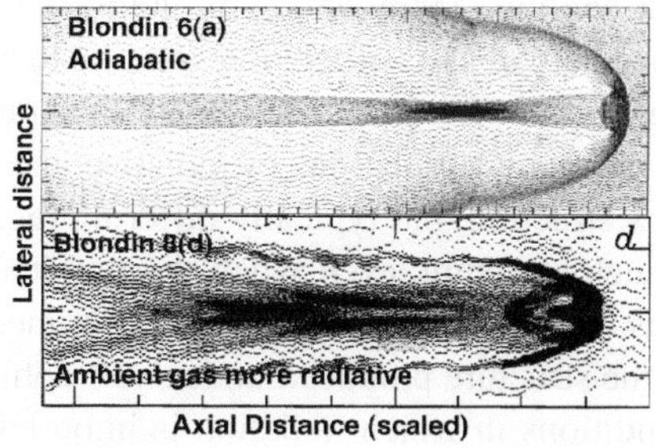

Fig. 1 Radiative jets, adapted from the paper by Blondin et al. *Top*: shows an adiabatic case without radiative cooling. *Bottom*: a case with much stronger cooling in the ambient gas than in the jet, like the proposed experiment

Figure 1 shows an example of such shocks, from the classic Astrophysical Journal paper by Blondin et al. (1990). The bottom portion of this figure shows the case that we propose to produce and study, in which strong radiative effects are present in the shocked, ambient gas. The radia-

R.P. Drake (✉)
University of Michigan, Ann Arbor, USA
e-mail: rpdrake@umich.edu

J.P. Knauer
Laboratory for Laser Energetics, University of Rochester, Rochester, USA

tive cooling collapses the blast wave around the jet and creates more complicated spatial structure through the action of various instabilities, notably including Kelvin-Helmholtz instabilities because of shear between the jet flow and the ambient medium. The structures along the edges of the jet in Fig. 1 were attributed to Kelvin-Helmholtz instabilities. Such structures have not been seen in non-radiative jet experiments (Foster et al. 2005; Rosen et al. 2005). When there is strong cooling of the shocked gas, Blondin et al. attribute some of the structure near the head of the jet to the Vishniac instability (Vishniac 1983). They attribute other structure, behind the head of the jet, to conical shocks launched dynamically. A large number of subsequent papers have explored jet structure under various assumptions, including for example the effect of Kelvin-Helmholtz in three dimensions (Xu et al. 2000), the effects of pulsations in the jet source combined with magnetic fields (Gardiner et al. 2000; O'Sullivan and Ray 2000), and the consequences of different types of cooling (Moraghan et al. 2006). Such simulations do not easily obtain sufficient resolution to resolve the details (Plewa 1993; Rosen and Smith 2004). At the present time there are many simulations of radiative jets, all showing complex structure, but typically differing regarding the details of such structure, even when the nominal physical assumptions are the same. The role for experiments is thus to observe the structure produced under well-defined experimental conditions in which radiation is important. This is the need that the present design addresses.

The experiment design discussed here follows a substantial body of jet-related research in the high-energy-density laboratory astrophysics community during the past decade. The first laser-driven jet experiments could produce jets in isolation and observe their radiative collapse using laser-ablated plasma (Farley et al. 1999), or could produce jets whose length did not greatly exceed their diameter by inverting a pin in a target (Foster et al. 2002). Designs for extended, laser-driven, supersonic jets that can penetrate an ambient medium involve the expansion and acceleration of a shocked plasma down a collimating tube These were developed by E.C. Harding and R.P. Drake in a memorandum in 2000, with the first multi-dimensional design simulations being discussed in a memorandum by A.M. Khokhlov in 2001. This led over time to a large team effort, culminating in publications of experimental results by Foster et al. (2005) and Rosen et al. (2005). In work contemporary with this, S. Sublett employed a similar approach and some others in doctoral thesis work (Sublett 2007; Sublett et al. 2007). Similar approaches have since been employed in experiments on the National Ignition Facility (Blue et al. 2005) and in Europe (Loupias et al. 2007a, 2007b). In addition, two alternative approaches to jet formation have been developed. These are the use of wire-ablation plasmas from Z-pinch devices to produce jets (Ciardi et al. 2002; Lebedev et al. 2002, 2004) and the use of ring-shaped laser irradiation of planar surfaces to produce jets (Kasperczuk et al. 2006; Nicolai et al. 2006, 2007). We will refer to the latter objects as "ring jets". Both of these techniques can also drive the jet into an ambient medium.

In both the Z-pinch and the ring jets the bow shocks must be driven into gas at densities that can be accommodated within a vacuum system, creating a physical system suitable for diagnosis by optical techniques such as interferometry and Schlieren. In contrast, the experiments discussed here would use denser gas within an enclosed target, enabling them to diagnose the dense shocked gas by radiography. This is thus a complementary approach, with some potential advantages in spatial resolution and in the ability to see structure within very dense regions.

2 Experiment design

In order to create a radiative bow shock it is necessary that the energy flux due to thermal radiative losses from an optically thick layer of shocked material would exceed the energy flux entering the shocked material. One can express this condition as $R_r > 1$, where

$$R_r = \frac{\sigma T_{init}^4}{\rho_o u_s^3/2} = \frac{32(\gamma-1)^4}{(\gamma+1)^8}\frac{\sigma}{c_v^4}\frac{u_s^5}{\rho_o}, \tag{1}$$

in which u_s is the shock velocity, σ is the Stefan-Boltzmann constant, ρ_o is the mass density of the unshocked material, γ is the polytropic index appropriate to the shock transition and the post-shock electron temperature, T_{init} is (Drake 2005)

$$T_{init} = \frac{2(\gamma-1)}{(\gamma+1)^2}\frac{u_s^2}{c_v}, \tag{2}$$

in which c_v is the specific heat at constant volume of the post shock material, approximately equal to $(Z+1)k_B/(Am_p)$ for a fully ionized gas. Here the post-shock level of ionization is Z, the atomic mass of the shocked material is A, the proton mass is m_p, and the Boltzmann constant is k_B. To exceed the threshold of (1), a shock wave in Xe at 1 atm pressure must propagate above about 50 km/s.

Figure 2 shows the schematic design of the experiment. Its operation is as follows. Three beams of Omega EP irradiate a thin plastic layer on the surface of a mid-Z material (Al or Ti). The resulting ablation pressure shocks and accelerates this material, which moves and expands down a collimating tube, allowing the leading edge to accelerate, before emerging as a jet. This technique has been used previously to produce hydrodynamic jets in foam (Foster et al. 2005; Rosen et al. 2005; Sublett et al. 2007). In our case the jet will emerge into a large volume of Xe gas at 1 atmosphere pressure. So long as the leading edge of the jet is moving above

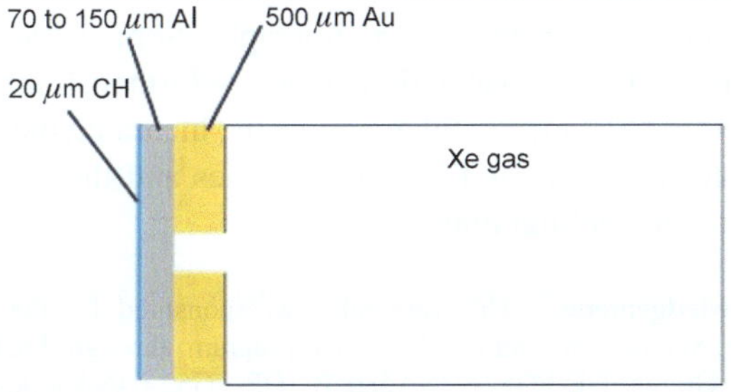

Fig. 2 Schematic of radiative blast wave target

50 km/s, the shock driven into the Xenon will be strongly radiative and will collapse spatially in consequence.

A principal and unknown limitation on such experiments is that for the immediate future at Omega EP, the available beam smoothing will be by Distributed Phase Plates (DPPs) only. This will produce a larger seed for Rayleigh-Taylor growth than has been present in past Omega-60 radiative shock experiments. The simulations discussed below indicate that the higher laser energy can make up for this, by enabling the acceleration of thicker blocks of material. The question of how much material can be accelerated and to what velocity remains significant for this experiment and will need to be addressed by measurements.

The options for irradiation range from 7.5 kJ at 1 ns to 19.5 kJ at 10 ns. Using a 300 μm spot at 1 ns would produce an irradiance of 10^{16} W/cm^2 while using a 500 μm spot at 10 ns would produce 10^{15} W/cm^2. These spot sizes were used to determine the laser irradiance for preliminary experimental design, although the availability of DPPs may alter the precise available spot sizes. We simulated these cases in 1D using HYADES, run with settings that reproduce other experiments. HYADES (Larsen and Lane 1994) is a Lagrangian code solving one set of continuity and momentum equations and three energy equations (for electron, ions, and radiation). It was run with multigroup-diffusion radiation transport and with flux-limited thermal conduction for electron transport, with an electron flux limiter of 0.05.

In both cases, a jet velocity above 100 km/s was obtained and endured for tens of ns. Correspondingly, a strongly radiative shock was produced in the Xe, which collapsed spatially into a thin, high-density layer. Figure 3 shows a result, for the lower irradiance at a time of 80 ns. After a period of initial evolution, the basic structure of the system, along the axis, is as follows, starting with the shock and moving toward the laser. There is a layer of dense, shocked Xe that has radiatively cooled and collapsed. A layer of low-density, ablated Al follows. The shocked Xe and ablated Al are at approximately the ram pressure of the shock. Behind this, a drifting clump of relatively dense Al follows. This supplies the momentum to sustain the shock. Its pressure decreases moving away from the shock, with a decrease in both density and temperature. Trailing this layer, beyond a temperature

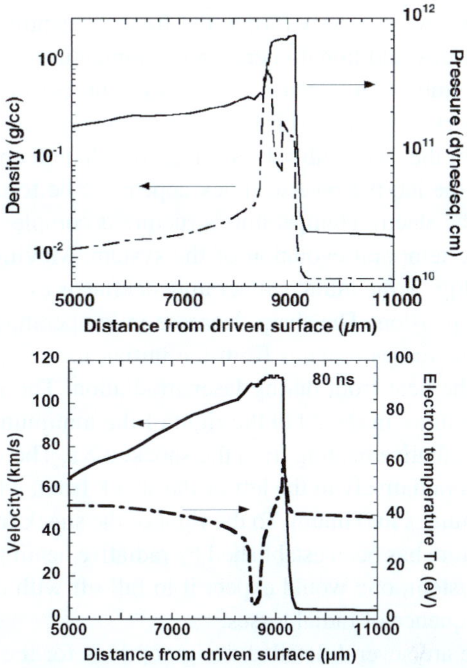

Fig. 3 Simulated structure of jet and blast wave at 80 ns for the 10 ns laser pulse case. (**a**) density and pressure (**b**) velocity and electron temperature

minimum, there is a transition to a lower-pressure, lower-density, freely expanding material that was originally heated and accelerated by the laser. The transition in density originated in the laser ablation front. Earlier in time, the pressure is larger in the laser-heated region; transition in pressure develops during post-laser-pulse cooling.

Some other details about the structure an evolution of the system are worth noting. The transition in pressure at the former ablation front develops quickly (within 10 ns and 1 mm) in the short-pulse case but takes much longer (about 40 ns and 4 mm) in the long-pulse case. The dense layer of shocked Xe is thicker in the figure than would actually be the case with lateral flow. The leading edge of the Al has opposed density and pressure gradients and thus is Rayleigh-Taylor unstable, as is the "working surface" of a typical astrophysical jet. The entire dense structure is about 1 mm thick, much larger than the diagnostic resolution of 10 to 30 μm (to be determined for backlighting on EP). The backlighting might be done either by using a 1 ns pulse with an x-ray framing camera or by using the 80 ps short-pulse beam. Experience gained as the facility is activated will determine the best choice.

The velocity shown in Fig. 3 increases linearly toward the right, as it should for a freely expanding system, and then levels out in the high-pressure region. In the long run, a system like this should develop a reverse shock, where the freely expanding material is slowed to the velocity of the dense material following the accumulated, shocked gas.

After the reverse shock forms, the freely expanding material provides additional energy and momentum to the jet. Enough ambient material must be swept up before this reverse shock will form. This has not happened yet in the case shown. In the case with a 1 ns drive, for which there is less mass in the jet, the reverse shock appears to be forming by 80 ns. The electron temperature structure is complex and reflects the temporal evolution of the system. Moving to the right in Fig. 3, the initial temperature decline is expected for a free expansion. The sharp decrease in temperature as the density increases remains from the initial, limited penetration of the heat front during laser irradiation. The increase in temperature in the Al to the right of the minimum is created by radiative heating from the shocked Xe. The shocked Xe cools radiatively to the left of the shock front, where the temperature is maximum. To the right of the shock front, the temperature has been established by radiative heating. In the actual system, one would expect it to fall off with distance in consequence of radial losses.

There are several detailed design options for the experiment. We tested whether to place a thin gas barrier on the output end of the collimating tube, so that the expansion can occur in vacuum, and found that such a barrier (1 μm plastic) has little effect The jet material might be Al, Ti, or some other mid-Z metal. Variation of this material or backlighter energy can be used to explore details of the jet morphology. The collimator could be Au as indicated, or another dense material such as W. Past experience has shown that some mass is stripped from the interior of such collimating tubes, and that this can limit the ability to diagnose the interior of the jet. One could line the collimating tube with a lower-Z material to reduce this effect. The precise hole diameter can also vary. The hole needs to stay open while the jet flows through it. With 1 ns laser pulses, 300 μm holes have proven to be sufficient in past experiments. When using a 10 ns laser pulse, a larger hole might be needed.

3 Conclusion

We have discussed the design of jet-driven, radiative-blast-wave experiments for a 10 kJ class pulsed laser facility. Using a 10 kJ class laser, it is possible to produce a jet that can drive a radiative bow shock in gas that is dense enough to permit diagnosis by x-ray radiography. Some issues must be addressed through experimentation or further simulations in order to field successful experiments, including the limits imposed by Rayleigh-Taylor instabilities and the optimum approach to backlighting.

Acknowledgements This research was sponsored by the Stewardship Science Academic Alliances program through DOE Research Grants DE-FG52-07NA28058, DE-FG52-04NA00064, and other grants and contracts.

References

Bally, J., Heathcote, S., Reipurth, B., Morse, J., Hartigan, P., Schwartz, R.: Astron. J. **123**, 2627 (2002)
Blondin, J.M., Fryxell, B.A., Konigl, A.: Astrophys. J. **360**, 370 (1990)
Blue, B.E., et al.: Phys. Rev. Lett. **94** (2005)
Boehly, T.R., et al.: Rev. Sci. Instr. **66**, 508 (1995)
Ciardi, A., Lebedev, S.V., Chittenden, J.P., Bland, S.N.: Laser Part. Beams **20**, 255 (2002)
Dal Pino, E.M.D.: In: Fundamentals of Space Environment Science, p. 908 (2005)
Drake, R.P.: Astrophys. Space Sci. **298**, 49 (2005)
Farley, D.R., et al.: Phys. Rev. Lett. **83**, 1982 (1999)
Foster, J.M., et al.: Phys. Plasmas **9**, 2251 (2002)
Foster, J.M., et al.: Astrophys. J. Lett. **634**, L77 (2005)
Gardiner, T.A., Frank, A., Jones, T.W., Ryu, D.: Astrophys. J. **530**, 834 (2000)
Kasperczuk, A., et al.: Phys. Plasmas **13** (2006)
Larsen, J.T., Lane, S.M.: J. Quant. Spectrosc. Radiat. Transfer **51**, 179 (1994)
Lebedev, S.V., et al.: Astrophys. J. **616**, 988 (2004)
Lebedev, S.V., et al.: Astrophys. J. **564**, 113 (2002)
Loupias, B., et al.: Astrophys. Space Sci. **307**, 103 (2007a)
Loupias, B., et al.: Phys. Rev. Lett. **99** (2007b)
Moraghan, A., Smith, M.D., Rosen, A.: Mon. Not. R. Astron. Soc. **371**, 1448 (2006)
Nicolai, P., Tikhonchuk, V.T., Kasperczuk, A., Pisarczyk, T., Borodziuk, S., Rohlena, K., Ullschmied, J.: Phys. Plasmas **13** (2006)
Nicolai, P., Tikhonchuk, V.T., Kasperczuk, A., Pisarczyk, T., Borodziuk, S., Rohlena, K., Ullschmied, J.: Astrophys. Space Sci. **307**, 87 (2007)
O'Sullivan, S., Ray, T.P.: Astron. Astrophys. **363**, 355 (2000)
Plewa, T.: Acta Astron. **43**, 235 (1993)
Reipurth, B., Bally, J.: Ann. Rev. Astron. Astrophys. **39**, 403 (2001)
Rosen, A., Smith, M.D.: Astron. Astrophys. **413**, 593 (2004)
Rosen, P.A., et al.: Astrophys. Space Sci. **298** (2005)
Sublett, S.: Physics and Astronomy. University of Rochester, Rochester (2007)
Sublett, S., Knauer, J.P., Igumenshchev, I.V., Frank, A., Meyerhofer, D.D.: Astrophys. Space Sci. **307**, 47 (2007)
Vishniac, E.T.: Astrophys. J. **274**, 152 (1983)
Xu, J.J., Hardee, P.E., Stone, J.M.: Astrophys. J. **543**, 161 (2000)

Astrophys Space Sci (2009) 322: 101–105
DOI 10.1007/s10509-008-9897-1

ORIGINAL ARTICLE

Laboratory experiments to study supersonic astrophysical flows interacting with clumpy environments

P.A. Rosen · J.M. Foster · B.H. Wilde · P. Hartigan · B.E. Blue · J.F. Hansen · C. Sorce · R.J.R. Williams · R. Coker · A. Frank

Received: 12 June 2008 / Accepted: 27 August 2008 / Published online: 20 September 2008

Abstract A wide variety of objects in the universe drive supersonic outflows through the interstellar medium which is often highly clumpy. These inhomogeneities affect the morphology of the shocks that are generated. The hydrodynamics are difficult to model as the problem is inherently 3D and the clumps are subject to a variety of fluid instabilities as they are accelerated and destroyed by the shock. Over the last two years, we have been carrying out experiments at the University of Rochester's Omega laser to address the interaction of a dense-plasma jet with a localised density perturbation. More recently, we have turned our attention to the interaction of a shock wave with a spherical particle. We use a 1.6-mm diameter, 1.2-mm length Omega hohlraum to drive a composite plastic ablator (which includes bromine to prevent M-band radiation from preheating the experiment). The ablator acts as a "piston" driving a shock into 0.3 $\mathrm{g\,cm^{-3}}$ foam containing a 0.5-mm diameter sapphire sphere. We radiograph along two orthogonal lines of sight, using nickel or zinc pinhole-apertured X-ray backlighters, to study the subsequent hydrodynamics. We present initial experimental results and two-dimensional simulations of the experiment.

Keywords Laboratory astrophysics · Supersonic astrophysical flows · Shocked clumps · Omega laser

P.A. Rosen (✉) · J.M. Foster · R.J.R. Williams
AWE Aldermaston, Reading, RG7 4PR, UK
e-mail: paula.rosen@awe.co.uk

B.H. Wilde · R. Coker
Los Alamos National Laboratory, Los Alamos, NM 87545, USA

P. Hartigan
Rice University, P.O. Box 1892, Houston, TX 77251-1892, USA

B.E. Blue
General Atomics, P.O. Box 85608, San Diego, CA 92186-5608, USA

J.F. Hansen · C. Sorce
Lawrence Livermore National Laboratory, Livermore, CA 94550, USA

A. Frank
University of Rochester, Rochester, NY 14627-0171, USA

1 Introduction

Much of the analysis of astrophysical flows relies upon models that make simplistic assumptions about the geometries of the flows. One of the most important of these assumptions, and one that is rarely realised in nature, is that of a pre-shock medium of uniform density. In real astrophysical situations, shock waves move into media that are often highly clumpy, which affects the morphology and spectra of material heated in the shocks, and also affects how the shocks propagate. It is difficult to model shocks that move into clumpy media in astrophysical situations because the problem is inherently 3D, and the clumps are subject to a variety of fluid instabilities as they are accelerated and destroyed by the shock (Poludnenko et al. 2002, 2004). Examples of shocked clumps in jets from young stars are shown in Fig. 1.

One way to shed some light on how shocks propagate into clumpy media is to recreate the situation in a laboratory plasma. In our previous NLUF-sponsored experiments (Foster et al. 2005 and Coker et al. 2007) at the Omega laser facility, we developed a well-characterised experimental 'test-bed' to study high-Mach-number, compressible, potentially turbulent, dense-plasma jets and their associated bow shocks, as they propagate through fluids that are either

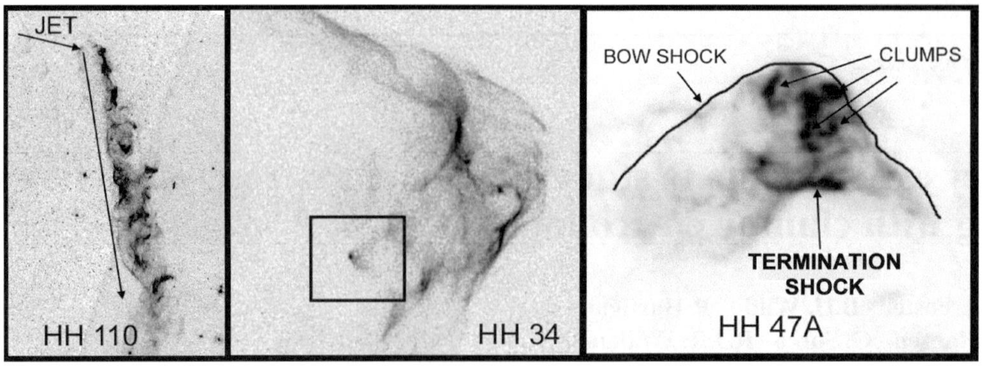

Fig. 1 Shocked clumps in jets from young stars. *Left* (**a**)—HST image of the HH 110 flow. The jet enters from the left, impacts a molecular cloud, and deflects downward, producing a spray of shocked gas. *Middle* (**b**)—HST images of the bow shock of protostellar jet HH 34. The large bow shock moves to the right in this image, and has overtaken a small clump shown in the box. *Right* (**c**)—An HST image of the HH 47A bow shock, showing a an inhomogeneous region. Images taken at a later time show that small clumps from the jet move upward through the Mach disk, and may survive all the way to the bow shock

homogeneous or contain a single, large-scale, density perturbation. These experiments were relevant to the astrophysical object HH110.

The new experiments, reported here, are addressing objects such as HH34 and HH47A, where small clumps become accelerated by much larger shocks. The experiment generates a strong, planar bow shock which collides with a spherical obstacle located at various distances from the axis of the flow.

In addition to the Omega experiments, astronomical observations have been carried out which measured velocity shear and turbulence in deflected and entrained flows. These used the telescope at Kitt Peak National Observatory (Hartigan et al. 2007a) and the Hubble Space telescope (Hartigan et al. 2007b).

Others have studied the interaction of a strong, near-planar shock with a single, spherical particle (Klein et al. 2000, 2003 and Robey et al. 2002). In the work by Klein and Robey, an ablative drive was used to launch a strong ($M = 10$) shock in a small (typically, 800 μm diameter) solid-polystyrene or foam-filled shock tube containing a single (typically, 100 μm diameter) spherical copper or aluminium particle. As in their experiments, we use X-ray backlighting radiography to record the hydrodynamic motion of this particle, in a succession of 'snapshot' images. These experiments are analogous to 'shock-bubble' experiments in conventional shock tubes (Haas and Sturtevant 1987; Jacobs 1993 and Ranjan et al. 2005), although significantly, the much higher Mach number available in the laser-driven case provides access to differences of hydrodynamic development for $M > 2$. In recent work (Hansen et al. 2007), the interaction of a strong shock with two adjacent aluminium spheres (120 μm diameter spheres in 0.3 $\mathrm{g\,cm^{-3}}$ density foam) has been studied.

Our present work builds on the experiments by Klein et al. Our longer-term goal is to study large shocks that overrun multiple small-diameter particles. That is, we will move from an experiment in which the jet/shock and density discontinuity scale sizes are comparable to one in which the scale sizes are significantly different (bow shock and planar shock propagation through small-scale clumpy medium).

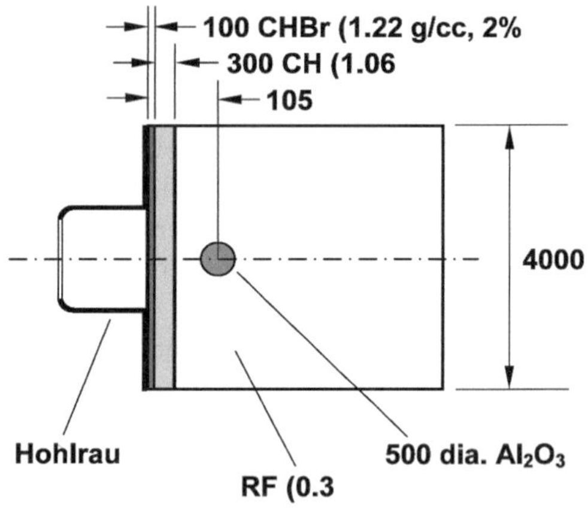

Fig. 2 The target design for the shock-sphere interaction experiment. All dimensions are in μm. The hohlraum measures 1.6 mm in diameter and is 1.2 mm long and the foam cylinder measure 4000 μm in length

2 Experiment configuration

The design for the experiments is shown in Fig. 2. Radiation drive is generated using a 1.6-mm-diameter, 1.2-mm-length cylindrical gold hohlraum target with a single 1.2-mm-diameter laser-entry hole. The experimental package is mounted over a 1.6-mm-diameter hole in the hohlraum, immediately opposite the laser entry hole. The hohlraum is heated by 12 beams of the Omega laser with a total energy of 6 kJ in a 1-ns duration, constant power laser pulse of 0.35-μm wavelength. The details and modelling

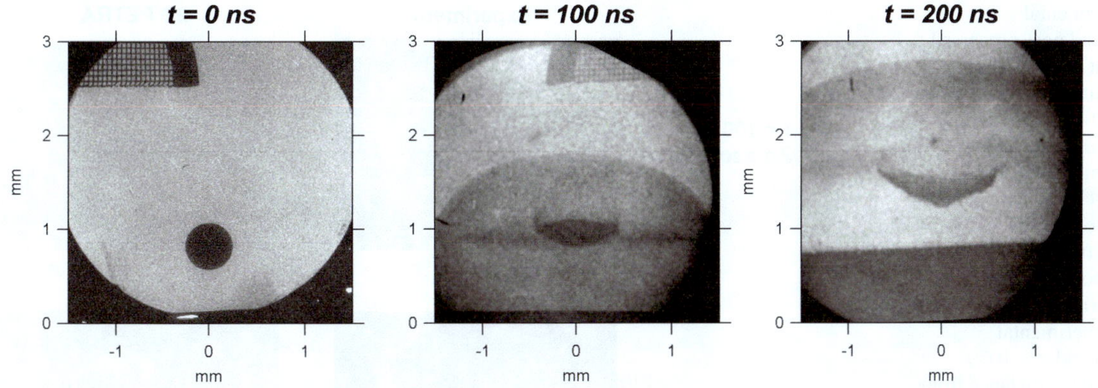

Fig. 3 Pre-shot radiograph ($t = 0$ ns) and experimental radiographs at $t = 100$ and 200 ns. The initial position (μm) of the ball (r, z) is (208, 1229), (160, 1114) and (166, 1195) respectively. The nominal position of the ball is (0, 1050) where $r = 0$ corresponds to a position on the cylindrically-symmetric axis of the experimental configuration and $z = 1050$ corresponds to the distance from the driven side of the ablator to the centre of the sapphire sphere. The radiographs are backlit with nickel

of this hohlraum are described in detail elsewhere (Foster et al. 2005). As the hohlraum used in the experiments described here has no end wall, the albedo is somewhat reduced and the resulting peak radiation temperature is lower (170–180 eV) than that in the previously referenced experiments. The hohlraum drives a composite plastic ablator consisting of a 100-μm thickness layer of CHBr (2% atomic Br, and 1.22 $\mathrm{g\,cm^{-3}}$ density) and a 300 μm thickness layer of CH (polystyrene of 1.06 $\mathrm{g\,cm^{-3}}$ density). The ablator acts as a 'piston' driving a shock into foam containing a 500-μm-diameter sapphire sphere. The foam comprises a 4-mm diameter, 6-mm length cylinder of resorcinol-formaldehyde ($C_{15}H_{12}O_4$) foam, of 0.3 $\mathrm{g\,cm^{-3}}$ density. Resorcinol-formaldehyde foam is used because of its very small (<1 μm) pore size. The centre of the sapphire sphere is placed 1050 μm from the driven surface of the ablator along the cylindrical axis of symmetry ($r = 0$). The precise position of the sapphire sphere is measured pre-shot, for each target, via dual-axis X-ray radiography. The precise position of the sphere inevitably varies from target to target. The size of the sapphire sphere is chosen such that the radius of curvature of the shock is much greater that the radius of curvature of the sphere (see Sect. 4 below).

The evolution of the shock-sphere interaction is diagnosed by X-ray backlighting radiography, using a pinhole-apertured, laser-produced-plasma, X-ray backlighting source to project an image of the sphere either directly onto radiographic film or via a gated micro-channel-plate detector onto film. Both nickel and zinc are (separately) used as backlighter materials. The He-like resonance lines of nickel and zinc are 7806 and 8999 eV respectively and lie above the K-absorption edge of aluminium (1560 eV). The backlighter foils are illuminated with three laser beams (~425 J per beam) in a 1-ns duration laser pulse (0.35-μm wavelength).

Figure 3 shows experimental radiographs, backlit with nickel, from two separate laser shots, backlit at $t = 100$ and 200 ns together with a pre-shot radiograph (equivalent to $t = 0$ ns). The bow shock and distorted sphere are clearly visible in the latter two radiographs.

3 Modelling

The experiment is modelled using the RAGE (Gittings 1992) and PETRA (Youngs, 1982, 1984) hydrocodes. Due to limitation of space, only the AWE (PETRA) calculations will be described in this paper. The experimental geometry was calculated, but the physical presence of the hohlraum was not included in the calculation. Instead, the hohlraum drive was approximated by a Planckian temperature-time profile input in a region of low-density gold immediately adjacent to the ablator. The code uses tabular LTE opacities which are calculated off-line using the IMP (Rose 1992) opacity code. Equation-of-state data are also input in tabular form using the SESAME (Lyon and Johnson 1992) database. Square, 5 μm zones are used in the central part of the problem (up to a radius of 800 μm), then the size of the radial zones increases geometrically, with an increase in cell size of 10% per cell (to limit the overall number of cells in the problem to make the calculation run more efficiently). The simulation is post-processed to produce a synthetic radiograph for comparison with the experimental data.

4 Hydrodynamic scaling

In order for these experiments to be directly applicable to astrophysical objects, a number of hydrodynamic criteria must be met. In particular (Ryutov et al. 1999), in order to

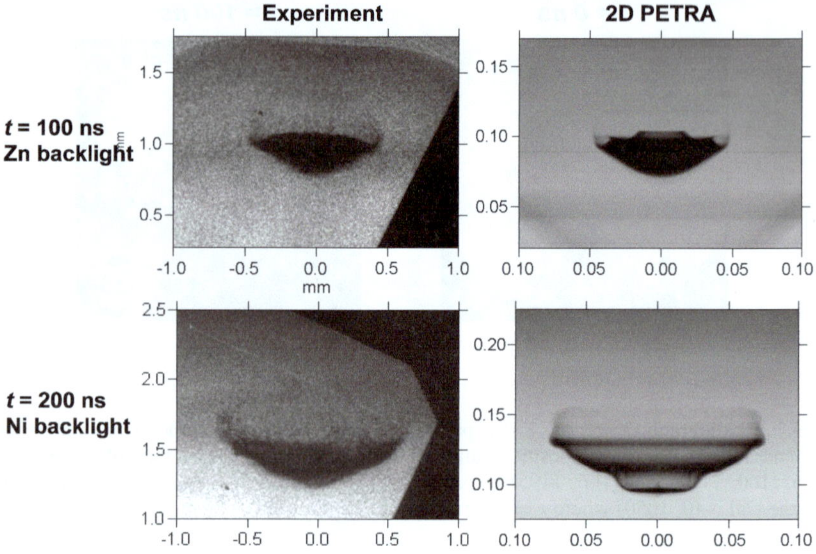

Fig. 4 Experimental radiographs and post-processed simulations at $t = 100$ and 200 ns, backlit with zinc and nickel respectively. The initial positions (μm) of the ball (r, z) in the experiment are (27, 1034) and (166, 1195) respectively. The position of the ball in the simulation is (0, 1050) where $r = 0$ corresponds to a position on the cylindrically-symmetric axis of the experimental configuration and $z = 1050$ corresponds to the distance from the driven side of the ablator to the centre of the sapphire sphere

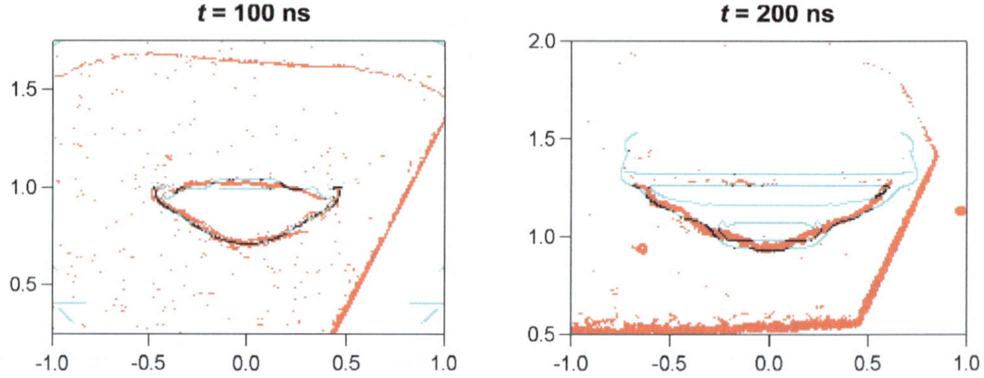

Fig. 5 Quantitative comparison of experimental data with post-processed simulation at $t = 100$ and 200 ns using an edge detection algorithm. Red represents experimental data and blue represents the PETRA simulation. Black denotes where the experiment and simulation agree precisely

scale, various dimensionless quantities must be comparable and both the laboratory and astrophysical systems must satisfy a number of dissipation criteria, such as negligible heat conduction, viscosity and energy loss by radiation. In practice, it has been found (Klein et al. 1994) that the behaviour of a strong shock running into an obstacle converges for Mach numbers above ~2.5; we obtain experimental internal isothermal Mach numbers of ~5. Another dimensionless number that must be matched is the density ratio η, of the obstacle to the surrounding medium. At present, the experiments have $\eta \sim 10$ while the astrophysical objects (cold clumps embedded in a warmer medium) probably have wide ranges of η, from ~3 to 100 or more.

There are three timescales in shock/obstacle problems: the time it takes for the shock to cross the obstacle (t_c), the time it takes to destroy the obstacle (t_d), and the evolution timescale of the shock itself (t_s). For $\eta \sim 10$ and treating the shock as a Sedov blast wave (Klein et al. 2003), we have $t_c \sim 2r_0/v_s$, $t_d \sim 3r_0/v_s$, and $t_s \sim 0.1r_s/v_s$, where r_0 is the radius of the obstacle, r_s is the radius of curvature of the shock, and v_s is the velocity of the shock. Thus, in order to be in the ideal 'small cloud' limit, we need the shock to have a radius of curvature $r_s \gg r_0$. Note this only applies to the part of the shock that interacts with the obstacle. If we do not meet this requirement, the shock evolves significantly in the time of the experiment, and the experiment would be more like a YSO jet interacting with ISM clumps (where the clumps are on the order of the size of the shock front) than a SN blast wave overrunning ejecta.

For the Ryutov Euler scaling to be valid, dissipative mechanisms must be unimportant. That is, the system must be collisional ($l_{\mathrm{coll}} \ll r_0$), thermal conduction must be small ($\mathrm{Pe} \gg 1$), radiative cooling should be small over the relevant timescale ($t_{\mathrm{cool}} \gg t_d$), and viscous dissipation should be negligible ($\mathrm{Re} \gg 1$). Using representative values of $n \sim 10^{23}\ \mathrm{cm}^{-3}$, $T \sim 1$ eV, $v \sim 10^6\ \mathrm{cm\,s}^{-1}$, $A \sim 10$ amu, and $Z \sim 3$, and solving the expressions in Ryutov et al. (1999), we find that $l_{\mathrm{coll}}/r_0 \sim 10^{-6}$, $\mathrm{Pe} \sim 10^7$, $t_d/t_{\mathrm{cool}} \sim 10^{-4}$, and $\mathrm{Re} \sim 10^5$. Thus, hydrodynamic scaling of these experiments to astrophysical objects should apply. It should be noted that we are not fully in the decoupled regime (regions of the experiment have $\Gamma \geq 1$).

5 Comparison of experimental data with simulation

Figure 4 shows a comparison of experimental data at 100 and 200 ns with a post-processed simulation. The gross hydrodynamic features are reproduced well by the simulation. The question arises of how to make a meaningful quantitative comparison between experiment and simulation. It was decided to use an edge detection algorithm (Canny 1986) to compare the experimental data to the simulation. Figure 5 shows a comparison at $t = 100$ and 200 ns.

6 Summary

We have carried out an astrophysically-relevant experiment to study the interaction of a shock with a spherical "clump". The simulations broadly agree with the gross features present in the experimental data, but differ in the finer-scale structure. Further experiments are planned which refine some of the experimental details (e.g. a lower-opacity ball to probe the structure within the sphere). These experiments lay the foundation for an experimental study into the interaction of a shock with multiple clumps.

Acknowledgements We would like to thank Peter Graham (AWE) for modifying his subroutine which calculates volume fractions in PETRA, and for shedding light on some of the idiosyncrasies of PETRA and the AWE supercomputer. We also extend our thanks to the teams at General Atomics for target manufacture and characterisation and the laser and operations staff at the Omega laser facility.

References

Canny, J.: IEEE Trans. Pattern Anal. Mach. Intell. **8**, 679 (1986)
Coker, R.F., et al.: Astrophys. Space Sci. **307**, 57–62 (2007)
Foster, J.M., et al.: Astrophys. J. Lett. **634**, L77 (2005)
Gittings, M.L.: Numerical Methods Symposium, April 28–30 (1992). Copies may be ordered from the Defence Nuclear Agency, 56801 Telegraph Road, Alexandria, VA 22310-3398
Haas, J.-F., Sturtevant, B.: J. Fluid Mech. **181**, 41 (1987)
Hansen, J.F., et al.: Astrophys. Space Sci. **307**, 147–152 (2007)
Hartigan, P., et al.: Proposal 06B-0182 (2007a)
Hartigan, P., et al.: Proposal 11179 (2007b)
Lyon, S.P., Johnson, J.D.: Sesame: the Los Alamos national laboratory equation of state database. Los Alamos National Laboratory, Los Alamos, NM, LA-UR-92-3407 (1992)
Jacobs, J.: Phys. Fluids A **5**, 2239 (1993)
Klein, R.I., McKee, C.F., Colella, P.: Astrophys. J. **420**, 213 (1994)
Klein, R.I., et al.: Astrophys. J. Suppl. Ser. **127**, 379–383 (2000)
Klein, R.I., et al.: Astrophys. J. **583**, 245 (2003)
Poludnenko, A., et al.: Astrophys. J. **576**, 832 (2002)
Poludnenko, A., et al.: Astrophys. J. **604**, 213 (2004)
Ranjan, D., Anderson, M., Oakley, J., Bonazza, R.: Phys. Rev. Lett. **94**, 184507 (2005)
Robey, H.F., et al.: Phys. Rev. Lett. **89**, 85001 (2002)
Rose, S.J.: J. Phys. B **25**, 1667 (1992)
Ryutov, D., Drake, R.P., Kane, J., Liang, E., Remington, B.A., Wood-Vasey, W.M.: Astrophys. J. **518**, 821 (1999)
Youngs, D.L.: In: Morton, K.W., Baines, M.J. (eds.) Numerical Methods for Fluid Mechanics. Academic Press, London (1982)
Youngs, D.L.: Physica D **12**, 32 (1984)

Astrophys Space Sci (2009) 322: 107–111
DOI 10.1007/s10509-009-9983-z

ORIGINAL ARTICLE

Scaling laws for radiating fluids: the pillar of laboratory astrophysics

E. Falize · S. Bouquet · C. Michaut

Received: 16 June 2008 / Accepted: 29 December 2008 / Published online: 16 January 2009

Abstract In this paper, we derive the scaling laws for different radiating fluids. The studied regimes are relevant for both laboratory astrophysics and High Energy Density Physics. Using Lie groups theory, we obtain scaling laws, the similarity properties and the number of free parameters to rescale experiments.

Keywords Scaling laws · Lie groups · Radiation hydrodynamics · Radiating fluids · Laboratory astrophysics · Optically thin plasmas · Optically thick plasmas

1 Introduction

The Universe is composed of a multitude of objects and phenomena for which radiation and matter are strongly coupled (Lowrie et al. 1999; Mihalas and Mihalas 1999; Drake 2006). From galaxy collisions to the matter ejected by young stellar objects, the radiative processes play a crucial role in the evolution of these objects. High Energy Density Laboratory Astrophysics (HEDLA) is a new approach for astronomers and astrophysicists to study the evolution of radiating plasmas (Remington et al. 2006). The spectacular development of laser facilities allow to bring the matter up to extreme states of density and temperature in laboratory. In laser targets, plasma densities and temperatures can be monitored respectively up to 100 g.cm^{-3} and 10^8 K which cover the range of variation of these quantities in typical stars like the Sun (Drake 2006). Although the spatial and temporal scales are very different, the astrophysical relevance of these experiments can be checked thanks to scaling laws provided that the physical system under study satisfies similarity properties (Ryutov et al. 1999, 2001; Ryutov and Remington 2002; Falize et al. 2008b). Thus, we must use a rigorous formalism in order to examine similarity properties and to establish scaling laws. In this paper we propose to use the Lie groups theory which appears as a systematical method in order to derive scaling laws and determine the similarity properties. In the first part, we discuss the similarity principle and remind elementary properties of Lie groups. In the second part, we present the scaling laws of hydrodynamics and optically thin radiative hydrodynamics systems. Using Drake's theoretical analogy we deduce scaling laws in the equilibrium diffusion approximation where the ingredients to describe radiation are the radiative flux. Finally, the equilibrium diffusion approximation including radiative pressure, radiative energy density and radiative flux are considered.

E. Falize (✉) · S. Bouquet
CEA-DAM-DIF, 91297 Arpajon cedex, France
e-mail: emeric.falize@obspm.fr

E. Falize · S. Bouquet · C. Michaut
Laboratoire Univers et Théories (LUTH), CNRS: UMR8102
INSU Observatoire de Paris Université Denis Diderot, Paris VII,
5 Place Jules Janssen, 92190 Meudon, France

2 Similarity, the homothetic symmetry and scaling laws

Laboratory astrophysics experiments can be divided into two fundamental classes which are connected to the type of similarity. The first class, also called *sameness* (Takabe 2001), includes the experiments which consists in reproducing thermodynamical conditions identical to those of the astrophysical media such as the opacity or equations of state experiments (Remington et al. 2006). More generally these experiments do not introduce the notions of temporal and spatial scales and thus they do not introduce the concept of scaling laws. The second class is the experiments which reproduce astrophysical objects, despite the fact that there exists 10–20 orders of magnitude in temporal and spatial scales and the different thermodynamical regimes (Ryutov et al. 1999, 2001; Ryutov and Remington 2002). This class of experiments corresponds to the scaled similarity experiments which verify structured relations (the scaling laws) which assure that the equations are rigorously scaled invariant. This type of experiments gathers, for instance, high Mach numbers plasmas and can be divided into two subclasses which depend on the type of equation invariance. Thus, we have the *absolute scaled similarity* which is obtained when the equation system is strictly invariant under the mathematical sense, *i.e.* conservation of the equation form added to conservation of all coefficients. In physics, usually these coefficients contain some microscopic physics as the ionization degree or magnetic field as explained below. In general this type of similarity is very constraining and it is very difficult to satisfy them. Consequently, it is more interesting to introduce the concept of *global scaled invariance* which is a more general notion and which only needs the invariance of the equation form (Ryutov et al. 2001; Drake 2006). This similarity type is less constraining and introduces additional free parameters in order to rescale astrophysical experiments as we will see in the next section. Symmetry provides tools to analyze the solutions of a problem without needing to solve the problem explicitly or it can also suggest methods to simplify it. The invariant transformation group theory elaborated by Lie is a very powerful formalism of theoretical physics in order to study the symmetry properties of partial differential equations (Coggeshall 1991; Coggeshall and Meyer-ter-Vehn 1992; Ibragimov 1999). This theory explains and unifies apparently disparate and ad hoc methods for obtaining solutions to differential equations. Among all Lie groups, one of them, namely the one-parameter homothetic group (HG) is frequently used, firstly because of its simplicity and, then, because it provides more general self-similar solutions than those derived from dimensional analysis. This property arises because the HG is a sub-group of the scaling transformation. Since the philosophy of HEDLA is to recreate, on short scales systems having astronomical size, it seems quite natural to use HG in order to study similarity properties, scaling laws and even self-similarity of these systems. In this paper we focus on two first points. Among the different important results that we can establish with this formalism, one of them is the fact that the invariance of equations by HG implies that the Rankine-Hugoniot relations are also invariant. In other words if the equation system admits a homothetic symmetry, then we assure that small scale shocks correspond to the homothetic structures of astrophysical shocks.

3 From optically thin to optically thick plasmas

Optically thin plasmas are a very important class of plasmas in astrophysics because they are easy to observe. When the cooling (or heating) characteristic time of a plasma gets close to its dynamical time, this should be considered in the modeling. Concerning optically thin plasmas, i.e. $\lambda_p \gg L$ (λ_p is the mean free path of photons and L is the characteristic plasmas of the system), a simple modeling of radiating losses (or heating) can be done simply by introducing a loss (or gain) of entropy. Consequently, the plasma is described by the following equations:

$$\frac{\partial \rho}{\partial t} + \vec{\nabla}.[\rho \vec{v}] = 0, \qquad dM = \rho dV, \qquad \frac{d\vec{v}}{dt} = -\frac{1}{\rho}\vec{\nabla} P_{th} \tag{1}$$

$$\frac{dP_{th}}{dt} - \gamma \frac{P_{th}}{\rho}\frac{d\rho}{dt} = -(\gamma - 1)\mathcal{L}(\rho, T) \tag{2}$$

where d/dt is the Lagrangian derivative and ρ, $\vec{v}$, P_{th}, γ, T and M are respectively the density, velocity, thermal pressure, polytropic index, temperature and the mass of the fluid. We can write $dM = \rho d^N x$ where N is the dimensionality of problem. The function $\mathcal{L}(\rho, T)$ writes $\mathcal{L}(\rho, T) = \mathcal{Q}_1(\rho, T) + \mathcal{Q}_2(\rho, T)$ where $\mathcal{Q}_1$ and $\mathcal{Q}_2$ are energy sources (or losses). We suppose that $\mathcal{Q}_i$-terms are given by:

$$\mathcal{Q}_i = \mathcal{Q}_{0,i}\,\rho^{\epsilon_i} P^{\zeta_i} x^{\theta_i} \tag{3}$$

where $\mathcal{Q}_{0,i}$, ϵ_i, ζ_i, θ_i are four constants (Drake 2006; Falize et al. 2008a). This form generalizes the optically thin case where $\theta = 0$. The arbitrarity of ϵ and ζ is motivated by the fact that $\mathcal{Q} \propto \kappa_P \sigma T^4$ where σ is the Stefan-Boltzmann constant and κ_P is the Planck opacity which can be modeled by a power law at high temperature and by the fact that several continue processes can be modeled exactly or approximatively by power laws. Experimentally, heating can represent the laser energy deposition by inverse bremsstrahlung although in astrophysical situation the heating can be due to stars or dust in interstellar medium. From the astrophysical viewpoint, this system of equations describes interstellar jets and bow shocks, radiating shocks (point C in Drake diagram, Drake 2006, Fig. 7.17) in Polars (Falize et al. 2008a, 2008b) or supernova remnants. In order to establish (2) we assumed a polytropic evolution, *i.e.* $P_{th} = (\gamma - 1)\rho e$ where e is the specific internal energy. Finally, we suppose that we have a Zeldovich-Raizer fluid (Zeldovich and Raizer 1967), thus the equation of state, which should be added to close the equations system, is given by $P_{th} = \varepsilon_0[Z]\rho^{\mu} T^{\nu}$ where $\varepsilon_0[Z]$ is only function of ionization and we have the constraint $\gamma(1-\nu) = (\mu - \nu)$ in order to assure the consistency of the thermodynamic description of the polytropic gas (Zeldovich and Raizer 1967). The relation between the typical quantities in astrophysical objects and laboratory experiments (that we note with $\sim$) are given by:

$x = a^{\delta_1}\tilde{x}$, $t = a^{\delta_2}\tilde{t}$, $\vec{v} = a^{\delta_3}\vec{\tilde{v}}$, $M = a^{\delta_4}\tilde{M}$, $\rho = a^{\delta_5}\tilde{\rho}$, $P_{th} = a^{\delta_6}\tilde{P}_{th}$, $T = a^{\delta_7}\tilde{T}$, $\gamma = a^{\delta_8}\tilde{\gamma}$, $\varepsilon_0 = a^{\delta_9}\tilde{\varepsilon}_0$, $\mathcal{Q}_1 = a^{\delta_{10}}\tilde{\mathcal{Q}}_1$, $\mathcal{Q}_2 = a^{\delta_{11}}\tilde{\mathcal{Q}}_2$, $\mathcal{Q}_{0,1} = a^{\delta_{12}}\tilde{\mathcal{Q}}_{0,1}$, $\mathcal{Q}_{0,2} = a^{\delta_{13}}\tilde{\mathcal{Q}}_{0,2}$ where a is the group parameter and δ_i are the homothetic exponants. Rescaling ε_0, and $\mathcal{Q}_i$ can absorb a modification of Z (global scaled invariance) from one system to the other (for example in bremsstrahlung cooling $\mathcal{Q}_0 \propto Z^2$). The invariance of equations under the HG gives the group invariants (Ibragimov 1999), namely: $I_1 = vt/x = St$ (Strouhal number), $I_2 = \gamma$, $I_3 = P_{th}t/\rho v x = Eu \times St = St/[\gamma\mathcal{M}^2]$ (Eu: Euler number, $\mathcal{M}$: Mach number), $I_4 = \mathcal{Q}_1 t/P_{th} \propto t/t_{\mathcal{Q}_1}$, $I_5 = \mathcal{Q}_2 t/P_{th} \propto t/t_{\mathcal{Q}_2}$, where $t_{\mathcal{Q}_i}$ is the characteristic time of the sources Q_i and $I_6 = M/[\rho x^N]$ (mass conservation). When we can neglect the cooling or heating effects (I_4 (or I_5) $\ll 1$) the equation system is equivalent to the hydrodynamics equations. This similarity properties are extensively studied in laboratory astrophysics as well as high density energy physics contexts. Using the homothetic group it is easy to recover all these results which are summed up in Table 1. In each column we give the conserved quantities. In Table 2, we propose to rescale the problem of radiative shock in polars (Chevalier and Imamura 1982; Mignone 2005; Falize et al. 2008a). From the second column in this table we present the global invariance results, where we have generally four free parameters (δ_1, δ_5, δ_6 and δ_9). If we preserve ionization, *i.e.* $\varepsilon_0/\tilde{\varepsilon}_0$ (and $\delta_9 = 0$), $\mathcal{Q}_{0,1}/\tilde{\mathcal{Q}}_{0,1} = 1$ and $\mathcal{Q}_{0,2}/\tilde{\mathcal{Q}}_{0,2} = 1$ only two (resp. one) exponent(s) remain(s) for a single source (resp. two sources). Since we have two more (resp. three more) constraints compared to the first column. The single cooling process correspond to bremsstrahlung, which constitutes the more efficient cooling process in the accretion column, with $\mathcal{Q}_1 \propto \rho^2 T^{1/2}$. In order to express $\mathcal{Q}_1$ in terms of ρ and P_{th}, we have taken an ideal equation of state. Therefore, we get $\epsilon_1 = 3/2$, $\zeta_1 = 1/2$ and $\theta_1 = 0$. In the fourth column, the second cooling process is the cyclotronic cooling (Saxton et al. 1998) with $\mathcal{Q}_2 \propto \rho^{0.15}T^{2.5}$ $\epsilon_2 = -2.35$, $\zeta_2 = 5/2$ and $\theta_2 = 0$ and (we also have an ideal gas). A detail discussion on this result is given in Falize et al. (2008a). Now if we consider that the source term is equivalent to radiative flux term, we have

$$\mathcal{Q} \propto \vec{\nabla}.\vec{F}_{rad} = \vec{\nabla}.[-\kappa_{rad}\vec{\nabla}T] \sim \rho^m T^{n+1} x^{-2} \tag{4}$$

where κ_{rad} is the radiative conductivity given by $\kappa_{rad} = \kappa_0 \rho^m T^n$. The last relation is obtained by dimensional con-

Table 1 Scaling for hydrodynamics system. We indicate the invariant quantities in the top of table. The generalized scaling laws are obtain by Ryutov et al. (1999), the scaling laws with s (entropy) invariant (for $\gamma \neq 1$) are derived by Birkhoff (1950), we extend to the isothermal case. The sixth and last column is obtain by Basko and Johner (1998)

Scaling ratio	Ryutov et al. (1999)	$s, \gamma \neq 1$	$s, \gamma = 1$	s, ρ, P, v	s, M	v, M
$x/\tilde{x}$	a^{δ_1}	a^{δ_1}	a^{δ_1}	a^{δ_1}	a^{δ_1}	a^{δ_1}
$t/\tilde{t}$	$a^{\delta_1-(\delta_6-\delta_5)/2}$	a^{δ_2}	a^{δ_1}	a^{δ_1}	$a^{[1+N(\gamma-1)/2]\delta_1}$	a^{δ_1}
$v/\tilde{v}$	$a^{(\delta_6-\delta_5)/2}$	$a^{\delta_1-\delta_2}$	1	1	$a^{-[N(\gamma-1)/2]\delta_1}$	1
$\rho/\tilde{\rho}$	a^{δ_5}	$a^{2(\delta_1-\delta_2)/(\gamma-1)}$	a^{δ_5}	1	$a^{-N\delta_1}$	$a^{-N\delta_1}$
$P_{th}/\tilde{P}_{th}$	a^{δ_6}	$a^{2\gamma(\delta_1-\delta_2)/(\gamma-1)}$	a^{δ_5}	1	$a^{-\gamma N\delta_1}$	$a^{-N\delta_1}$
$T/\tilde{T}$	$a^{(\delta_6-\mu\delta_5)/\nu}$	$a^{2(\gamma-\mu)(\delta_1-\delta_2)/[\nu(\gamma-1)]}$	$a^{[(1-\mu)/\nu]\delta_5}$	1	$a^{N(\mu-\gamma)\delta_1/\nu}$	$a^{N[(\mu-1)/\nu]\delta_1}$
$s/\tilde{s}$	$a^{\delta_6-\gamma\delta_5}$	1	1	1	1	$a^{N(\gamma-1)\delta_1}$
$M/\tilde{M}$	$a^{\delta_5+N\delta_1}$	$a^{(2/(\gamma-1)+N)\delta_1-[2/(\gamma-1)]\delta_2}$	$a^{N\delta_1+\delta_5}$	$a^{N\delta_1}$	1	1

Table 2 Scaling for optically thin plasmas for power law models of sources (second column) and application for plane geometry ($N = 1$) for the radiative shock problem for magnetic cataclysmic variables (third and fourth column). BC, CC correspond respectively to bremsstrahlung cooling and cyclotronic cooling

Physical ratio	Global invariance	Ideal fluids + BC	Ideal fluids + BC + CC
$x/\tilde{x}$	a^{δ_1}	$a^{\delta_6-2\delta_5}$	$a^{-3\delta_5/40}$
$t/\tilde{t}$	$a^{\delta_1+(\delta_5-\delta_6)/2}$	$a^{(\delta_6-3\delta_5)/2}$	$a^{-43\delta_5/80}$
$v/\tilde{v}$	$a^{(\delta_6-\delta_5)/2}$	$a^{(\delta_6-\delta_5)/2}$	$a^{37\delta_5/80}$
$M/\tilde{M}$	$a^{\delta_5+N\delta_1}$	$a^{\delta_6-\delta_5}$	$a^{37\delta_5/40}$
$\rho/\tilde{\rho}$	a^{δ_5}	a^{δ_5}	a^{δ_5}
$P_{th}/\tilde{P}_{th}$	a^{δ_6}	a^{δ_6}	$a^{77\delta_5/40}$
$T/\tilde{T}$	$a^{(\delta_6-\delta_9-\mu\delta_5)/\nu}$	$a^{\delta_6-\delta_5}$	$a^{37\delta_5/40}$
$s/\tilde{s}$	$a^{\delta_6-\gamma\delta_5}$	$a^{\delta_6-\gamma\delta_5}$	$a^{(77/40-\gamma)\delta_5}$
$\mathcal{Q}_{0,1}/\tilde{\mathcal{Q}}_{0,1}$	$a^{(3/2-\zeta_1)\delta_6-(\epsilon_1+1/2)\delta_5-(\theta_1+1)\delta_1}$	1	1
$\mathcal{Q}_{0,2}/\tilde{\mathcal{Q}}_{0,2}$	$a^{(3/2-\zeta_2)\delta_6-(\epsilon_2+1/2)\delta_5-(\theta_2+1)\delta_1}$	0	1
$\varepsilon_0/\tilde{\varepsilon}_0$	a^{δ_9}	1	1

Table 3 General scaling laws (global invariance) for optically thick plasmas (*Column 2*), ideal gas, $\mu = \nu = 1$ (*Column 3*) and dust plus ideal gas with $\kappa_{rad} \propto \rho^{-1} T$ (*Column 4*)

Physical ratio	Global invariance	Ideal fluids	Ideal gas + Dust opacity
$x/\tilde{x}$	$a^{\delta_{16}+([n-5]/[4-\nu])\delta_9+([m+1/2]+\mu[n-5]/[4-\nu])\delta_5}$	$a^{([m+1/2]+[n-5]/3)\delta_5}$	$a^{(-11/6)\delta_5}$
$t/\tilde{t}$	$a^{\delta_{16}+([n-7]/[4-\nu])\delta_9+(m+1+\mu[n-7]/[4-\nu])\delta_5}$	$a^{(m+1+[n-7]/3)\delta_5}$	$a^{(-2)\delta_5}$
$v/\tilde{v}$	$a^{(2/[4-\nu])\delta_9+([4\mu+\nu-4]/[8-2\nu])\delta_5}$	$a^{\delta_5/6}$	$a^{\delta_5/6}$
$\rho/\tilde{\rho}$	a^{δ_5}	a^{δ_5}	a^{δ_5}
$P_{th}/\tilde{P}_{th}$	$a^{(4/[4-\nu])\delta_9+(4\mu/[4-\nu])\delta_5}$	$a^{(4/3)\delta_5}$	$a^{(4/3)\delta_5}$
$T/\tilde{T}$	$a^{(1/[4-\nu])\delta_9+(\mu/[4-\nu])\delta_5}$	$a^{\delta_5/3}$	$a^{\delta_5/3}$
$E_{rad}/\tilde{E}_{rad}$	$a^{(4/[4-\nu])\delta_9+(4\mu/[4-\nu])\delta_5}$	$a^{(4/3)\delta_5}$	$a^{(4/3)\delta_5}$
$F_{rad}/\tilde{F}_{rad}$	$a^{(6/[4-\nu])\delta_9+([12\mu-4+\nu]/[8-2\nu])\delta_5}$	$a^{(3/2)\delta_5}$	$a^{(3/2)\delta_5}$
$P_{rad}/\tilde{P}_{rad}$	$a^{(4/[4-\nu])\delta_9+(4\mu/[4-\nu])\delta_5}$	$a^{(4/3)\delta_5}$	$a^{(4/3)\delta_5}$
$s/\tilde{s}$	$a^{(4/[4-\nu])\delta_9+([4\mu]/[4-\nu]-\gamma)\delta_5}$	$a^{(4/3-\gamma)\delta_5}$	$a^{(4/3-\gamma)\delta_5}$
$\mathcal{Q}_{rad}/\tilde{\mathcal{Q}}_{rad}$	$a^{-\delta_{16}+([11-n]/[4-\nu])\delta_9+(\mu[11-n]/[4-\nu]-[m+1])\delta_5}$	No source	No source
$\kappa_0/\tilde{\kappa}_0$	$a^{\delta_{16}}$	1	1
$\varepsilon_0/\tilde{\varepsilon}_0$	a^{δ_9}	1	1

siderations (Drake 2006). Thanks to this relation we see that choosing $\theta = -2$ we obtain scaling laws in optically thick regimes which are established by Murakami and Iida (2002). This type of modeling allows to describe a list of the top astrophysical situations such as the ionized fronts HII regions and the evolution of strong supernova explosions. Strong UV radiations emitted by massive main sequence stars (OB stars) interact with molecular clouds and photoevaporate it in thin layers at the surface of clouds. Thus, an ablative front begins and a shock propagates into the cloud. Several self-similar studies were performed about this radiative regime and different Lie symmetries are obtained (Coggeshall 1991; Coggeshall and Meyer-ter-Vehn 1992).

4 Optically thick radiative pressure regime

Generally, including radiative flux, F_{rad}, in laboratory experiments is enough and for that reason, researches about scaling laws in this regime have been made in ICF, mainly. Here, we add the energy density and pressure of radiation, E_{rad} and P_{rad}, ($o(v^2/c^2)$ effects see Mihalas and Mihalas 1999) in the diffusion approximation at LTE. This radiative regime is commonly referred to as the equilibrium diffusion limit. In experiments, LTE is usually satisfied (Castor 2007) and this regime will be achieved on LMJ and NIF experiments. In Astrophysics, radiation pressure and energy play a key role in stars, supernovae and in evaporation phenomena in clumps (Konigl 1984). The plasma evolution is then governed by the equations:

$$\rho \frac{d\vec{v}}{dt} = -\vec{\nabla}[P_{th} + P_{rad}], \tag{5}$$

$$\frac{dE_{tot}}{dt} - \frac{E_{tot} + P_{th} + P_{rad}}{\rho} \frac{d\rho}{dt} = -\vec{\nabla}.\vec{F}_{rad} - \mathcal{Q}, \tag{6}$$

where $E_{tot} = \rho e + E_{rad}$ and where $\mathcal{Q}$ is the energy source term which can model optically thick processes. In the application we will consider that $E_{rad} = a_R T^4$, where a_R is the first radiative constant $P_{rad} = E_{rad}/3$, $\vec{F}_{rad} = -\kappa_{rad}\vec{\nabla}T$ where κ_{rad} is the radiative conductivity given by $\kappa_{rad} = \kappa_0 \rho^m T^n$ as in the previous section. In addition to the optically thin case we add the radiative relations: $\mathcal{Q} = a^{\delta_{10}}\tilde{\mathcal{Q}}$, $\vec{F}_{rad} = a^{\delta_{14}}\vec{\tilde{F}}_{rad}$, $\kappa_{rad} = a^{\delta_{15}}\tilde{\kappa}_{rad}$, $\kappa_0 = a^{\delta_{16}}\tilde{\kappa}_0$, $E_{rad} = a^{\delta_{17}}\tilde{E}_{rad}$, $P_{rad} = a^{\delta_{18}}\tilde{P}_{rad}$. As before I_1, I_2, I_3, I_4 (or I_5) and I_6 are five invariants. The additional ones are $I_7 = P_{rad}t/(\rho v r) = Eu_{rad} \times St$ (Eu_{rad}: radiative Euler number), $I_8 = E_{rad}/P_{th} \propto 1/R$ (R: Mihalas number), $I_9 = tF_{rad}/(P_{th}r) = 1/Bo$ (Bo: Boltzmann number). We recover the standard dimensionless numbers (Mihalas and Mihalas 1999) which describe these radiating fluids. The scaling laws are presented in Table 3. We see that for an ideal gas with an ionization conservation state, we have a single (δ_5) parameter to rescale the experiments, but if the ionization is not preserved, we have at least three free parameters (δ_5, δ_9, δ_{16}). Finally, we find that s (entropy) is invariant for $\gamma = 4/3$, which corresponds to a dominant photon regime. Notice that if we set $P_{rad} = E_{rad} = 0$, we find the scaling laws of optically thick radiating fluid of the previous section.

5 Discussion and conclusion

We have presented the general scaling for different radiative regimes and we have determined the number of free parameters which depends on the structure of the model and the similarity criteria that we impose. In view of the established results we can see that the more phenomena we add, the harder it is to rescale an experiment. However, since if there exists a homothetic symmetry the Rankine-Hugoniot relations are conserved, a large variety of astrophysical radiative shocks can be reproduced in laboratory.

References

Basko, M.M., Johner, J.: Nucl. Fusion **38**, 1779 (1998)

Birkhoff, G.: Hydrodynamics: A Study in Logic, Fact, and Similitude. Princeton University Press, Princeton (1950), pp. 129–130

Castor, J.: Astrophys. Space Sci. **307**, 207 (2007)

Chevalier, R.A., Imamura, J.N.: Astrophys. J. **261**, 543 (1982)

Coggeshall, S.V.: Phys. Fluids A **3**, 757 (1991)

Coggeshall, S.V., Meyer-ter-Vehn, J.: J. Math. Phys. **33**, 3585 (1992)

Drake, R.P.: High Energy Density Physics: Fundamentals, Inertial Fusion and Experimental Astrophysics. Springer, Berlin (2006)

Falize, E., et al.: Astrophys. Space Sci., Proc. of HEDLA'08 held in St Louis, MO, USA, 12–15 April 2008, in this issue (2008a)

Falize, E., Bouquet, S., Michaut, C.: J. Phys. Conf. Ser. **112**, 042016 (2008b)

Ibragimov, N.H.: Elementary Lie Group Analysis and Ordinary Differential Equations. Pasadena (1999)

Konigl, A.: Astrophys. J **284**, 303 (1984)

Lowrie, R.B., Morel, J.E., Hittinger, J.A.: Astrophys. J. **521**, 432 (1999)

Mignone, A.: Astrophys. J **626**, 373 (2005)

Mihalas, D., Mihalas, B.: Foundations of Radiation Hydrodynamics. Dover, New York (1999)

Murakami, M., Iida, S.: Phys. Plasma **9**, 2745 (2002)

Remington, B.A., Drake, R.P., Ryutov, D.D.: Rev. Mod. Phys. **78**, 755 (2006)

Ryutov, D.D., Remington, B.A.: Plasma Phys. Control. Fusion **44**, B407 (2002)

Ryutov, D.D., Drake, R.P., Kane, J., Liang, E., Remington, B.A., Wood-Vasey, W.M.: Astrophys. J. **518**, 821 (1999)

Ryutov, D.D., Remington, B.A., Robey, H.F., Drake, R.P.: Phys. Plasma **8**, 1804 (2001)

Saxton, C.J., Wu, K., Pongracic, H., Shaviv, G.: Mon. Not. R. Astron. Soc. **299**, 862 (1998)

Takabe, H.: Progress Theor. Phys. Suppl. **143**, 202 (2001)

Zeldovich, Y.B., Raizer, Y.P.: Physics of Shock Waves and High-temperature Hydrodynamic Phenomena. Academic Press, New York (1967)

Astrophys Space Sci (2009) 322: 113–116
DOI 10.1007/s10509-008-9963-8

ORIGINAL ARTICLE

Opacity effects on the polarization of line emissions in astrophysical plasmas

Peter Hakel · Roberto C. Mancini

Received: 13 June 2008 / Accepted: 1 December 2008 / Published online: 23 December 2008

Abstract Polarization-based line spectroscopy is a valuable tool in determining the characteristics of electron distribution functions in anisotropic plasmas. For instance, directional electrons can unevenly populate the magnetic sublevels of atomic energy levels resulting in partial polarization of the emitted spectral lines. The large dimensions of astrophysical sources raise the possibility of non-negligible self-absorption effects on spectral line properties. Alternatively, the high densities characteristic of laser-produced laboratory plasmas can also result in substantial optical depth values. We present a modeling study of He-like Fe line emissions in which we investigate the effects of radiation transport on the polarization of selected spectral lines under corona conditions.

Keywords Polarization · Spectral lines · Opacity

1 Introduction

Spectroscopic analysis of line emissions is an important diagnostic technique for both the astrophysical and laboratory plasmas. For instance, plasma temperature can be determined by measuring the ratios of intensities of certain spectral lines. Plasma density can be inferred from the Stark broadening of lines. Polarization is a degree of freedom through which anisotropic characteristics of the emitting plasma can be revealed. This anisotropy can assume a number of forms, be it a beam of directional energetic particles or a strong magnetic field; anything that breaks the isotropic nature of the plasma can in principle lead to polarized line emissions. Polarized-sensitive spectroscopy is therefore of great utility in a number of astrophysical applications, e.g., solar magnetometry, modeling of anisotropic radiation transport in stellar atmospheres (Landi Degl'Innocenti and Landolfi 2004), and in radiation emissions from accretion disks (Chen et al. 1997; Matt 1993; Gnedin et al. 2006).

In this paper we expand on our previous work in the area of polarized line emissions driven by beams of energetic electrons. Via anisotropic excitation these beams can place unequal populations into the magnetic sublevels of a given fine-structure atomic level. This condition is called *alignment* and can manifest itself in partial polarization of spectral lines originating from this excited level. We present a model study in which we investigate the potential of opacity (i.e., self-absorption) effects to alter the polarization degrees of spectral lines.

P. Hakel (✉) · R.C. Mancini
University of Nevada, Reno, USA
e-mail: phakel@unr.edu

2 Model

2.1 Magnetic-sublevel atomic kinetics

Magnetic sublevels are the fundamental electronic quantum states of an atom or an ion. They are characterized by a collection of quantum numbers, which include the total angular momentum J and its projection M along the quantization axis. To calculate the line emission characteristics we have used the collisional-radiative model for magnetic sublevels *POLAR* (Hakel et al., 2004, 2007) which we have augmented to account for opacity effects. *POLAR* is a two-step model in which the level populations are calculated first and are then postprocessed by a spectral model which extracts the line properties such as intensity and polarization degree.

This is based on the assumption that the kinetics is collisionally dominated and that radiation-induced processes do not significantly alter the level populations and hence affect the line formation only through radiation transport. This assumption allows us, for now, to avoid a much more complicated and computationally intensive task of solving the atomic kinetics and radiation transport self-consistently, which would otherwise be necessary in low-density, photoionized plasmas often encountered in astrophysical applications. Therefore the kinetics portion of the model is the same as in Hakel et al. (2004). We apply *POLAR* under steady-state conditions and thus we solve,

$$\bar{\bar{\mathbf{A}}}\mathbf{f} = \mathbf{0} \tag{1}$$

for the sublevel populations $\mathbf{f}$ based on the rates of atomic processes organized into the matrix $\bar{\bar{\mathbf{A}}}$. The rates are calculated for given plasma conditions such as temperature and density using a database of atomic and collisional data (energy level structure, cross sections, etc.). This database can be created by any suite of atomic structure and collisional codes. In this work we used the Flexible Atomic Code (Gu 2007) for this purpose.

2.2 Electron distribution function

The processes controlling the sublevel populations in our model are collisional excitation and deexcitation, and spontaneous radiative decay. The collisional processes are driven by electron impact and it is the anisotropic nature of the electron distribution that drives alignment and hence line polarization. We characterize the electron distribution by four parameters: N_e: electron number density, α: non-thermal electron fraction, T_M: temperature of the Maxwellian component, T_h: "temperature" of the non-thermal component. We model the polarization-driving non-thermal component as a beam of electrons directed along the quantization axis, of density αN_e, with energy distribution,

$$f_B(E, T_h) = kT_h \exp\left(-\frac{E}{kT_h}\right) \tag{2}$$

The thermal component of the electron distribution function has the density $(1-\alpha)N_e$ and its energy distribution is characterized by a Maxwellian of temperature T_M.

2.3 Line emission and transport

We account for opacity effects in the second (spectral) part of the model, in a manner analogous to our *CRAK* model in Andiel et al. (2002). Therefore, in addition to the spontaneous radiative decay process, we add the transition rates representing photoexcitation and stimulated emission. These three radiative processes are represented by their respective Einstein coefficients (Mihalas and Weibel-Mihalas 1999).

For spontaneous emission rate from an upper magnetic sublevel u to a lower one l we have (Takács et al. 1996)

$$A(J_u M_u \to J_l M_l) = A(J_u \to J_l) \times (J_l q M_l - \Delta M \mid J_u M_u)^2 \tag{3}$$

where $A(J_u \to J_l)$ is the spontaneous decay rate between fine-structure levels, q is the multipolarity of the transition, $\Delta M \equiv M_l - M_u$, and ( ) is the Clebsch-Gordan coefficient.

The coefficient for stimulated emission $B(J_u M_u \to J_l M_l)$ is given by,

$$A(J_u M_u \to J_l M_l) = \frac{2h\nu^3}{c^2} B(J_u M_u \to J_l M_l) \tag{4}$$

and is the same as the photoexcitation coefficient $B(J_l M_l \to J_u M_u)$, since the statistical weights of magnetic sublevels are $g_u = g_l = 1$. Hence we have,

$$B(J_l M_l \to J_u M_u) = B(J_u M_u \to J_l M_l) \equiv B \tag{5}$$

This allows us to construct the polarization-state-dependent emissivities j and opacities κ for a selected spectral line as,

$$j_{||,\perp} = \frac{h\nu}{4\pi} N_i \sum_u \sum_l f_u A(J_u M_u \to J_l M_l)\, MI_{||,\perp}(\Delta M), \tag{6}$$

$$\kappa_{||,\perp} = \frac{h\nu}{4\pi} N_i \sum_u \sum_l (f_l - f_u) B\, MI_{||,\perp}(\Delta M) \tag{7}$$

where f are fractional (dimensionless) magnetic sublevel populations, N_i is the total ion number density, and MI's are the relative multipole intensities for a given line of sight (usually at 90° with respect to the quantization axis) (Hakel et al. 2007).

The polarization-dependent total line intensities $I_{||,\perp}$ satisfy the radiation transport equation,

$$\frac{dI_{||,\perp}}{ds} = j_{||,\perp} - \kappa_{||,\perp} I_{||,\perp} \tag{8}$$

which for a uniform slab of thickness L has the solution,

$$I_{||,\perp} = \frac{j_{||,\perp}}{\kappa_{||,\perp}} [1 - \exp(-\tau_{||,\perp})] \tag{9}$$

where

$$\tau_{||,\perp} \equiv \kappa_{||,\perp} L \tag{10}$$

is the polarization-state-dependent optical depth.

The polarization degree P of a line is defined as the relative difference between the intensities of the line components linearly polarized in the directions parallel and perpendicular to the axis of quantization, i.e.,

$$P \equiv \frac{I_{||} - I_{\perp}}{I_{||} + I_{\perp}} \tag{11}$$

The line of sight is taken to be perpendicular to the quantization axis.

We note that the use of the Einstein B coefficients to describe the photoexcitation and stimulated emission processes is really only justified for isotropic (i.e., angle-independent) radiation fields. In our case the radiation field is inherently anisotropic since we are considering transitions between magnetic sublevels. For example, a given spontaneous transition has a characteristic angular intensity pattern (e.g., $\propto \sin^2\theta$ for electric dipole) whereas photons created via stimulated emission copy the angular distribution of the ambient radiation field. The above-presented model is therefore a simplified one and only serves to illustrate what we may expect when a polarized line becomes sufficiently optically thick. A more detailed radiation transport or a photon-kinetics treatment coupled with a *POLAR*-type magnetic sublevel model would be needed to obtain more accurate results.

3 Results

3.1 Limiting cases

With the earlier stated assumption that radiation-driven processes do not notably affect atomic kinetics, the population distribution is fully determined by the results of (1). This then translates into uniquely determined emissivities and opacities defined in (6) and (7). In this case the only parameter left to affect the polarization of a given line (11) is the physical length of the plasma L through which the radiation must work its way out toward an observer. Varying values of L result in optical depth values (10) ranging from 0 (marking the optically thin limit) to infinity (i.e., the optically thick limit). At present we do not consider situations with $\tau < 0$ characteristic of systems with population inversion (Rybicki and Lightman 1985) (also see (7) with $f_l < f_u$ resulting in negative κ). Since the intensities of the two polarized line components $I_{||,\perp}$ are given by (9), for the two limiting cases we obtain,

$$I_{||,\perp} = j_{||,\perp} L \quad (\tau \to 0), \tag{12}$$

$$I_{||,\perp} = \frac{j_{||,\perp}}{\kappa_{||,\perp}} \quad (\tau \to \infty) \tag{13}$$

Therefore we expect the line polarization P, as a function of optical depth, to also gradually transition between two values characteristic of these two limits.

3.2 Application to He-like Fe lines

Iron is an element whose line emissions are of great utility in solar physics and astrophysics. For instance, polarized EUV and X-ray emissions from Fe can be driven by jets of energetic particles during solar flares (Shlyaptseva et al. 1981). In Figs. 1 and 2 we present our calculations of the polarization of the $1s\,3d\ ^1D_2 \to 1s\,2p\ ^1P_1$ line in He-like Fe, which is an $h\nu = 881$ eV ($\lambda = 1.4$ nm) transition. The plasma conditions are typical of the solar corona at $N_e = 10^9$ cm^{-3} and $T_M = 100$ eV. For the non-thermal electrons we select $\alpha = 0.2$ and two values for T_h: 10 keV and 100 keV. As the two polarization-state-dependent optical depths are similar but not identical in value, the horizontal axis on all plots represents their average value defined by $\tau \equiv (\tau_{||} + \tau_{\perp})/2$. In order to calculate the needed sublevel populations we constructed a three-level model for *POLAR* containing the two levels defining the transition of interest and the ground state of He-like Fe, $1s^2\ ^1S_0$. A more extensive population model

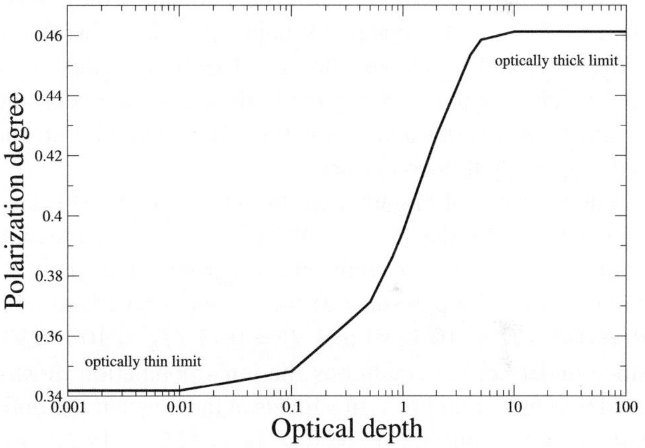

Fig. 1 Polarization of the $1s\,3d\ ^1D_2 \to 1s\,2p\ ^1P_1$ line in He-like Fe; $T_h = 10$ keV

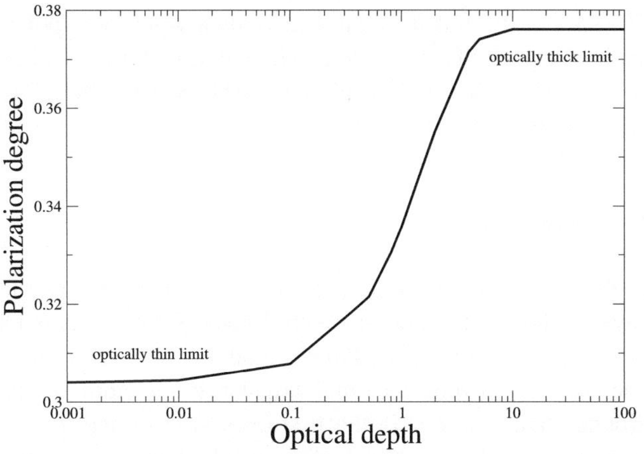

Fig. 2 Polarization of the $1s\,3d\ ^1D_2 \to 1s\,2p\ ^1P_1$ line in He-like Fe; $T_h = 100$ keV

is not necessary in this case since under these low-density (i.e., "corona") conditions most of the ion's population resides in its ground state and multilevel (i.e., "cascade") kinetics effects are therefore negligible.

Both Figs. 1 and 2 show the expected behavior of the polarization P transitioning between the two limits. The values of the limits in the two figures are different owing to the change in atomic rates corresponding to the two values of T_h. This illustrates how the observed polarization degree can be used to infer the characteristics of the anisotropic electron distribution—in this case the value of T_h. (A more detailed discussion of the diagnostic value of *POLAR* modeling is presented in Hakel et al. (2004), see Fig. 11 therein.) In the optically thin limit the sublevel populations of the line's upper level f_u completely determine the polarization of the outgoing radiation (6 and 12). As opacity effects become more important with increasing L (and hence τ) the plasma absorption increasingly attenuates the outgoing line intensities until an equilibrium between emission and absorption is achieved for large values of τ (13). Since the 3 sublevels of the lower (i.e. photoabsorbing) level ($1s\ 2p\ ^1P_1$) are also unequally populated, they absorb the two families of polarized photons at different rates. This opacity effect changes the relative difference between the $I_{||}$ and $I_{\perp}$ intensities and hence P compared to what it was under optically thin conditions.

The presence of alignment in the $1s\ 2p\ ^1P_1$ level is also demonstrated by the fact that the spectral line originating from this level and decaying into the ground-state $1s^2\ ^1S_0$ (known as the He-α resonance line) is itself polarized with $P = 0.48$ ($T_h = 10$ keV) and $P = 0.11$ ($T_h = 100$ keV). Since under corona conditions the ion's population mostly resides in the ground state this line is in fact much more optically thick than our line of interest, $1s\ 3d\ ^1D_2 \to 1s\ 2p\ ^1P_1$. However, while opacity effects can severely reduce the *total* line intensity $I \equiv I_{||} + I_{\perp}$, they will affect its polarization only if there is alignment in the lower level of the transition. The ground state of a He-like ion has only one sublevel ($J = M_J = 0$), thus its alignment is always zero and polarization degrees of lines such as He-α will be unaffected by opacity effects. The polarization of the He-α line is therefore independent of optical depth.

4 Conclusions

In this work we demonstrated the effects that opacity can have on the polarization of spectral lines. We constructed a magnetic-sublevel atomic kinetic model arising from three fine-structure levels; the total angular momentum of the ground state was $J = 0$, the first excited state had $J = 1$, and the second had $J = 2$. This resulted in a magnetic-sublevel atomic kinetics model consisting of 9 sublevels; one from the ground state ($M = 0$), three from the first excited state ($M = 0, \pm 1$), and five from the second excited state ($M = 0, \pm 1, \pm 2$). We solved this kinetics model for sublevel populations under solar corona conditions and post-processed the results into polarization degrees with a spectral model that accounts for the effects of radiation transport.

We applied this model to the study of two lines in He-like Fe. The He-α line $((J = 1) \to (J = 0))$, while being a prime candidate of being optically thick due to a large concentration of population in its lower level, is unaffected by opacity effects as far as its polarization degree is concerned. This is caused by the fact that the sole lower magnetic sublevel is equally suited to absorb photons of either polarization state. On the other hand, the aligned upper level of the He-α line serves as the lower (photoabsorbing) level of the $(J = 2) \to (J = 1)$ line. Here the two polarization-state-dependent opacities are different reflecting the unequally populated sublevels of the $(J = 1)$ level. As a result the polarization degree of this line changes with the increasing optical depth (i.e., the physical dimension of the plasma). In the presented model the sublevel populations are completely determined by the plasma conditions and therefore the change in polarization is due to radiation transport effects.

This paper illustrates the need for developing a new generation of spectroscopic models. Such models will combine the multilevel kinetics with the treatment of opacity effects. This way we will be able to better understand the underlying mechanisms that give rise to observed polarization properties of spectral lines and hence have a more accurate diagnostic tool available. Such modeling capability will find applications on both astrophysics and laboratory plasma research.

Acknowledgements This work was supported by the University and Community College System of Nevada.

References

Andiel, U., et al.: Europhys. Lett. **60**(6), 861 (2002)
Chen, K., et al.: Astrophys. J. **483**, 194 (1997)
Gnedin, Yu.N., et al.: Astrophys. J. **32**(1), 39 (2006)
Gu, M.F.: http://kipac-tree.stanford.edu/fac/ (2007)
Hakel, P., et al.: Phys. Rev. E **69**, 056405 (2004)
Hakel, P., et al.: Phys. Rev. A **76**, 012716 (2007)
Landi Degl'Innocenti, E., Landolfi, M.: Polarization in Spectral Lines. Kluwer, Dordrecht (2004)
Matt, G.: Mon. Not. R. Astron. Soc. **260**(3), 663 (1993)
Mihalas, D., Weibel-Mihalas, B.: Foundations of Radiation Hydrodynamics. Dover, New York (1999)
Rybicki, G.B., Lightman, A.P.: Radiative Processes in Astrophysics. Wiley, New York (1985)
Shlyaptseva, A.S., et al.: P.N. Lebedev Physical Institute Report No. 194 (1981) (unpublished)
Takács, E., et al.: Phys. Rev. A **54**(2), 1342 (1996)

Astrophys Space Sci (2009) 322: 117–121
DOI 10.1007/s10509-008-9930-4

ORIGINAL ARTICLE

Modelling, design and diagnostics for a photoionised plasma experiment

I.M. Hall · T. Durmaz · R.C. Mancini · J.E. Bailey · G.A. Rochau · M.J. Rosenberg · D.H. Cohen · I.E. Golovkin · J.J. MacFarlane · M.E. Sherril · J. Abdallah · R.F. Heeter · M.E. Foord · S.H. Glenzer · H.A. Scott

Received: 16 June 2008 / Accepted: 10 October 2008 / Published online: 31 October 2008

Abstract Photoionised plasmas are common in astrophysical environments and new high resolution spectra from such sources have been recorded in recent years by the Chandra and XMM-Newton satellites. These provide a wealth of spectroscopic information and have motivated recent efforts aimed at obtaining a detailed understanding of the atomic-kinetic and radiative characteristics of photoionised plasmas. The Z-pinch facility at the Sandia National Laboratories is the most powerful terrestrial source of X-rays and provides an opportunity to produce photoionised plasmas in a well characterised radiation environment. We present modelling work and experimental design considerations for a forthcoming experiment at Sandia in which X-rays from a collapsing Z-pinch will be used to photoionise low density neon contained in a gas cell. View factor calculations were used to evaluate the radiation environment at the gas cell; the hydrodynamic characteristics of the gas cell were examined using the Helios-CR code, in particular looking at the heating, temperature and ionisation of the neon and the absorption of radiation. Emission and absorption spectra were also computed, giving estimates of spectra likely to be observed experimentally.

Keywords Photoionised plasma · Laboratory Astrophysics · Hydrodynamic simulation · Atomic-kinetic modelling · Plasma spectroscopy

1 Introduction

Photoionised plasmas are common in astrophysical environments, typically occurring near strong sources of accretion powered X-rays such as active galactic nuclei and X-ray binaries. In recent years, the Chandra and XMM-Newton satellites have recorded high resolution spectra from astrophysical photoionised plasmas. These contain a wealth of spectroscopic information and have motivated recent efforts aimed at obtaining a detailed understanding of the atomic-kinetic and radiative characteristics of photoionised plasmas. However, models used to interpret spectra from astrophysical plasmas are complex and require many parameters to be known including the geometry of the system, gas composition and density distribution, and the energy distribution of the ionising X-ray source. Uncertainties inherent in these parameters will affect the validity of the results obtained from the spectral analysis.

In photoionised plasmas, the radiation field plays a dominant role in the atomic-kinetic behaviour, and collisional processes do not strongly influence the ionisation balance. In contrast, the population distribution in most laboratory produced plasmas is largely determined by collisional processes. Conditions of strong ionising radiation field and low collisionality are difficult to achieve experimentally, and

I.M. Hall (✉) · T. Durmaz · R.C. Mancini
University of Nevada, Reno, USA
e-mail: ihall@unr.edu

J.E. Bailey · G.A. Rochau
Sandia National Laboratories, Albuquerque, USA

M.J. Rosenberg · D.H. Cohen
Swarthmore College, Swarthmore, USA

I.E. Golovkin · J.J. MacFarlane
Prism Computational Sciences, Madison, USA

M.E. Sherril · J. Abdallah
Los Alamos National Laboratory, Los Alamos, USA

R.F. Heeter · M.E. Foord · S.H. Glenzer · H.A. Scott
Lawrence Livermore National Laboratory, Livermore, USA

there have been relatively few studies of terrestrial photoionised plasmas (Bailey et al. 2001; Foord et al. 2004; Van Hoof et al. 2005). The Z-pinch facility at Sandia National Laboratories is currently the most powerful terrestrial source of X-rays, and is capable of producing a photionised plasma.

In this paper, we present modelling work and experimental design considerations for a forthcoming experiment at Sandia in which X-rays from a collapsing Z-pinch plasma will be used to photoionise low density neon contained in a gas cell. The aim is to produce a photoionised plasma with a simple geometry, known ion number density and quantified ionising X-ray flux. Absorption spectra from this well characterised plasma will be obtained, using the Z-pinch as a backlighter. This work builds on a previous 'proof of principle' experiment (Bailey et al. 2001). In this way, we hope to obtain a dataset which can be used to test the accuracy of atomic data and simulation codes. The photoionised nature of the plasma makes this data likely to be useful to the astrophysics community.

2 Simulations

The model of the gascell behaviour was constructed in an incremental fashion using several simulation codes to analyse the radiation environment at the gascell, simulate the hydrodynamic behaviour of the gascell, and model the atomic kinetic behaviour to produce synthetic emission and absorption spectra. The results of the modelling help us to gain an understanding of the behaviour of the system and allow us to investigate the role of key experimental parameters, guiding the experimental design.

2.1 Radiation environment

The first part of the model is to characterise the X-ray flux incident on the gascell from the collapsing Z-pinch. The total ionising flux has contributions from both the Z-pinch itself and re-radiated emission from parts of the experimental setup such as the return current cannister and anode plate. This part of the model is a key element, since the gascell is heated solely by the radiation field and the hydrodynamic and atomic kinetic behaviour are directly dependent on its characteristics.

The radiation environment is modelled using the Visrad view factor code (MacFarlane 2003). Based on the specified geometry and material emission properties, the code calculates the radiation re-emitted from each surface of the setup in response to a user specified X-ray radiation source. Each surface radiates with a Planckian energy distribution. The geometry used in the model is based on the experimental parameters as far as possible. The Z-pinch radiates

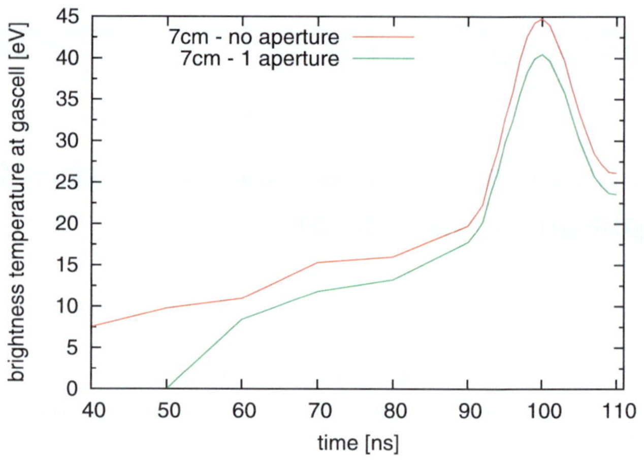

Fig. 1 Brightness temperature at the gascell front surface with and without aperture in place

as a black-body with a Planckian energy distribution, and a radius-power time history inferred from an earlier experiment (shot #z541). The Z-pinch emission is characterised by a 60 ns run-in phase, reaching a peak power of 168 TW at 100 ns with a colour temperature of 220 eV and a pinch diameter of 1.1 mm; the majority of the X-ray emission occurs between 95–105 ns. A detector was positioned at 7 cm from the pinch axis to represent the front surface of the gascell. The X-ray flux at the gascell comprises both the directly observed Z-pinch and re-radiated emission, giving a net energy distribution that is non-Planckian. A shielding aperture between pinch and gascell was used in one of the model configurations. This allows us to reduce the contribution of re-radiated X-ray components to the total flux arriving at the gascell, making the energy distribution closer to being Planckian. This is desirable since for each Z-pinch shot, the pinch energy spectrum is measured and by allowing the gascell to see only the pinch we can use this information directly to assist analysis of spectra from the gascell.

The total net flux arriving at the centre of the gascell front surface as a function of time is shown in Fig. 1. The flux is expressed in terms of radiation brightness temperature, T_B, using the Stefan-Boltzmann law $F_{tot} = \sigma T_B^4$ where F_{tot} ($\mathrm{W\,cm^{-2}}$) is the total energy-integrated flux arriving at the surface. The model predicts peak values of T_B of around 40 and 45 eV (corresponding to 2.5×10^8 and 4.5×10^8 $\mathrm{W\,cm^{-2}}$ of X-ray power) with and without the aperture respectively.

Figure 2 shows the energy resolved flux components from the pinch, anode and return current cannister with and without aperture at 100 ns. With the aperture in place, anode and return current cannister contributions are reduced by a factor of ~3 with the overall flux reduced by a factor of around 2. In each case, there is almost the same contribution from the pinch. These results show that carefully designed shielding may be used to reduce the contribution of

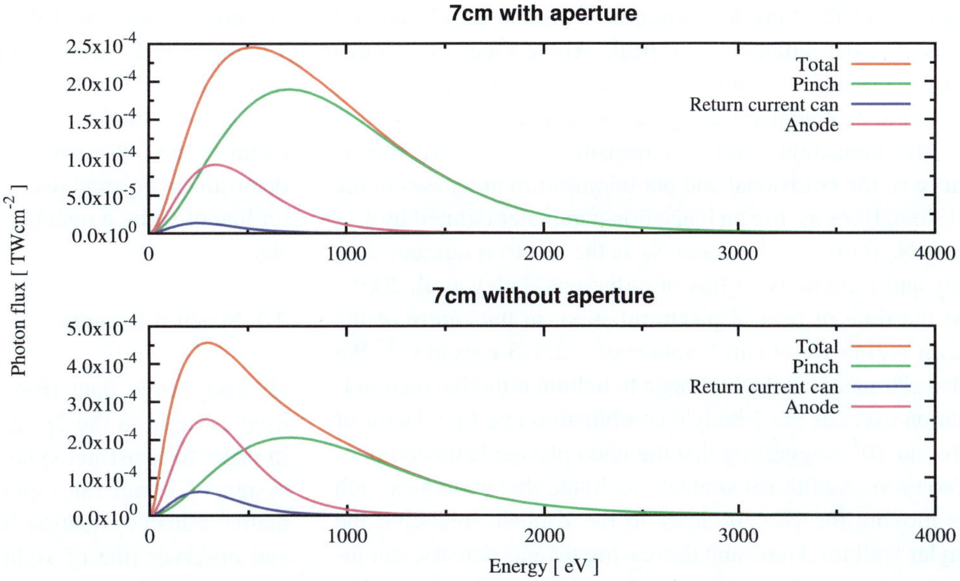

Fig. 2 Composition of the total X-ray flux, showing directly observed pinch radiation and the main re-radiated contributions from return current cannister and anode

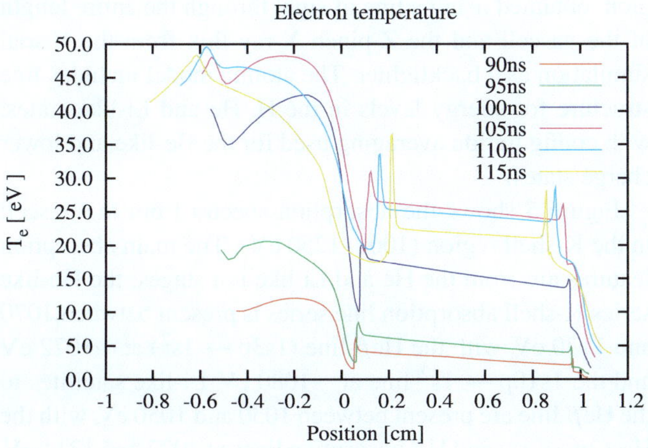

Fig. 3 Space resolved electron temperature in the gascell.

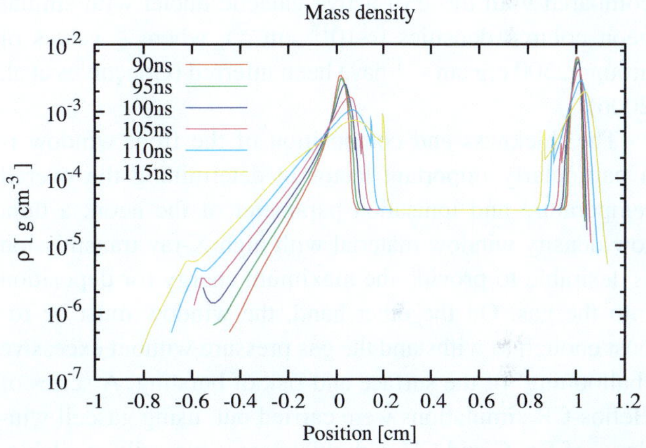

Fig. 4 Space resolved mass density in the gascell

re-radiated X-rays at the gascell at the expense of reducing the overall flux by a factor of ∼2.

2.2 Hydrodynamic behaviour

The time history of the energy resolved ionising X-ray flux was used to drive a hydrodynamic simulation of the gascell using the Helios-CR code (MacFarlane et al. 2006). A 1-dimensional model was used, with 1.4 μm Mylar walls enclosing a 1 cm length of neon gas with an ion density of 10^{18} cm^{-3} (30 Torrs). The X-ray flux was applied to one side of the system. LTE opacity lookup tables and Propaceos equation of state data were used for the neon and Mylar regions. A multiangle radiation transport model was used to account for the radiation propagation charateristics. Results presented here correspond to a gascell with a front surface at 7 cm from the Z-pinch without an aperture in place.

Figures 3 and 4 show space resolved temperature and mass density behaviour from the model for times near to the peak of the X-ray flux (100 ns). The Mylar walls are initially at 0 and 1 cm with the gas in between. Approximately 70% of the incident radiation field is absorbed by the first Mylar window, leading to heating and ablation expansion of the material into the surrounding vacuum and into the gas region. A smaller proportion of the unattenuated radiation is absorbed by the neon gas and by the rear Mylar window. Shock waves are launched into the neon gas at each Mylar-neon interface, leading to a compression of the gas and shock induced heating (Fig. 4). The central region of the neon is unperturbed by the expanding Mylar near the peak of the radiation drive, with the density remaining largely uniform and the velocities negligible. The temperature profile in the neon is also quite uniform, typically varying less than 2 eV along the full length of the unperturbed region and reaching a maximum of around 25 eV around 5 ns after the radiation drive peak. At the times of spectroscopic interest, we therefore expect the centre of the neon region to be characterised by largely uniform conditions.

The ionisation parameter measures the relative importance of the collisional and photoionisation processes in the plasma. Here we use an ionisation parameter defined by $\xi = 4\pi I/N_e$ (erg cm s^{-1}) where N_e is the electron number density and I the ionising flux of radiation (Bailey et al. 2001). At the time of peak Z-pinch radiation, in the centre of the neon region we obtain ξ values of ~2.5–3 erg cm s^{-1}. We also estimate that the hydrogen to helium radiative recombination exceeds the 3-body recombination rate by a factor of around 10^4, suggesting that the neon plasma is indeed photoionised. Additional simulations found that strategies such as moving the gascell closer to the Z-pinch, reducing the mylar wall thickness and decreasing the gas density, can increase the ionisation parameter by a factor of 3 or more. We note that the degrees of photoionisation found here are small compared with those of active galactic nuclei with similar neon column densities (~10^{18} cm^{-2}), where ξ values of around 2500 erg cm s^{-1} have been inferred (Gonçalves et al. 2006).

The thickness and composition of the front window is a particularly important factor in determining the overall temperature and ionisation parameter of the neon; a thin, low density window material with high X-ray transmission is desirable to provide the maximum energy for deposition into the gas. On the other hand, the window must be robust enough to withstand the gas pressure without excessive 'ballooning' of the surface and risk of bursting. A series of Helios-CR simulations were carried out, using gascell windows of Be, C and CH with thicknesses typically available; bursting pressure calculations for these materials were also performed, using a thin membrane model (Timoshenko and Woinowsky-Krieger 1959). For the fill pressure of 30 Torrs, 1.4 µm of Mylar was found to provide a reasonable compromise between X-ray attenuation, material strength and thickness and availability. As a practical note, it is important to quantify the deformation of the gascell windows since this determines the areal density of the gas along the spectrometer line of sight, a quantity required for spectroscopic analysis.

2.3 Spectral features

Hydrodynamic data from the Helios-CR simulation were used as input to the Spect3d code (MacFarlane et al. 2007) in order to construct synthetic spectra. The code solves the system of atomic rate equations self consistently with the radiative transfer equation for a specified plasma distribution and observer line of sight. Solutions may be obtained using either a time-dependent or steady-state approximation. Here, we present absorption spectra in the neon K-shell region, obtained using a line of sight through the entire length of the gascell and the Z-pinch X-ray flux from the Visrad simulation as a backlighter. The atomic model includes fine structure for energy levels in the H, He and Li-like states, with configuration averaging used for the Be-like and lower charge states.

Figure 5 shows the absorption spectra from the gascell in the K-shell region (1000–1250 eV). The main absorption features are from the He and Li like ion stages; the He-like series K-shell absorption line series is present between 1070 and 1170 eV, with the Heβ line (1s3p $\rightarrow$ 1s^2) at ~1072 eV and the 1s10p $\rightarrow$ 1s^2 line at ~1080 eV. Li-like satellites to the Heβ line are present between 1030 and 1050 eV, with the Hα (2p $\rightarrow$ 1s) and Hβ absorption lines at 1022 and 1211 eV. The overall signal strength follows the strength of the backlighting Z-pinch continuum, peaking at 100 ns.

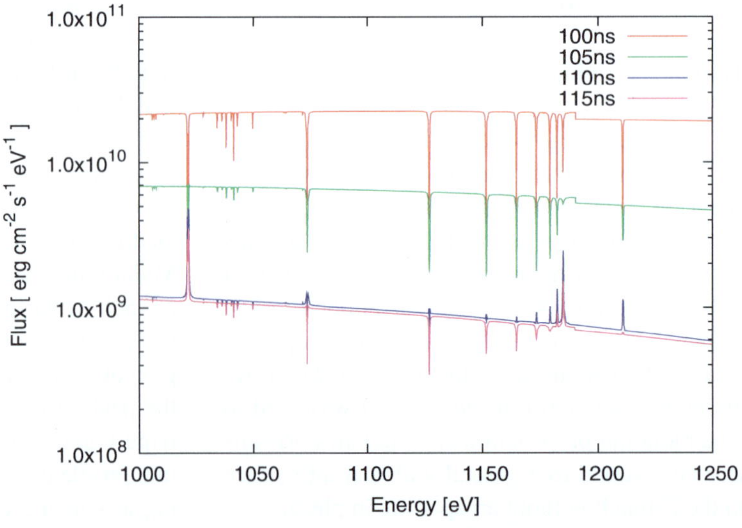

Fig. 5 Neon K-shell absorption spectra from Spect3d

The energy region shown in Fig. 5 contains features from three ionisation stages that can be used to assist diagnosis of the plasma from measured spectra. For example, the relative strengths of features from different ion stages are valuable for inferring the charge state distribution whilst comparison of the detailed Li-like satellite structure with computed results provides information the accuracy of the atomic model. We note that these features have been observed experimentally (Bailey et al. 2001; Bailey 2008) and provide the basis for an analysis of the plasma properties currently in progress.

3 Conclusions

We have presented results from a modelling study of a photoionised plasma with the aim of aiding the design and understanding of a forthcoming experiment using the Z-pinch facility at Sandia National Laboratories. The aim of the experiment is to produce a well characterised system that can be used to gain insight into atomic-kinetic and radiative properties of photoionised plasmas and provide a test bed for model calculations. The modelling work presented helps us to identify and quantify important experimental parameters such as the radiation environment and the hydrodynamics of the gascell and the interplay between them. Knowledge gained from this study will be valuable for optimising the experimental design and interpreting experimental spectra. This work is supported by DOE/SSAA Grant DE-FG52-06NA27587.

References

Bailey, J.E.: Private communication (2008)

Bailey, J.E., et al.: Neon photoionization experiments driven by z-pinch radiation. J. Quant. Spectrosc. Radiat. Transfer **71**, 157–168 (2001)

Foord, M.E., et al.: Charge-state distribution and Doppler effect in an expanding photoionized plasma. Phys. Rev. Lett. **93**, 055002 (2004)

Gonçalves, A.C., et al.: A new model for the warm absorber in NGC 3783: a single medium in total pressure equilibrium. Astron. Astrophys. **451**, L23–L26 (2006)

MacFarlane, J.J.: Visrad—a 3-d viewfactor code and design tool for high-energy density physics experiments. J. Quant. Spectrosc. Radiat. Transfer **81**, 287–300 (2003)

MacFarlane, J.J., Golovkin, I.E., Woodruff, P.R.: Helios-CR—a 1-d radiation-magnetohydrodynamics code with inline atomic kinetics modeling. J. Quant. Spectrosc. Radiat. Transfer **99**, 381–397 (2006)

MacFarlane, J.J., Golovkin, I.E., Wang, P., Woodruff, P.R., Pereyra, N.A.: Spect3d—a multi-dimensional collisional-radiative code for generating diagnostic signatures based on hydrodynamics and PIC simulation output. High Energy Density Physics **3**, 181–190 (2007)

Timoshenko, Woinowsky-Krieger: Theory of Plates and Shells. McGraw-Hill, New York (1959)

Van Hoof, P.A.M., et al.: Modeling x-ray photoionized plasmas produced at the Sandia Z-facility. Astrophys. Space Sci. **298**, 147–153 (2005)

Astrophys Space Sci (2009) 322: 123–127
DOI 10.1007/s10509-008-9948-7

ORIGINAL ARTICLE

To see the inside of a planet in a drop of deuterium

D. Saumon

Received: 30 June 2008 / Accepted: 13 November 2008 / Published online: 22 November 2008

Abstract We briefly review the close connection between research on the equation of state of dense hydrogen and helium and models of the structure of Jupiter, Saturn and exoplanets and how they inform the process of planet formation. Recent progress in equation of state experiments and theory is high lighted.

Keywords Jupiter · Saturn · Equation of state · Hydrogen · Helium · Exoplanets

1 Introduction

Since the discovery of the first exoplanet orbiting a solar-type star (Mayor and Queloz 1995), nearly 300 exoplanets have been found with no sign of abatement in the discovery rate. The planets are primarily being discovered with the radial velocity technique, which detects periodic variations in the radial velocity of the host star that betray the gravitational influence of a planetary companion. Other techniques such as transit photometry and gravitational microlensing now routinely uncover new planets. Approximately 7% of solar-type stars are known to harbor planets (Marcy et al. 2005).

The current crop of planets and planetary systems is remarkable by its diversity. The masses of the planets range from 0.013 to over 10 Jupiter masses (M_J), their orbital periods range from 1.1 day to 14 years, and their orbits go from circular to highly elongated with eccentricities up to $e = 0.92$.[1] By comparison, Jupiter orbits the Sun in 11.9 years in a nearly circular orbit ($e = 0.05$). Strong biases in each of the discovery methods artificially enhance the unusual appearance of extrasolar planetary systems when compared to our own. Improved sensitivity and search techniques are expected to reveal many systems more similar to ours in the next decade. Nevertheless, the fact remains that planetary systems show much unexpected diversity.

The radial velocity technique provides the orbital period, eccentricity and a lower limit on the mass of the planet, because only the radial component of the orbital velocity can be measured, which leaves the inclination of the orbital plane to the line of sight undetermined. For a small fraction of systems, the orbit is small enough and the inclination such that the planet transits the disk of the star every orbit, causing a dimming of the star's light by ~1–2%. This fractional dimming gives the ratio of the surface area of the planet to that of the star. This fortuitous circumstance allows a determination of both the mass and the radius of a transiting planet. These are the first important clues to its global composition. Over 30 such transiting systems are known (Fig. 1). All have radii comparable to that of Jupiter and Saturn, indicating that their bulk composition is dominated by hydrogen and helium. Rocky planets would have much smaller radii for a given mass (Seager et al. 2007). Again, the range of radii is rather surprising, revealing much diversity in the planetary properties, in addition to the variety of their orbital parameters. For a given mass, the radius decreases if elements heavier than H and He (so-called "metals", such as C, O, Si, Fe) are present, or if the planet's interior is colder (i.e. lower specific entropy). For instance, the planet

[1]From The Extrasolar Planets Encyclopedia, http://exoplanet.eu/index.php

D. Saumon (✉)
Los Alamos National Laboratory, PO Box 1663, MS F663, Los Alamos, NM 87545, USA
e-mail: dsaumon@lanl.gov

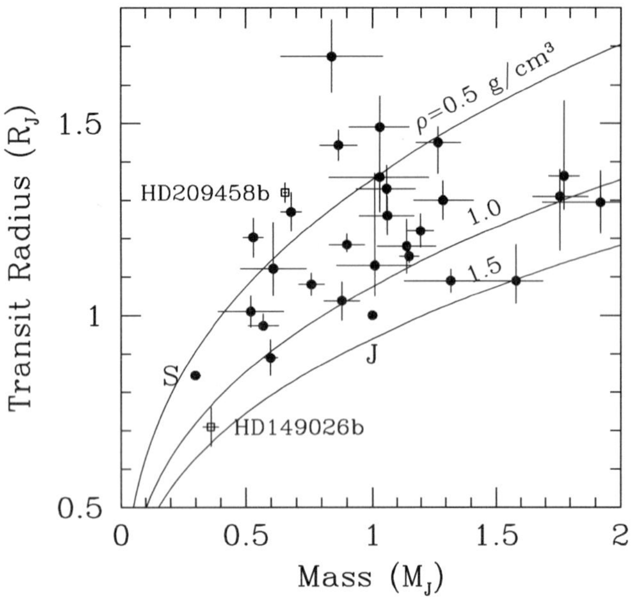

Fig. 1 Measured radii and masses of transiting planets, in units of the Jupiter radius (R_J) and Jupiter's mass (M_J). Jupiter and Saturn are included for comparison, with their symbols labeled "J" and "S", respectively. *Lines* of constant average density are shown. Two planets of particular interest, HD149026b and HD209458b are shown with *open squares*. Data compiled by F. Pont (http://www.inscience.ch/transits)

HD 149026b has a small radius (Fig. 1) that indicates that ~55–80% of its mass is composed of heavy elements (Fortney et al. 2006). For comparison, Jupiter and Saturn respectively have 4–13% and 13–35% of their masses in "metals" (Saumon and Guillot 2004). At the other extreme, the large radius of planet HD209458b cannot be reproduced even if it is composed only of H and He. Clearly some external process has maintained its interior entropy at a relatively high value, but its nature remains unknown (Burrows et al. 2007). The data shown in Fig. 1 raises fundamental questions about the planet formation process: What determines the amount of metals in a giant planet? What determines the mass of the planet?

2 Planet formation

The diversity of planetary systems and planetary properties poses a challenge to planet formation theories. Solar system studies have led to the widely accepted core accretion model of giant planet formation (Pollack et al. 1996; Lissauer and Stevenson 2007). Solid bodies in the protosolar system (known as planetesimals whose present-day analogs are asteroids and comet nuclei) orbited in a gas disk left over from the formation of the Sun. The gas was primarily composed of H and He, most of the other relatively abundant elements having condensed in various compounds. Multiple collisions between planetesimals led to the formation of a core of heavy elements. After a few $\sim 10^5$ years, the protoplanet's mass reaches ~10–15 $M_\oplus$, where $M_\oplus$ is the mass of the Earth, and starts accreting the gas from the surrounding disk. After ~1 Myr, the mass has grown sufficiently to cause runaway accretion of the gas where the planet accumulates the bulk of its mass in a very short time, perhaps 10^3–10^4 years. The newly formed planet gradually cools and contracts by radiating the heat of accretion to space. This formation mechanism thus predicts that the fully formed planet should have a core of heavy elements, a massive H/He envelope, and some admixture of "metals" in the envelope. This formation scenario is reasonably successful at accounting for the gross properties of the giant planets of the solar system, and perhaps of extrasolar planetary systems too. On the other hand, we have very little understanding of what happens to the core during the phase of gas accretion and of the final distribution of metals in the planet. The remarkable data compiled in Fig. 1 is rather minimal for any given planet. On the other hand, Jupiter and Saturn have been scrutinized by several spacecrafts. These two planets are the best studied outcomes of the giant planet formation process and serve as end point boundary conditions to the formation process. Can our theories of planet formation account for the present structure of Jupiter and Saturn?

3 Interior models of Jupiter and Saturn

The interior can be shown to be fully convective and thus characterized by a constant specific entropy (Hubbard 1968), which simplifies the calculation of their structure. Observations of the giant planets from the Earth give their mass, radius, rotation period, atmospheric composition, and surface entropy. Since they are rapidly rotating fluid masses in hydrostatic equilibrium, their shape is visibly oblate. The response of the planet to the rotational perturbation depends on its radial density profile and is described by the theory of rotating fluids (Zharkov and Trubitsyn 1978). Departures from sphericity are reflected in the gravitational potential of the planet which has been derived from spacecraft trajectories. These deviations, characterized as "gravitational moments" are moments of the radial density profile $\rho(r)$ of the planet. While these integral constrains are not sufficient to derive a unique density profile, simple and reasonable assumptions for the structure—inspired in part from the core accretion mechanism of formation—allow the construction of models that fit all the observations with a minimal number of parameters (Hubbard and Marley 1989; Chabrier et al. 1992; Guillot et al. 1994; Saumon and Guillot 2004 and references therein). The principal outcome of this exercise is a determination of the mass of the core and of the heavy elements mixed in the overlying gaseous envelope. These two numbers provide a crude, step function description of the distribution of heavy elements in the planet. The

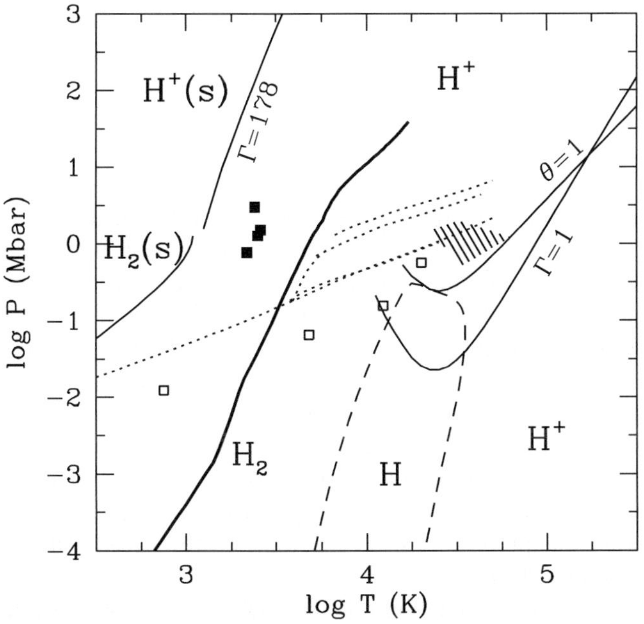

Fig. 2 Phase diagram of hydrogen. Physical regimes are indicated with solid lines showing the plasma coupling parameter ($\Gamma = 1$) and the electron degeneracy parameter ($\theta = 1$). For $\Gamma > 178$, the plasma freezes into a bcc Coulomb solid. The melting curve of H_2 is also indicated for $\log T \leq 3$. The *dashed curve* shows the dissociation and ionization boundaries in the low density gas. Above $\log P \sim 0.7$, hydrogen is fully ionized. The Jupiter isentrope is shown by the *heavy solid line*. The regions probed by single, double and triple shock experiments on deuterium are indicated with *dotted lines*. *Filled squares* show the near isentropic compression data of Fortov et al. (2008) suggesting a PPT in hydrogen. The single and double shock helium points of Nellis et al. (1984) are indicated with *open squares*. Finally, the hashed region outlines the locus of the shocked states achieved by Eggert et al. (2008) from pre-compressed He targets

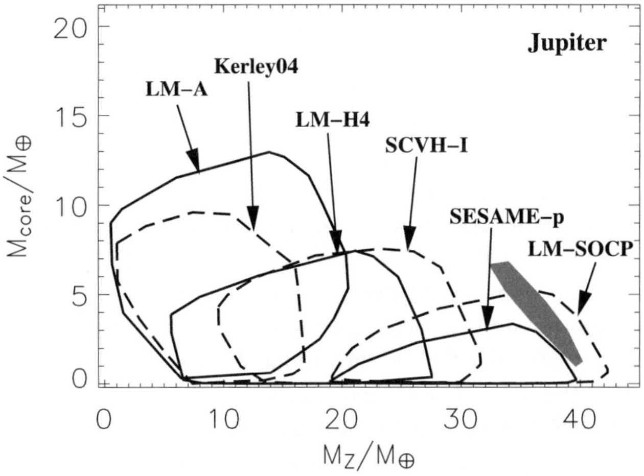

Fig. 3 Core mass and mass of heavy elements in the envelope of Jupiter in units of the Earth's mass ($M_\oplus$). Each box represents the ensemble of solutions that fits the observational constraints for a given hydrogen EOS. The *gray area* shows the solutions obtained by Nettelmann et al. (2008) using *ab initio* EOS for H, He and H_2O. Adapted from Fig. 9 of Saumon and Guillot (2004)

limited amount of data and their accuracy as well as the unknown distribution of He and the internal rotation profile of the giant planets preclude a more detailed level of modeling of their internal composition.

The density profile of an isentropic object in hydrostatic equilibrium is determined by its composition, its entropy and the equation of state (EOS) of the fluid mixture. The conditions in Jupiter's envelope are shown in Fig. 2. Most of the mass of Jupiter is found in the multi-Mbar pressure regime. The central conditions are $P \sim 70$ Mbar and $T \sim 17000$ K. In the inner 70% of the radius, the plasma coupling parameter Γ and the electron degeneracy parameter θ are

$$\Gamma = \frac{e^2}{ak_BT} \sim 20\text{–}30, \qquad \theta = \frac{T}{T_F} \sim 0.02, \tag{1}$$

where e is the electron charge, a the interparticle distance, k_B the Boltzmann constant, and T and T_F are the temperature and the Fermi temperature, respectively. The plasma is thus strongly non-ideal and the electrons are strongly degenerate. The Jupiter isentrope crosses the region of pressure ionization of hydrogen, which occurs at $P \sim 1$–5 Mbar and $T \lesssim 10^4$ K. This physical regime, where hydrogen goes from an insulating molecular fluid to a conducting, dissociated fluid, is particularly difficult to model. The past decade has seen intensive experimental work on shock-compressed deuterium (Da Silva et al. 1997; Collins et al. 1998; Knudson et al. 2004; Boriskov et al. 2005) but helium has, until recently, received much less attention (Fig. 2). The conditions achieved in shock experiments reproduce those found in Jupiter up to 0.7 Mbar for H and ~0.04 Mbar for He. The highest pressures achieved in EOS experiments of H and He are ~1–3 Mbar, but occur at temperatures that are ~10 times hotter than found at the corresponding pressures along Jupiter's isentrope. The EOS of H and He in Jupiter's interior thus remains poorly constrained by experimental data.

Considering the difficulty in modeling the pressure ionization of hydrogen and the paucity of data in the regime of interest, there remains considerable uncertainty in the EOS of hydrogen inside Jupiter. The situation for helium is similarly unsatisfactory but this is less critical due to its lower abundance (~10–15% by number). Saumon and Guillot (2004) performed an extensive study of the range of structures for Jupiter and Saturn that can fit all the data, taking into account the uncertainties in the measurements, the uncertainties arising from the assumptions that underly the simple interior model, and considering 7 different plausible EOS for H and 2 different He EOS. The results for the distribution of heavy elements in Jupiter, based on the computation of about 25000 different structures, are summarized in Fig. 3. Each polygonal box encloses the values of the core mass $M_{\rm core}$ and the mass of heavy elements mixed in the envelope M_Z of models that fit all the observational constraints for a given hydrogen EOS. Each box corresponds to the projection of a multi-dimensional error box onto the (M_Z, M_{core}) plane. While the boxes overlap to some extent, the choice of the hydrogen EOS largely determines the range of values of M_Z, and, to a lesser degree, of M_{core}. Saumon and Guillot (2004) found $M_{core} = 0–12\ M_\oplus$ and $M_Z = 1–39\ M_\oplus$, representing an enrichment in heavy elements of 1.5 to 6 times the solar value.[2] An enrichment in heavy elements is expected in the core accretion formation model, but the possibility that Jupiter may have no core is more problematic. For Saturn, they found $M_{core} = 9–22\ M_\oplus$ and $M_Z = 0–8\ M_\oplus$. Heavy elements are 6–14 times more abundant in Saturn than in the Sun. The choice of the hydrogen EOS affects primarily M_Z, but its effect is comparable to the other sources of uncertainty. An important conclusion from this study is that the interiors of Jupiter and Saturn appear to be qualitatively different; a fact that theories of planet formation must eventually explain (Alibert et al. 2005).

While the Saumon and Guillot (2004) study of interior models uncertainties was quite extensive, it is possible that an accurate EOS for H/He mixtures lies outside the range considered and that it could lead to models of Jupiter and Saturn that are more consistent with each other and with the core accretion formation model. Before reliable interior models of the giant planets can be computed, however, several long-standing EOS problems need to be solved. Does the pressure ionization of hydrogen occur gradually as the pressure increases or is there an associated first order phase transition (the so-called "plasma phase transition")? What is the EOS of mixtures of H and He at Mbar pressures? Is there a phase separation in H/He mixtures at Mbar pressures? If so, what is its phase diagram? Finally, what is the effect of the trace heavy elements (e.g. C, O, Si, Fe) mixed with H and He on the total EOS?

4 Recent developments

Models of the interiors of giant planets, in the absence of new measurements of their gravity field, are largely driven by progress in EOS experiments and models. Developments on several fronts during the last two years have invigorated this field of research. Multiple shock compression (which approximates isentropic compression) can reach Mbar pressures while keeping the temperature relatively low. In multi-shock experiments on deuterium (Fig. 2), Fortov et al. (2008) find a density discontinuity around 1.5 Mbar that suggests a first order phase transition—the long-hypothesized plasma phase transition. The reported discontinuity hinges on the validity of one data point out of total of five, however. While two independent quantities must be measured to obtain an EOS point in a shock experiment, this experiment measures only the density of the compressed deuterium. Their analysis uses EOS models to provide the missing information. The effect of such a discontinuity in the EOS would decrease M_Z and would raise M_{core} in Jupiter (Guillot 1999). This very intriguing experimental result should be independently corroborated.

After a 24 year hiatus, new shock experiments have been conducted on helium. Eggert et al. (2008) have shocked gaseous He targets that were pre-compressed from 1 to 3.3 times the liquid density (Fig. 2). While the temperatures of the shocked states are very high compared to planetary interiors ($\sim 10^5$ K), this new data provides much needed constraints on EOS models for He (Saumon et al. 1995; Militzer 2006; Ross et al. 2007). It will be rather difficult to compress helium to (P, T) conditions along Jupiter's isentrope that probe deep in its envelope, however (Militzer and Hubbard 2007).

Ab initio simulations of the EOS are a very valuable complement to experimental data in regimes where experiments are not yet available of simply not feasible. Such simulations have recently focused on the regime of the isentropes of Jupiter and Saturn using Quantum Molecular Dynamics (QMD) simulations coupled with the Density Functional Theory (DFT). Extensive simulations of the EOS of H, He, H/He mixtures and H_2O have been performed (Militzer 2006; Vorberger et al. 2007; Nettelmann et al. 2008). Interestingly, the simulations of the hydrogen EOS do not find evidence of a plasma phase transition. Overall, those simulations reproduce experimental results fairly well (Desjarlais 2003; Eggert et al. 2008). For the first time, the excess thermodynamics of mixing H and He has been quantified without prior assumptions as to their ionization state (Vorberger et al. 2007). For mixing ratios of interest for planetary interiors, the excess pressure and internal energy amount to a significant 5–10% of the total values.

The new *ab initio* EOS tables, which represent a significant improvement over previously available EOS, have been applied to modeling the interior of Jupiter. Nettelmann et al. (2008) have computed EOS tables for H, He and H_2O, where the latter is used as a proxy for all heavy elements. An EOS for mixtures of H/He/H_2O is obtained by a composition interpolation scheme known as the additive volume rule. They find relatively small cores ($M_{core} = 1–7\ M_\oplus$) and large amounts of heavy elements mixed in the H/He envelope ($M_Z = 33–40\ M_\oplus$), which are in agreement with the more extreme values found by Saumon and Guillot (2004) (Fig. 3). On the other hand, Militzer et al. (these proceedings), using their own simulations of H/He mixtures, find a very low value of M_Z and a core mass closer to the values typically found for Saturn. This result falls somewhat beyond the opposite limit found by Saumon and Guillot (2004). To summarize, the two latest sets of interior models based on *ab initio* EOS find qualitatively very different

[2] The total mass is 317.82 $M_\oplus$ for Jupiter and 95.147 $M_\oplus$ for Saturn.

results for the amount and distribution of heavy elements in Jupiter and thus have very different implications for its formation. It would be very interesting to have corresponding models of Saturn to see if the qualitative differences found by Saumon and Guillot (2004) persist with the new *ab initio* EOS.

5 Conclusions

The bewildering variety of planets and planetary systems discovered so far poses many challenges to the theory of planet formation. While the information available on any given planet remains rather rudimentary, it is somewhat compensated by the large and growing sample of objects. An important requirement of theories of formation is to account for the present day structure of Jupiter and Saturn, which are relatively well observed. The history of their formation is imprinted primarily in the distribution of heavy elements in their interior. Heavy elements are supplied as solid bodies while the massive H/He envelopes of giant planets accumulate through gas accretion. The relative importance of the two accretion processes during the formation of the planet is closely tied to the formation process and the surrounding environment.

Experiments, which are typically conducted on very small amounts ("a drop") of deuterium or helium provide vital EOS data to model the interior of giant planets. Because shock experiments typically reach very high temperatures at Mbar pressures, there is very little data taken under the conditions found inside Jupiter and Saturn, however. Off-Hugoniot, isentropic and additional pre-compressed target experiments on H and He are very much needed to solve long-standing EOS puzzles.

The recent application of *ab initio* EOS simulations to the conditions found inside giant planets is an important development in the field and should produce more reliable EOS tables than have been previously available, as well as better models of the interiors. The latest two modeling efforts based on *ab initio* EOS computed independently but with essentially the same method give very different results for the amount and distribution of heavy elements in Jupiter. The reason(s) for this discrepancy is discussed by Militzer and Hubbard (these proceedings).

The two main goals of the JUNO mission to Jupiter (to be launched in 2011) is to obtain very accurate measurements of the planet's gravity field and of the surface oxygen abundance. The latter should be the most abundant element after H and He but the current determination remains problematic and inconsistent with heavy element abundances derived from interior models. This mission will gather new data that will provide tight constraints on interior models and on the EOS of H/He/metals inside Jupiter.

Acknowledgement This work was supported by the United States Department of Energy under contract DE-AC52-06NA25396.

References

Alibert, Y., Mousis, O., Mordasini, C., Benz, W.: Astrophys. J. **626**, 57 (2005)

Boriskov, G.V., Bykov, A.I., Il'kaev, R.I., Selemir, V.D., Simakov, G.V., Trunin, R.F., Urlin, V.D., Shuikin, A.N., Nellis, W.J.: Phys. Rev. B **71**, 092104 (2005)

Burrows, A., Hubeny, I., Budaj, J., Hubbard, W.B.: Astrophys. J. **661**, 502 (2007)

Chabrier, G., Saumon, D., Hubbard, W.B., Lunine, J.I.: Astrophys. J. **391**, 817 (1992)

Collins, G.W., Da Silva, L.B., Celliers, P., Gold, D.M., Foord, M.E., Wallace, R.J., Ng, A., Weber, S.V., Budil, K.S., Cauble, R.: Science **281**, 1178 (1998)

Da Silva, L.B., Celliers, P., Collins, G.W., Budil, K.S., Holmes, N.C., Barbee, T.W., Hammel, B.A., Kilkenny, J.D., Wallace, R.J., Ross, M., et al.: Phys. Rev. Lett. **78**, 483 (1997)

Desjarlais, M.P.: Phys. Rev. B **68**, 064204 (2003)

Eggert, J., Brygoo, S., Loubeyre, P., McWilliams, R.S., Celliers, P.M., Hicks, D.G., Boehly, T.R., Jeanloz, R., Collins, G.W.: Phys. Rev. Lett. **100**, 124503 (2008)

Fortney, J.J., Saumon, D., Marley, M.S., Lodders, K., Freedman, R.S.: Astrophys. J. **642**, 495 (2006)

Fortov, V.E., Ilkaev, R.I., Arinin, V.A., Burtzev, V.V., Golubev, V.A., Iosilevskiy, I.L., Khrustalev, V.V., Mikhailov, A.L., Mochalov, M. A, Ternovoi, V.Y., et al.: Phys. Rev. Lett. **99**, 185001 (2008)

Guillot, T.: Planet. Space Sci. **47**, 1183 (1999)

Guillot, T., Chabrier, G., Morel, P., Gautier, D.: Icarus **112**, 354 (1994)

Hubbard, W.B.: Astrophys. J. **152**, 745 (1968)

Hubbard, W.B., Marley, M.S.: Icarus **78**, 102 (1989)

Knudson, M.D., Hanson, D.L., Bailey, J.E., Hall, C.A., Asay, J.R., Deeney, C.: Phys. Rev. B **69**, 144209 (2004)

Lissauer, J.J., Stevenson, D.J.: In: Reipurth, B., Jewitt, D., Keil, K. (eds.) Protostars & Protoplanets V. p. 591 (2007)

Marcy, G.W., Butler, R.P., Fisher, D., Vogt, S., Wright, J.T., Tinney, C.G., Jones, H.R.A.: Prog. Theor. Phys. Supp. **158**, 24 (2005)

Mayor, M., Queloz, D.: Nature **378**, 355 (1995)

Militzer, B.: Phys. Rev. Lett. **97**, 175501 (2006)

Militzer, B., Hubbard, W.B.: In: Elert, M., Furnish, M.D., Chau, R., Holmes, N.C., Nguyen, J. (eds.) Shock Compression of Condensed Matter, p. 1395. AIP, Melville (2007)

Nellis, W.J., Holmes, N.C., Mitchell, A.C., Trainor, R.J., Governo, G.K., Ross, M., Young, D.A.: Phys. Rev. Lett. **53**, 1248 (1984)

Nettelmann, N., Holst, B., Kietzmann, A., French, M., Redmer, R., Blaschke, D.: Astrophys. J. **683**, 1217 (2008)

Pollack, J.B., Hubickyj, O., Bodenheimer, P., Lissauer, J.J., Podolak, M., Greenzweig, Y.: Icarus **124**, 62 (1996)

Ross, M., Rogers, F., Winter, N., Collins, G.: Phys. Rev. B **76**, 020502 (2007)

Saumon, D., Chabrier, G., Van Horn, H.M.: Astrophys. J. Suppl. Ser. **99**, 713 (1995)

Saumon, D., Guillot, T.: Astrophys. J. **609**, 1170 (2004)

Seager, S., Kuchner, M., Hier-Majumder, C.A., Militzer, B.: Astrophys. J. **669**, 1279 (2007)

Vorberger, J., Tamblyn, I., Militzer, B., Bonev, S.A.: Phys. Rev. Lett. **75**, 024206 (2007)

Zharkov, V.V., Trubitsyn, T.V.: In: Hubbard, W.B. (ed.) Physics of Planetary Interiors. Pachart, Tucson (1978)

Astrophys Space Sci (2009) 322: 129–133
DOI 10.1007/s10509-008-9958-5

ORIGINAL ARTICLE

Comparison of Jupiter interior models derived from first-principles simulations

B. Militzer · W.B. Hubbard

Received: 23 June 2008 / Accepted: 25 November 2008 / Published online: 5 December 2008

Abstract Recently two groups used first-principles computer simulations to model Jupiter's interior. While both studies relied on the same simulation technique, density functional molecular dynamics, the groups derived very different conclusions. In particular estimates for the size of Jupiter's core and the metallicity of its hydrogen-helium mantle differed substantially. In this paper, we discuss the differences of the approaches and give an explanation for the differing conclusions.

Keywords Equation of state · First-principles simulations · Density functional theory · Hydrogen-helium mixtures · Giant planet interiors

The characterization of the interior structure of giant planets in our solar system is crucial for identifying their formation mechanism and for understanding the evolution of the solar system. Establishing the history of our solar system will help interpreting the observed similarities and differences between our and other solar systems. The unexpected diversity among the over three hundred recently discovered extra solar planets has challenged existing theories of planetary formation and migration.

The planets in our solar system have been studied in great detail with observations and space missions but many questions about their interior structures have remained unanswered. Jupiter is predicted to have a relatively small rocky core of between zero and seven Earth masses (Saumon et al. 1995; Saumon and Guillot 2004), which is surprising because similar theories predicted between 10 and 25 Earth masses for the core in Saturn (Saumon and Guillot 2004). This prediction for Jupiter has lent support to core-accretion theories with comparatively small cores (Pollack et al. 1996), late-stage core erosion scenarios (Guillot 2004), or suggesting that Jovian planets are able to form directly from gases without a triggering core (Boss 2007).

The small core hypothesis for Jupiter has now been challenged in a paper by Militzer, Hubbard, Vorberger, Tamblyn, and Bonev (MHVTB) (Militzer et al. 2008) who used first-principles computer simulations of hydrogen-helium mixtures to compute the equation of state (EOS) in the interior of Jupiter. This work predicts a large core of 14–18 Earth masses for Jupiter, which is in line with estimates for Saturn and suggests that both planets may have formed by core-accretion. The paper further predicts small fraction of planetary ices in Jupiter's envelope suggesting that the ices were incorporated into the core during formation rather than accreted along with the gas envelope. Jupiter is predicted to have an isentropic and fully convective envelope that is of constant chemical composition. In order to match the observed gravitational moment J_4, the authors suggest that Jupiter may not rotate as a solid body and predicted the existence of deep winds in the interior leading to differential rotation on cylinders.

The first-principles simulations used in the MHVTB paper are a major difference compared to chemical EOS models developed by Saumon, Chabrier, and van Horn (Saumon and Chabrier 1992; Saumon et al. 1995). With first-principles simulations one simulates a fully interacting quantum system of over a hundred electrons and nuclei and therefore

B. Militzer (✉)
Departments of Earth and Planetary Science and of Astronomy, University of California, Berkeley, CA 94720, USA
e-mail: militzer@berkeley.edu

W.B. Hubbard
Lunar and Planetary Laboratory, The University of Arizona, Tucson, AZ 85721, USA

avoids a number of approximations used in chemical models. In chemical models one for example describes hydrogen as an ensemble of stable molecules, atoms, free electrons, and protons and is then required to make additional approximations to treat their interactions. These approximations are dependent on the material under consideration and may also depend on the temperature and pressure. First-principles methods describe the interactions on a fundamental level. One is also required to make some approximations to solve the many-body Schrödinger equation but those are very different in nature, are not material specific and do not depend on the T and P. Therefore one expects EOSs derived from first-principles to be more accurate than chemical models unless they have been fit to experiments. However, no experimental EOS data are available for the deep interiors of giant gas planets.

The MHVTB paper described significant differences between the first-principles EOS and chemical models. However these were not solely responsible for the different predictions for Jupiter's interior. In a different paper, Nettelmann, Holst, Kietzmann, French, Redmer and Blaschke (NHKFRB) Nettelmann et al. (2008) also used first-principles method to study Jupiter's interior but came to very different conclusions. Very much in line with earlier models, they predict a small core for Jupiter and a large amount of heavy elements in the envelope.

In this paper, we will objectively discuss the differences between the MHVTB and NHKFRB approaches in order to explain how such different conclusions were derived, although we encourage the reader to compare the two original papers also. The differences between the two papers can be sorted into three categories: (1) differences in DFT-MD simulations, (2) differences in the subsequent construction of adiabats, and (3) different assumptions in the models for Jupiter's interior. We will go through these differences in the following three sections and demonstrate that the main difference arises from point (3).

1 Comparison of simulation parameters

In this section we discuss the differences in the first-principles simulations performed by the two groups. Although there are differences in the level of accuracy, they are unlikely to be the main reason for the differences in the Jupiter models. Both groups used density functional molecular dynamics (DFT-MD) simulations and should derive identical equations of state for hydrogen, helium, and their mixtures. However, computational details are important for accuracy of the derived EOS.

The NHKFRB group used exclusively VASP code while MHVTB used CPMD as well as the VASP code. MHVTB verified that both codes yield identical results when run accurately.

The MHVTB group performed simulations of hydrogen-helium mixtures using a mixing ratio that closely resembles Jupiter's composition within the limitations of a finite simulation cell. The NHKFRB instead relied on the ideal mixing approximation using EOSs of pure hydrogen and helium. This approximation neglects all interactions between hydrogen and helium. In Vorberger et al. (2007a, 2007b), it was shown that this interaction leads to significant corrections in the derived pressures and energy but, more importantly, the presence of helium increases that stability of the hydrogen molecules for given T and P. These interaction effects also have an impact on the derived adiabat that will be discussed in the next section. However, linear mixing approximation does not affect the predicted core mass very much. In fact, the MHVTB group used their own set of simulations for pure hydrogen and helium (Militzer 2008) in a separate Jupiter model and derived a core mass that differed by only 2 Earth masses. MHVTB use this independent calculation to estimate the uncertainty of the predictions.

The NHKFRB group performed DFT-MD simulations with 500–2000 time steps. MHVTB performed simulations for 2 picoseconds with 5000 time steps. Longer simulations lead to more accurate averages for thermodynamic variables such as pressure and internal energy.

Concerning the size of the simulation, the NHKFRB paper states that simulations with between 32 and 162 atoms for hydrogen, helium, and water were performed. A conservative estimate of the precision of the EOS 5% is given. Conference presentations by this group showed results with 64 hydrogen atoms and 32 helium atoms (64 electrons). The MHVTB group used simulations with 128 electrons throughout, which are more accurate.

The NHKFRB used $(\frac{1}{4}, \frac{1}{4}, \frac{1}{4})$ Baldereschi point to sample the Brillouin zone for all simulations. The MHVTB group analyzed the k-point dependence using simulation with up to $4 \times 4 \times 4$ k-points and then performed simulation with $2 \times 2 \times 2$ in the metallic regime and using the Γ point at lower densities. It was shown that Γ point only simulation with 128 electrons overestimate the pressure near Jupiter's core-mantle boundary by 1.6%.

2 Derivation of the adiabats

Convection dominates the heat transport in giant gas planets. This leads to an adiabatic temperature-pressure profile for the planet's interior with the exception of a small low pressure region near the surface where radiative heat transport takes over. It is not clear if the cores of giant planets themselves are convective but the following planetary models are insensitive to the temperature profile in the core. On the other hand, the temperature profile in the hydrogen-helium rich envelope and the required derivation of adiabats are very important to characterize the interiors of giant planets. However, neither Monte Carlo nor molecular dynamics methods can compute the entropy directly because both techniques save orders of magnitude of computer time by generating only a representative sample of configurations instead of integrating over the whole configuration space that would be needed to derive the entropy.

One typically derives the entropy by thermodynamic integration from a known reference state. This can be very computationally demanding and is also not needed to map out planetary interiors. The absolute value of the entropy is not important as long as one is able to construct (T, P) curves of constant entropy. This can be achieved by using the pressure and the internal energy from first-principles simulations at different (T, V) conditions. Using Maxwell's relations, one finds,

$$\left.\frac{\partial T}{\partial V}\right|_S = -\frac{\left.\frac{\partial S}{\partial V}\right|_T}{\left.\frac{\partial S}{\partial T}\right|_V} = -T\frac{\left.\frac{\partial P}{\partial T}\right|_V}{\left.\frac{\partial E}{\partial T}\right|_V}. \tag{1}$$

By solving this ordinary differential equation, (V, T)-adiabats can be constructed as long as a sufficiently dense mesh of high-quality EOS points are available to make the required interpolation and differentiation of E and P with respect to temperature satisfactorily accurate.

One drawback of formula (1) is that it is not necessarily thermodynamically consistent if pressures and internal energies are interpolated separately. This is the primary reason why we developed a fit for the free energy (Militzer 2008) that is thermodynamically consistent by construction. The free energy, $F(V, T)$, is fit in such a way that first-principles pressures and internal energies are reproduced by,

$$P = -\left.\frac{\partial F}{\partial V}\right|_T \quad \text{and} \quad E = F - T\left.\frac{\partial F}{\partial T}\right|_V. \tag{2}$$

The details of this method are described in Militzer (2008) where a helium EOS derived from first-principles simulations is presented for a large temperature and pressure interval.

We apply this fit across the insulator-to-metal transition in fluid hydrogen-helium mixtures (Chabrier et al. 2006). According to predictions from the best simulation methods currently available, quantum Monte Carlo (Delaney et al. 2006) and DFT-MD (Vorberger et al. 2007a) this transition is expected to occur gradually. However, for pure hydrogen, the dissociation transitions gives raise to a region of negative $\partial P/\partial T|_V$ (Bagnier et al. 2000; Vorberger et al. 2007a), which leads to a negative Grüneisen parameter and might introduce a barrier into Jupiter's convection. However, the MHVTB work demonstrated that in a hydrogen-helium mixture the region of $\partial P/\partial T|_V < 0$ is shifted to lower temperatures than occur in Jupiter. Significant effort went into analyzing this transition within the DFT-MD method. The conclusion is that Jupiter's interior is fully convective and this is main reason for constructing a two-layer model.

The two papers under consideration both use the Born-Oppenheimer MD (BOMD) technique where the electronic wave function is converged at every ionic step to obtain instantaneous ground state. Earlier DFT-MD simulations (Scandolo 2003; Bonev et al. 2004) instead used the Car-Parrinello (CP) method (Car and Parrinello 1985) to propagate the electronic wave functions. The key idea of the CP method is the construction of a fictitious dynamical system that couples ionic and electronic degrees of freedom to allow for very efficient updates of the wave function. It relies on a clear separation of electronic and ionic frequencies. For many materials, the method has been shown to give results that are in very good agreement with the more accurate but also more expensive BOMD method. However, dense hydrogen near the molecular-to-metallic transition is a special case where CPMD and BOMD differ. The CPMD method (Scandolo 2003; Bonev et al. 2004) predicts an abrupt dissociation transition while BOMD predicts it to be a continuous process.

At high pressure, Pauli exclusion leads the delocalization of electronic charge (Militzer 2005) and the electrons can no longer provide a sufficient binding force for the protons. In such an event the shape of the electronic orbitals is tightly coupled to the motion of the nuclei, and a clear separation of electronic and ionic frequencies no longer exists. That is reason why CPMD has difficulties predicting the dynamical properties of dense hydrogen near the dissociation transition. The fictitious dynamics is prone to deviate from the Born-Oppenheimer surface and to predict dissociation to occur too soon. One should rely on the BOMD technique instead.

The dissociation transition leads to a region where $\partial P/\partial T|_V$ is small but positive. Following (1), the temperature along the adiabat raise very little in this region. As a result, the predicted temperatures for the metallic regime reaching all the way down to core-mantle boundary are significantly lower than predicted by the NHKFRB model. There is no information in the NHKFRB paper on how the entropy was derived. At this point, one cannot verify whether (1) is fulfilled. The NHKFRB adiabats agree well with adiabat from the SCvH EOS model for lowest and highest pressures in Jupiter's envelope. A temperature of 18 600 K is predicted for the core-mantle boundary.

The MHVTB adiabats, on the other hand agree with SCvH only in the low density limit. In the region of molecular dissociation, the slope of the adiabats, $\partial T/\partial\rho|_S$, is reduced, which leads to a lower temperature profile for the regime of metallic hydrogen. The MHVTB model predicts a temperature of 13 300 K for the core-mantle boundary. This temperature is substantially lower than the NHKFRB prediction and more work will be needed to resolve the discrepancy. At Jupiter's core-mantle boundary the pressures are not yet high enough for the hydrogen-helium mixtures would be weakly interacting. To obtain an independent estimate for the entropy and therefore temperature, one would need to construct another path for thermodynamic integration.

3 Comparison of Jupiter interior models

The main difference between the two Jupiter's model under consideration arise from the modeling assumptions. The MHVTB group derived a new type of Jupiter model that has only *two layers*: a dense core and a completely isentropic (consistent with full convection and no phase transitions) mantle composed of mostly hydrogen and helium. Because there is no freedom in this type of model to redistribute mass by invoking chemical discontinuities in the mantle, the group could only match the gravity moment J_4 and all other constraints by invoking differential rotation with deep winds in Jupiter's envelope instead of the conventional solid-body rotation with minor surface winds. Substantial differential rotation in Jupiter, potentially detectable by the forthcoming Juno orbiter, was one of main predictions of the MHVTB paper.

The NHKFRB group used a *three layer model* that is much closer to previous models by Saumon and Guillot (2004). Besides a solid rock core, NHKFRB model assumes the mantle to be composed of two layers with differing composition. The flexibility to distribute helium and heavier elements unevenly in the two mantle layers and the associated redistribution of mass allows NHKFRB group to match J_4 without additional assumptions such as differential rotation. *This is main difference between the two models.* The redistribution of mass in the NHKFRB model to match J_4 also reduces the predicted core mass and leads to a larger amount of heavier elements in the envelope.

In the NHKFRB model, the inner layer of Jupiter is helium rich (24.5%) and contains metallic hydrogen while the outer layer is comparatively helium poor (23.3%) and contains molecular hydrogen. However, the NHKFRB model also predicts a very large difference in the concentration of heavier elements: 2.1% in the outer and 16.6% in inner layer. An explanation how such large concentration differences can arise during Jupiter's evolution remains to be given. In this regard, the MHVTB is much simpler. It assumes a fully convective mantle of constant chemical composition.

NHKFRB used the standard theory of figures to derive Jupiter's gravity field. In contrast, MHVTB used two independent approaches to derive the gravity field. The first approach used the theory of figures, while the second approach used the self-consistent-field method incorporating arbitrary differential rotation (Hubbard et al. 1975; Hubbard 1982) to monitor numerical errors in the gravity field calculations and to confirm that the J_4 mismatch for solid-body rotation is not a numerical artifact. Because NHKFRB used a standard solution for the uniformly-rotating polytrope of index one to test their code, it seems unlikely that their gravity calculations are affected by errors either. However, as a minor point, NHKFRB fitted their models to an older value for J_4 with larger error bars instead of using (Jacobson 2003).

Summarizing one can say that MHVTB predicted a large core in Jupiter of 14–18 Earth masses using a new two-layer model with a fully convective envelope. The NHKFRB work predicted a much smaller core based on a three-layer model that is very similar to earlier Jupiter models. The crucial difference lies in the treatment of the molecular-to-metallic transition in dense fluid hydrogen and more work is needed to characterize this transition with different experimental and theoretical techniques. MHVTB analyzed this transition within density functional theory and concluded that this transition is continuous leading to an Jupiter envelope of constant chemical composition. Instead of analyzing this transition, NHKFRB follow previous models and made the additional assumption of two chemically different mantle layers. Besides the discussed accuracy differences in the computed EOS, we attribute this extra assumption to be the primary reason with the difference in the predicted core masses for Jupiter.

Acknowledgements We thank J. Vorberger and D. Stevenson for comments.

References

Bagnier, S., Blottiau, P., Clarouin, J.: Phys. Rev. E **63**, 015301 (2000)
Bonev, S.A., Militzer, B., Galli, G.: Phys. Rev. B **69**, 014101 (2004)
Boss, A.P.: Astrophys. J. **661**, L73 (2007)
Car, R., Parrinello, M.: Phys. Rev. Lett. **55**, 2471 (1985)
Chabrier, G., Saumon, D., Potekhin, A.Y.: J. Phys. A **39**, 4411 (2006)
Delaney, K.T., Pierleoni, C., Ceperley, D.M.: Phys. Rev. Lett. **97**, 235702 (2006)
Guillot, T., et al.: In: Bagenal, F., Dowling, W.M.T.E. (eds.) Jupiter. Cambridge Univ. Press, Cambridge (2004) Chap. 3
Hubbard, W.B.: Icarus **52**, 509 (1982)
Hubbard, W.B., Slattery, W.L., DeVito, C.L.: Astrophys. J. **199**, 504 (1975)
Jacobson, R.A.: JUP230 orbit solution. http://ssd.jpl.nasa.gov/?gravity_fields_op (NASA-JPL document) (2003)
Militzer, B.: J. Low Temp. Phys. **139**, 739 (2005)
Militzer, B.: Submitted to Phys. Rev. B. arXiv:0805.0317 (2008)
Militzer, B., Hubbard, W.H., Vorberger, J., Tamblyn, I., Bonev, S.A.: Astrophys. J. Lett. **688**, L45 (2008)
Nettelmann, N., Holst, B., Kietzmann, A., French, M., Redmer, R., Blaschke, D.: Astrophys. J. **683**, 1217 (2008)

Pollack, J., Hubickyj, O., Bodenheimer, P., Lissauer, J.J., Podolak, M., Greenzweig, Y.: Icarus **124**, 62 (1996)
Saumon, D., Chabrier, G.: Phys. Rev. A **46**, 2084 (1992)
Saumon, D., Guillot, T.: Astrophys. J. **609**, 1170 (2004)
Saumon, D., Chabrier, G., Horn, H.M.V.: Astrophys. J. Suppl. **99**, 713 (1995)
Scandolo, S.: Proc. Natl. Acad. Sci. **100**, 3051 (2003)
Vorberger, J., Tamblyn, I., Militzer, B., Bonev, S.: Phys. Rev. B **75**, 024206 (2007a)
Vorberger, J., Tamblyn, I., Bonev, S., Militzer, B.: Contrib. Plasma Phys. **47**, 375 (2007b)

Astrophys Space Sci (2009) 322: 135–139
DOI 10.1007/s10509-009-0034-6

ORIGINAL ARTICLE

The role of high-pressure experiments on determining super-Earth properties

Diana Valencia · Richard J. O'Connell · Dimitar D. Sasselov

Received: 16 July 2008 / Accepted: 20 March 2009 / Published online: 12 May 2009

Abstract Super-Earths are the newest class of extra-solar planets with a mass range between about 1–10$M_{\oplus}$. With their large masses, they experience very large internal pressures. The central pressure scales proportionately with mass, reaching values that require us to extend our understanding of rock and H_2O behavior to such extreme conditions. Pressure also constrains the power law relationship between mass and radius of solid planets $R \sim M^{\beta}$. The value for the exponent is $0.262 \leq \beta \leq 0.274$ as constrained by the different internal structure models for super-Earths, while it is 0.3 for planets between 5–50% the mass of Earth. Despite uncertainties in planetary composition, temperature structure and equation of state, the mass-radius relationship is robust, and thus, useful for inferring the expected signal in transit searches.

In the next few years many super-Earths will be discovered and their masses and radii will be known with some uncertainty. Even without errors in both the data and structure models, a large number of compositions can fit the same average density. However, the follow-up observations with space telescopes will yield very precise radius measurements and even probe the atmospheres of super-Earths. This radius uncertainty will then be comparable to the current error derived from the equation of state used by the structure models. Thus, there is a need for accurate equations of state of solid planetary materials. Furthermore, information on the structure, such as the size and state of the core, crucially depends on the exact behavior of super-Earth materials (i.e. silicates, iron, iron alloy and ices) at high pressures and temperatures. In addition, information about the atmospheric composition of these planets may prove useful in constraining their interiors. Ultimately any inference on the structure of super-Earths, including information from atmospheres, depends on the precision of interior models, which in turn require accurate equations of state.

D. Valencia (✉) · R.J. O'Connell
Earth and Planetary Sciences Dept., Harvard University,
20 Oxford Street, Cambridge, MA 02138, USA
e-mail: valencia@eps.harvard.edu

D.D. Sasselov
Harvard-Smithsonian Center for Astrophysics, 60 Garden Street,
Cambridge, MA 02138, USA

Present address:
D. Valencia
Observatoire de la Cote d'Azur, BP 4229, 06304 Nice, France

Keywords Equation of state · Super-Earths · Internal structure · Exoplanets

1 Introduction

Since the discovery of the first super-Earth four years ago (Rivera et al. 2005), the initiative to better understand the physical properties of these planets has intensified. There has been considerable effort to detect them more readily with space missions CoRoT (Borde et al. 2003) and the recently-launched *Kepler* (Borucki et al. 2003) as well as projects from the ground. Up to date, there have been around a dozen discovered super-Earths (Rivera et al. 2005; Beaulieu et al. 2006; Udry et al. 2007; Lovis et al. 2006; Bennett et al. 2008; Bouchy et al. 2009; Mayor et al. 2009). All but one of them is transiting. That is, we know their minimum mass and orbital parameters, except for the recently discovered transiting super-Earth named CoRoT-7b (Leger et al. 2009, in press). While this planet has the smallest measured radius (of $1.72 \pm 0.13\ R_{\oplus}$), radial velocity follow-ups have yet to yield its mass.

Super-Earths are planets that have a mass larger than Earth and do not harbour substantial amounts of H/He. Formation models (Ida and Lin 2004) have suggested a value of about $10M_\oplus$ to differentiate the planets that are massive enough to capture H/He during formation and become gas giants, from the smaller ones that are unable to retain a substantial envelope. Even though this value is model dependent, it sets a good limit in planetary mass to help differentiate the planets that are mostly solid or liquid (i.e. super-Earths) from the planets dominated by an atmosphere (Jovians). Clearly, there is a fuzzy line between the large super-Earths and the mini-Neptunes (planets with an incipient H/He atmosphere). There is also a continuum in composition ranging from planets that are completely dry to H_2O-rich planets. The former planets are termed 'rocky' or 'terrestrial' following the fact that Earth has very little water in terms of its mass (at most about 0.1% of its mass is water). The latter planets, with more than a few percent water/ices are termed 'icy' or 'ocean' planets.

Apart from the microlensing planets (OGLE-2005-390Lb, MOA-2007-BLG-192-Lb), which are at large orbital radii (2.1 AU, 0.61 AU respectively) around M-dwarfs and hence, can safely be expected to be icy, it is unclear what the dominant components of the other discovered super-Earths are. In other words, we do not know their planetary type (rocky vs. icy).

In a parallel track, there have been several studies that examine the internal structure of these planets. The goal is that in combination with observations, we can constrain the different possibilities for the internal structure of super-Earths and hence, their composition and physical properties.

This conference proceeding pertains to the role of high-pressure experiments and ab-initio calculations of equations of state (EOS) in improving internal structure predictions and thus, in helping characterize super-Earths. For our arguments, we mostly rely on the detail structure model we developed, its results and limitations and refer to the capabilities of other models as explained in the literature.

2 Internal structure models

The first internal structure models proposed mass-radius relationships (Valencia et al. 2006, 2007a; Fortney et al. 2007; Sotin et al. 2007; Seager et al. 2007) in an effort to help transit studies with a prediction of radius (R) given planetary mass (M), and to perhaps be used to infer planetary composition provided a measured mass and radius. The models have different degrees of sophistication and use different EOS, but they all predict a similar exponent β in the power law relationship $M \sim R^\beta$. The different models predict an exponent in the range $0.26 \leq \beta \leq 0.274$. That is, from the different exponents the radius of a 10 earth-mass planet would be predicted with a 3% uncertainty. The reason of the agreement, despite the different treatments, is that the exponent depends mostly on pressure (P) and is quite insensitive to temperature (T), and exact composition.

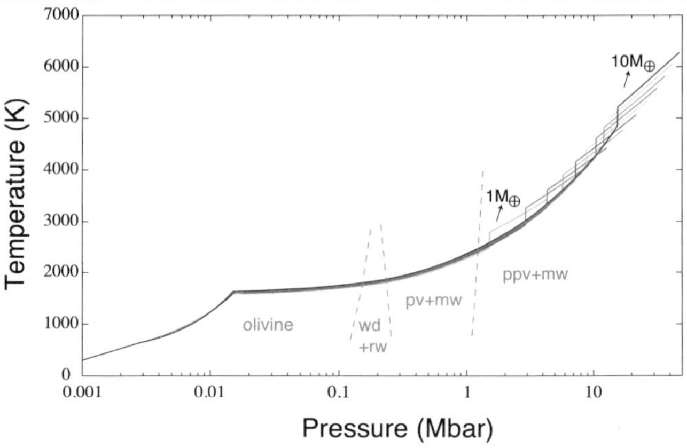

Fig. 1 P–T profiles of terrestrial super-Earths. The family of planets, from 1 to $10M_\oplus$, have a similar Fe/Si ratio as Earth. The highest internal P is 1.56 Mbar. The different phase transitions in the mantle are shown in *dashed lines* ranging from olivine (ol), wadsleyite (wd) and ringwoodite (rw), perovskite (pv) and magnesiowusite (wu), and post-perovskite (ppv) and wu. The discontinuities are caused by the boundary layers at the top and bottom of the mantle. The mantles of super-Earths with masses larger than $\sim 4M_\oplus$ are mostly composed of ppv + wu, compared to the dominance of pv + wu on Earth

The central P that terrestrial super-Earths achieve varies almost linearly with M (see Fig. 1). Unavoidably, the models that consider the internal structure of super-Earths extrapolate their EOS of state to untested regimes, or consider simple treatments. The Vinet EOS (used by Valencia et al. 2007a) and third order Birch-Murnaghan EOS (BM3) (used by Sotin et al. 2007) are adequate in the low P regime but are extrapolated beyond a safe pressure of $\sim$3 Mbars; while the Thomas Fermi Dirac (TFD) is adequate in the high P regime ($\gtrsim$10 Mbars) beyond the central P of most super-Earths. Seager et al. (2007) adopted a combination of Vinet/BM3 and modified TFD within the intermediate pressures. Fortney et al. (2007) used the well known EOS Sesame and ANEOS (which are based on a TFD approach) for ice/olivine/iron and have limitations in the P regime of super-Earths. In addition, some experiments indicate that the Vinet EOS extrapolates better to larger P than the BM3 (Hama and Suito 1996; Cohen et al. 2000), but no experiment has been conducted out at to the P and T of super-Earths.

Furthermore, Valencia et al. (2006, 2007a); Sotin et al. (2007) consider thermal corrections to the EOS throughout and phase transitions within the planets, while Fortney et al. (2007) considers isothermal planets with the exception of the icy ones where they adopt a simple temperature treatment, and Seager et al. (2007) only considers isother-

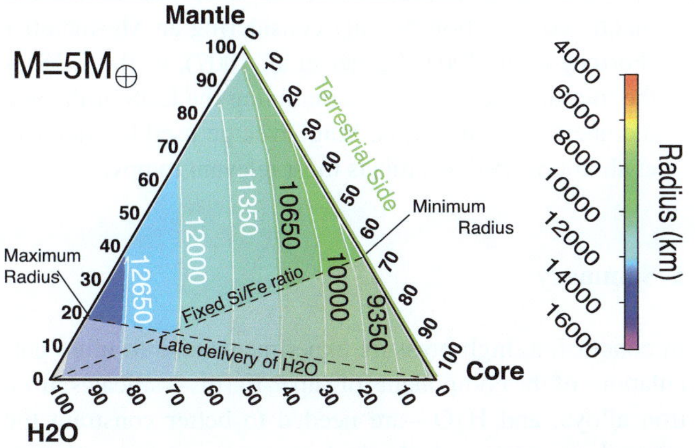

Fig. 2 Ternary diagram for a $5M_{\oplus}$-planet. The radius of each mixture is shown in *color* with the *color bar* spanning the radius of the smallest (a $1M_{\oplus}$ pure Fe planet with $R = 5400$ km) and largest (a $10M_{\oplus}$-pure H_2O planet with $R = 16000$) super-Earth. The *shaded region* shows the unlikely compositions that can conform a super-Earth from solar nebula condensation and secondary accretion constraints. A ternary diagram exists for every planetary mass value

mal planets. In addition Fortney et al. (2007); Seager et al. (2007) consider pure compositions or partial mixtures (compositions depicted only on the sides connecting H_2O-mantle and mantle-core of the ternary diagram of Fig. 2).

Nevertheless, as shown by the agreement of the different internal structure models, all these differences are not crucial when deriving a M–R relationship. However, an adequate EOS throughout the whole P–T range is crucial for determining details in the internal structure that soon will be relevant, thanks to the improvements in the accuracy of observations.

3 Inferring composition

Even though the sizes of the discovered super-Earths are not known, in the near future we expect to have dozens of transiting super-Earths with measured M and R. This information will not unambiguously yield the composition of the planets. Valencia et al. (2007b, 2007a) showed the extent of compositional degeneracy in planets that are composed of mantle silicates, iron cores and H_2O. Adams et al. (2008) showed an additional extension of degeneracy when considering another end-member component, that of an atmosphere of H and He. This additional degeneracy will become more important for the massive super-Earths and Neptune-like planets.

Figure 2 shows the ternary diagram derived from our model for a $5M_{\oplus}$-planet, such as GJ 581c. Each vertex shows the case of 100% composition of core, mantle or H_2O. The opposite side shows 0% of the corresponding end-member, and lines parallel to this zero-value side show increasing amounts of the component. Every point in the ternary diagram corresponds to a unique mixture of mantle, core and ices material.[1] Each composition yields a radius that is shown in colour. Such a diagram exists for every value in planetary mass, so that the full diagram is a triangular prism. In this prism the isolines shift to accommodate larger planets, rotate and curve slightly with increasing mass (Valencia et al. 2007a). It is clear, that for a given M (such as $5M_{\oplus}$) and a measured R, there is a wide range in compositions that can fit the data. To help constrain planetary composition we need additional information. The shaded region of the ternary diagram shows the compositions that are unlikely to conform a planet given constraints from solar nebula composition and secondary accretion processes (i.e. collisions, late delivery, etc.).

4 High-pressure experiments and ab-initio calculations

The uncertainty in R from the internal structure model is 2–3%. The error comes from uncertainties in the EOS parameters and temperature profiles (including viscosity treatment that affects the convective profile). The expected radius uncertainty with *Kepler* is ~4% on average, but follow up measurements with Warm Spitzer or the James Webb Space Telescope (JWST) will yield better precision. With really accurate radius measurements, precise EOS will be needed to help determine the composition of super-Earths.

However, even before the accuracy in R calls for a precise EOS, we already need high-P experiments and ab-initio calculations to resolve questions about the behavior of materials inside these planets, which will feed into the models. The following is a list (in order of importance) describing some of the necessary information that will considerably improve the internal structure models:

- Experiments at the 10 Mbar-level on silicates, ices and iron would illuminate the extrapolation qualities and deficiencies of the different EOS used by the models (Vinet, Birch-Murnaghan, TFD, ANEOS). Even a few experiments will be extremely helpful in this matter.
- What is the stability field of post-perovskite and are there other higher pressure silicate phases? This newly found silicate phase is stable beyond 1.26 Mbars of P (Murakami et al. 2004), which means that for the Earth it is stable for only 0.1 Mbars before it is truncated by the core (the P at the core-mantle boundary on Earth is 1.36 Mbar). However, super-Earths reach a P of 1.26 Mbars at much shallower radii, so that most of their mantles are composed of post-perovskite (see Fig. 1). It is unclear if this is the last mantle silicate phase stable all the way up to

[1] For more information on how the ternary diagram works we refer the reader to Valencia et al. (2007a).

16 Mbars, which corresponds to the P at the core-mantle boundary of a of $10M_\oplus$ terrestrial super-Earth (with an Earth-like Mg/Fe ratio); or, if there is an additional silicate phase. Mashimo et al. (2006) showed that the transition to a virtually incompressible oxide $Gd_3Ga_5O_{12}$ happened after 1.20 Mbars and suggested that "similar quasi-incompressible oxide phases composed of Si, Fe, Mg, and other elements with relatively high natural abundances, rather than Gd and Ga, might exist in the deep mantles of large extrasolar rocky planets". It is important to know which phases are present in planets and their EOS for internal structure models to include them. This means that at most, the radii obtained with internal structure models are an upper limit.

- What is the melting curve and EOS of H_2O within the P–T space of hot super-Earths. The discovery of CoRoT-7b has ignited the question of the possibility of 'steam' planets. That is, planets with a surface temperature above that of the sublimation curve and in some cases above the critical point of water. Due to the bias in detection to short-period planets, these hot-H_2O worlds might be discovered first. To model this cases it is imperative to not only have an adequate EOS at this high temperatures, but also to accurately model the transition to high-pressure forms of ices for which the melting curve is needed. Even though there has been efforts to constrain the melting curve of H_2O to high P and T, there is still a large discrepancy between the different studies (Lin et al. 2005; Schwegler et al. 2008).
- How much iron can post-perovskite accommodate? Mao et al. (2004) showed that post-perovskite (ppv) can accommodate large quantities of iron in its structure. For the same Si/Fe ratio, a super-Earth with a larger amount of content in ppv would have a smaller core. This would have implications on the state of the core (liquid versus molten) and to some extent on the planet's radius.
- What are the thermodynamic properties of all mantle materials? This knowledge is fundamental for correctly assessing the thermal state of a planet. In particular, a recent study (Cizkova et al. 2008) argues for the viscosity of ppv to be ~2–3 orders magnitude less than perovskite (pv). This would have a significant effect on the convection state and thermal evolution of a planet. In addition, it is important to know with accuracy the Gruneisen parameter (coefficient that relates internal energy to thermal pressure) of ppv (including the effects of iron) to correctly determine the state of the core.
- What are the EOS of compositional mixtures? Most of the equations of state available correspond to end-member components MgO, FeO, $MgSiO_3$, etc., but the minerals are a mixture $(Mg_{1-x},Fe_x)O$ of the Mg and Fe members, where x determines the molar fraction of Fe to Mg. Therefore, the models either simplify the treatment of the mantle composition by only considering an Mg-member (Fortney et al. 2007; Seager et al. 2007), or the EOS of the mixture is obtained by combining the EOS of the end members (such as with a Voigt average used by Valencia et al. 2006). Perhaps this is most relevant to ppv.

5 Summary

In conclusion, high pressure experiments and ab-initio calculations of the components of super-Earths—silicates, iron, iron alloys, and H_2O—are needed to better constrain the internal structure and physical properties of super-Earths. These extrasolar planets achieve large pressures, up to ~60 Mbars where current equations of state are being extrapolated without any experimental data to constrain them. Even though the mass-radius relationship for these planets is robust even to the choice of EOS, information on the profile structure, such as the size and state of the core, or information affecting the total radius, such as the existence of additional phases or amount of iron intake by ppv, crucially depend on experiments and calculations of materials to high pressures and temperatures.

In addition, with the new generation of telescopes (including JWST to be launched in 2013—Clampin 2008), the precision in radius for super-Earths will call for a smaller error in internal structure models. A source of error in these models comes from the extrapolation of the equation of state to pressures where the different EOS used differ by several tens of GPa.

As the occurrence of discovered super-Earths increases, and precision improves, high-pressure physics will be more relevant for determining the properties of these planets.

References

Adams, E., Seager, S., Elkins-Tanton, L.: Ocean planet or thick atmosphere: On the mass-radius relationship for solid exoplanets with massive atmospheres. Astrophys. J. **673**, 1160–1164 (2008)

Beaulieu, J.P., Bennett, D.P., Fouque, P., Williams, A., Dominik, M., Jorgensen, U.G., Kubas, D., Cassan, A., Coutures, C., Greenhill, J., Hill, K., Menzies, J., Sackett, P.D., Albrow, M., Brillant, S., Caldwell, J.A.R., Calitz, J.J., Cook, K.H., Corrales, E., Desort, M., Dieters, S., Dominis, D., Donatowicz, J., Hoffman, M., Kane, S., Marquette, J.B., Martin, R., Meintjes, P., Pollard, K., Sahu, K., Vinter, C., Wambsganss, J., Woller, K., Horne, K., Steele, I., Bramich, D., Burgdorf, M., Snodgrass, C., Bode, M., Udalski, A., Szymanski, M., Kubiak, M., Wieckowski, T., Pietrzynski, G., Soszynski, I., Szewczyk, O., Wyrzykowski, L., Paczynski, B., Collaboration, T.M.: Discovery of a cool planet of 5.5 earth masses through gravitational microlensing. Nature **439**, 437–440 (2006)

Bennett, D.P., Bond, I.A., Udalski, A., Sumi, T., Abe, F., Fukui, A., Furusawa, K., Hearnshaw, J.B., Holderness, S., Itow, Y., Kamiya, K., Korpela, A.V., Kilmartin, P.M., Lin, W., Ling, C.H., Masuda, K., Matsubara, Y., Miyake, N., Muraki, Y., Nagaya, M., Okumura,

T., Ohnishi, K., Perrott, Y.C., Rattenbury, N.J., Sako, T., Saito, T., Sato, S., Skuljan, L., Sullivan, D.J., Sweatman, W.L., Tristram, P.J., Yock, P.C.M., Kubiak, M., Szymanski, M.K., Pietrzynski, G., Soszynski, I., Szewczyk, O., Wyrzykowski, L., Ulaczyk, K., Batista, V., Beaulieu, J.P., Brillant, S., Cassan, A., Fouque, P., Kervella, P., Kubas, D., Marquette, J.B.: A low-mass planet with a possible sub-stellar-mass host in microlensing event moa-2007-blg-192 (2008)

Borde, P., Rouan, D., Leger, A.: Exoplanet detection capability of the corot space mission. Astron. Astrophys. **405**, 1137 (2003)

Borucki, W.J., Koch, D., Basri, G., Brown, T., Caldwell, D., Devore, E., Dunham, E., Gautier, T., Geary, J., Gilliland, R., Gould, A., Howell, S., Jenkins, J.: Kepler mission: a mission to find earth-size planets in the habitable zone. In: Fridlund, M., Henning, T. (eds.) Proceedings of the Conference on Towards Other Earths: DARWIN/TPF and the Search for Extrasolar Terrestrial Planets, ESA SP-539, Noordwijk, pp. 69–81, 2003

Bouchy, F., Mayor, M., Lovis, C., Udry, S., Benz, W., Bertaux, J.-L., Delfosse, X., Mordasini, C., Pepe, F., Queloz, D., Segransan, D.: The HARPS search for southern extra-solar planets. XVII. Super-Earth and Neptune-mass planets in multiple planet systems HD 47 186 and HD 181 433. Astron. Astrophys. **496**, 527–531 (2009). doi:10.1051/0004-6361:200810669

Cizkova, H., Oganov, A.R., Yuen, D.A., Cadek, O., Matyska, C.: Why post-perovskite should have a low viscosity, its dynamical consequences. In: AGU Fall Meeting Abstracts, A1742+, 2008

Clampin, M.: The James webb space telescope (jwst). Adv. Space Res. **12**, 1983–1991 (2008)

Cohen, R., Gulseren, O., Hemley, R.J.: Accuracy of equations of state formulations. Am. Mineral. **85**, 338–344 (2000)

Fortney, J.J., Marley, M.S., Barnes, J.W.: Planetary radii across five orders of magnitude in mass and stellar insolation: application to transits. Astrophys. J. **659**(2), 1661–1672, Part 1 APR20 (2007)

Hama, J., Suito, K.: The search for a universal equation of state correct up to very high pressures. J. Phys. Condens. Matter **8**, 67–81 (1996)

Ida, S., Lin, D.: Toward a deterministic model of planetary formation. I. A desert in the mass and semimajor axis distributions of extra-solar planets. Astrophys. J. **604**, 388–413 (2004)

Lin, J., Gregoryanz, E., Struzhkin, V., Somayazulu, M., Mao, H., Hemley, R.: Melting behavior of H_2O at high pressures and temperatures. Geophys. Res. Lett. **32**, 11306 (2005)

Lovis, C., Mayor, M., Pepe, F., Alibert, Y., Benz, W., Bouchy, F., Correia, A.C.M., Laskar, J., Mordasini, C., Queloz, D., Santos, N.C., Udry, S., Bertaux, J., Sivan, J.: An extrasolar planetary system with three Neptune-mass planets. Nature **441**, 305–309 (2006)

Mao, W.L., Shen, G., Prakapenka, V.B., Meng, Y., Campbell, A.J., Heinz, D.L., Shu, J., Hemley, R.J., Mao, H.-K.: Ferromagnesian postperovskite silicates in the D$''$ layer of the Earth. Proc. Natl. Acad. Sci. (USA) **101**(45), 15867–15869 (2004)

Mashimo, T., Chau, R., Zhang, Y., Kobayoshi, T., Sekine, T., Fukuoka, K., Syono, Y., Kodama, M., Nellis, W.J.: Transition to a virtually incompressible oxide phase at a shock pressure of 120 GPa (1.2 mbar): Gd3ga5o12. Phys. Rev. Lett. **96**, 105504:1–4 (2006)

Mayor, M., Udry, S., Lovis, C., Pepe, F., Queloz, D., Benz, W., Bertaux, J.-L., Bouchy, F., Mordasini, C., Segransan, D.: The HARPS search for southern extra-solar planets. XIII. A planetary system with 3 super-Earths (4.2, 6.9, and 9.2 M-Earth). Astron. Astrophys. **493**, 639–644 (2009). doi:10.1051/0004-6361:200810451

Murakami, M., Hirose, K., Kawamura, K., Sata, N., Ohishi, Y.: Post-perovskite phase transition in $MgSiO_3$. Science **304**(5672), 855–858 (2004)

Rivera, E.J., Lissauer, J.J., Butler, R.P., Marcy, G., Vogt, S.S., Fischer, D.A., Brown, T.M., Laughlin, G., Henry, G.W.: A 7.5 earth-mass planet orbiting the nearby star. gj 876. Astrophys. J. **634**, 625–640 (2005)

Schwegler, E., Sharma, M., Gygi, F., Gallir, G.: Melting of ice under pressure. Proc. Natl. Acad. Sci. (USA) **105**, 14779–14783 (2008)

Seager, S., Kuchner, M., Hier-Majumder, C., Militzer, B.: Mass-radius relationships for solid exoplanets. Astrophys. J. **669**, 1279–1297 (2007)

Sotin, C., Grasset, O., Mocquet, A.: Mass-Radius curve for extrasolar Earth-like planets and ocean planets. Icarus **191**, 337–351 (2007)

Udry, S., Bonfils, X., Delfosse, X., Forveille, T., Mayor, M., Perrier, C., Bouchy, F., Lovis, C., Pepe, F., Queloz, D., Bertaux, J.-L.: The HARPS search for southern extra-sola planets. XI. Super-Earths (5 and 8 M-Earth) in a 3-planet system. Astron. Astrophys. **469**, L43–L47 (2007)

Valencia, D., O'Connell, R.J., Sasselov, D.D.: Internal structure of massive terrestrial planets. Icarus **181**, 545–554 (2006)

Valencia, D., Sasselov, D.D., O'Connell, R.J.: Detailed models of super-earths: How well can we infer bulk properties? Astrophys. J. **665**, 1413 (2007a)

Valencia, D., Sasselov, D.D., O'Connell, R.J.: Radius and structure models of the first super-earth planet. Astrophys. J. **656**, 545–551 (2007b)

Astrophys Space Sci (2009) 322: 141–146
DOI 10.1007/s10509-009-9980-2

ORIGINAL ARTICLE

Using the Rayleigh-Taylor instability for *in situ* measurements of thermal conductivity of warm dense matter

D.D. Ryutov · B.A. Remington

Received: 16 July 2008 / Accepted: 29 December 2008 / Published online: 31 January 2009

Abstract The Rayleigh-Taylor (RT) instability of a material with stratified density, temperature, and composition is considered. The variation of composition gives rise to the appearance of modes whose growth rate is directly related to the finite thermal conductivity. It is suggested that this effect could be used for *in situ* measurements of thermal conductivity of warm dense matter. Expressions for the RT growth rate for the general equation of state are derived and the modes most suitable for measurements are identified. A desired perturbation can be introduced by machining the package or by using masks during the surface deposition process. To visualize the evolution of the embedded perturbation, higher-Z tracers can be used. A concept for a laser-driven experiment where this approach could be realized is presented.

Keywords Warm dense matter · Thermal conductivity · Equation of state

1 Introduction

The properties of warm dense matter (WDM) are of a significant interest for a astrophysics (Remington et al. 2006) and inertial confinement fusion (Lindl 1995). A lot of effort is being spent on the experimental measurements of the equation of state of WDM (e.g., Knudson et al. 2003; Lorenz et al. 2005; Koenig et al. 2005, and references therein). A related problem is determining transport coefficients (electrical conductivity, viscosity, and thermal conductivity). The latter is a particularly difficult task, given that the WDM samples typically occupy microscopic volumes and exist for only fleetingly short time (see, e.g., Celliers et al. 2004; Widmann et al. 2004 regarding measurements of electrical conductivity).

D.D. Ryutov (✉) · B.A. Remington
Lawrence Livermore National Laboratory, Livermore, CA 94551, USA
e-mail: ryutov1@llnl.gov

In this paper, we consider a possible approach to measuring the thermal diffusivity and some thermodynamic properties of matter by exploiting the destabilizing effect of the thermal conductivity on the RT instability of a medium with a spatially-varying composition (Ryutov 2000). As we show, one can use this approach to measure the thermal conductivity and some thermodynamic properties of a material of given composition.

The concept of the corresponding experiment is illustrated by Fig. 1. The ablative drive is from below, the experimental package is accelerated in the upward direction, so that the effective force of gravity in the co-moving frame is directed downward. We assume that the drive is tailored in such a way as to generate, initially, a compressional wave (shock wave) that heats the matter and creates the desired temperature distribution. This phase is followed with a prolonged stage of constant (ideally) or slowly varying acceleration. The desired distribution of composition (graded composition) will be created at the manufacturing stage. Initial perturbations will also be introduced at the manufacturing stage. Conceptually, this experimental setting is not dramatically different from those successfully used in a number of studies of "conventional" Rayleigh-Taylor (RT) instability (Lorenz et al. 2005; Remington et al. 1995; Budil et al. 1996; Glendinning et al. 1997).

A difference is in that we are going to study the instability in a graded material, where there is no discontinuity in composition. The length-scales of the perturbations will be much smaller than the length-scale of the density and temperature distribution. So, the growth of the pre-imposed perturbation

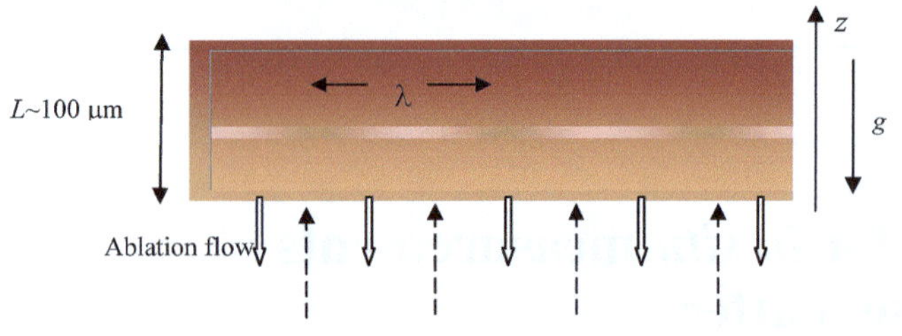

Fig. 1 Schematic of a proposed experiment with a graded composition varying in the z direction and pre-imposed lateral internal perturbation. The dashed arrows show the direction of the incident radiation and block arrows show the ablation outflow. The package is accelerated in the upward direction; the effective gravity acceleration g (in the co-moving frame) is directed downward. The wavelength of the perturbation is strongly exaggerated

will occur in the vicinity of its initial location, at least during the linear stage which we are interested in for the evaluation of the growth rate.

The RT instability is both strong and pervasive, so, the question may arise as to how one can be sure that the observed perturbations grow due to the presence of a finite thermal conductivity and not due to the standard mechanism. A critical point here is that the instability modes that we discuss in this paper grow only in the presence of a finite thermal conductivity, with the growth rate being a function of the conductivity; we identify this functional dependence for specific cases.

We are concerned with matter at a temperature of order $T \sim 1$ eV or higher, where the matter is ionized. Under such circumstances, thermal heat conduction occurs much faster than the interspecies diffusion or viscous velocity damping, that is, the thermal diffusivity is much greater than kinematic viscosity and particle diffusivity. The difference is significant, usually a couple of orders of magnitude. This is due to the much smaller velocity of heavy particles compared to the electrons, whereas mean free path lengths are either comparable or longer for the electrons. In the rest of the paper we neglect viscosity and diffusion. The latter means that the composition of any parcel of the medium does not change. The matter is considered a fluid.

The purpose of this paper is to provide a general framework for the assessment of the possibilities of the proposed technique. We leave detailed analysis of numerous design-specific issues for future work.

2 Basic equations

We characterize the matter by an equation of state

$$p = p(\rho, T, C) \tag{1}$$

and an equation for the internal energy (per unit volume),

$$\varepsilon = \varepsilon(\rho, T, C). \tag{2}$$

Here p, ρ, T, and ε are the pressure, density, temperature, and internal energy, respectively. The dimensionless parameter C characterizes the composition of the medium. For a two-component medium C can be considered as a relative atomic concentration of one of the two components, so that $0 < C < 1$. The generalization to a multi-component medium is straightforward.

As seen from (1), (2) we take as basic thermodynamic variables ρ, T, and C, and express other parameters as functions of these three. So, when having to introduce partial derivatives of thermodynamic parameters, for example $\partial\varepsilon/\partial T$, we do not specify what parameters are held constant (in the case of $\partial\varepsilon/\partial T$, for example, we assume that ρ and C are held constant). This allows us to use convenient short-hand for derivatives, e.g., $\partial\varepsilon/\partial T \equiv \varepsilon_T$, $\partial p/\partial\rho \equiv p_\rho$, etc.

At a given density ρ and temperature T, the change of the parameter C, generally speaking, leads to the change of the pressure p. Consider, as an example, the two-component system, with atomic weights A_1 and A_2 and relative atomic concentrations of the components $C_1 \equiv C$ and $C_2 = 1 - C$. Then, at a given mass density ρ, the atomic densities of the components will be $n_1 = (\rho/m_p)C[A_1C + A_2(1-C)]^{-1}$ and $n_2 = (\rho/m_p)(1-C)[A_1C + A_2(1-C)]^{-1}$, where m_p is the proton mass. Accordingly, even if the density ρ and the temperature T do not change, the pressure will depend on C due to the composition change. In other words, the derivative $\partial p/\partial C \equiv p_C \neq 0$. Interestingly, this thermodynamic parameter can be found on the basis of the proposed set of measurements.

For definiteness, we will denote by C the concentration of the heavier component, i.e. assume that $A_1 > A_2$. Then, for the systems whose equation of state does not differ too much from that of an ideal gas, one has $p_C < 0$. Indeed, for the ideal gas, $p \propto (n_1 + n_2) \propto (\rho/m_p)[(A_1 - A_2)C + A_2]^{-1}$, and $p_C < 0$. In a general case, the sign of p_C can be both positive and negative.

Unperturbed equilibrium is just the state of hydrostatic equilibrium,

$$\frac{dp_0}{dz} = -\rho_0 g. \tag{3}$$

Here z is directed upward, and g is an absolute value of acceleration, $g > 0$.

With these assumptions made, one obtains the following linearized mass, composition and momentum equations for perturbations with $\exp\Gamma t$ time dependence, Γ being a complex RT growth rate (Ryutov 2000):

$$\Gamma\delta\rho + v_z\rho_0' = -\rho_0\nabla\cdot\mathbf{v}, \tag{4}$$

$$\Gamma\delta C + v_z C_0' = 0, \tag{5}$$

$$\Gamma\rho_0\mathbf{v} = -\nabla\delta p - g\delta\rho. \tag{6}$$

Here and below the prime means differentiation of unperturbed quantities over z. In the equation for C we neglect diffusion, and in the momentum equation (6) we neglect viscosity.

We will analyze perturbations with a scale-length $\lambda\!\!\!^{-} \equiv \lambda/2\pi$ much smaller than the global scale-length L. This allows us to use the eikonal approximation in the z-direction and look for perturbations of the following structure: $a = a_0 \exp i(kx + qz)$, with a representing perturbation of density, temperature, etc. In the experiment the length-scale is determined by the wavelength of pre-imposed perturbations. We, therefore, assume that this wavelength is small enough, $\lambda\!\!\!^{-}/L \ll 1$.

By eliminating the pressure perturbation from (6), we obtain:

$$\Gamma \rho_0 \mathrm{v}_z = \frac{q}{k} \Gamma \rho_0 \mathrm{v}_x - g \delta\rho. \tag{7}$$

Then, using the continuity equation (4), we express v_z as a function of $\delta\rho$:

$$\mathrm{v}_z = -\frac{gk^2}{\Gamma(k^2+q^2)} \frac{\delta\rho}{\rho_0}. \tag{8}$$

When deriving this equation, we neglected small terms of order of $\lambda\!\!\!^{-}/L$. Substituting this result into (5), one relates the concentration perturbation and the density perturbation:

$$\delta C = \frac{gk^2 C_0'}{\Gamma^2(k^2+q^2)} \frac{\delta\rho}{\rho_0}. \tag{9}$$

The energy equation in the case under consideration (no diffusion, no viscosity) reads as (Landau and Lifshitz 1987):

$$\frac{\partial \varepsilon}{\partial t} + (\mathrm{v}\cdot\nabla\varepsilon) + (p+\varepsilon)\nabla\cdot\mathrm{v} = \nabla(\kappa\nabla T), \tag{10}$$

where κ is thermal conductivity. Linearization of this equation yields:

$$\Gamma(\varepsilon_{0\rho}\delta\rho + \varepsilon_{0T}\delta T + \varepsilon_{0C}\delta C) + \mathrm{v}_z \varepsilon_0' - (p_0+\varepsilon_0)\frac{\Gamma\delta\rho + \rho_0' \mathrm{v}_z}{\rho_0} = -\kappa_0(k^2+q^2)\delta T. \tag{11}$$

We have dropped all the sub-dominant terms in the small parameter $\lambda\!\!\!^{-}/L$ in the right-hand side.

As was shown in Catto (1978), the pressure perturbation is small compared to the density and temperature perturbation

$$\delta p/p \sim (\lambda\!\!\!^{-}/L)^2(\delta\rho/\rho) \sim (\lambda\!\!\!^{-}/L)^2(\delta T/T). \tag{12}$$

On the other hand, the equation of state (1) yields:

$$\delta p = p_\rho \delta\rho + p_T \delta T + p_C \delta C. \tag{13}$$

According to (12), δp is of the second order in the small parameter $\lambda\!\!\!^{-}/L$. Hence, one can set the left-hand side of (13) to zero, thereby obtaining:

$$\delta T = -(p_\rho/p_T)\delta\rho - (p_C/p_T)\delta C. \tag{14}$$

Substituting δT from (14), δC from (9), and v_z from (8) into (11), we obtain a single linear homogeneous equation for $\delta\rho$, which yields the dispersion relation:

$$\frac{1}{\Gamma^2}\frac{gk^2}{k^2+q^2}\left(\frac{\rho'}{\rho} + \frac{g}{s_{AD}^2}\right) - 1 = \frac{\chi s_{IS}^2(k^2+q^2)}{s_{AD}^2\Gamma}\left(1 + \frac{p_C}{s_{IS}^2\rho}\frac{gC'}{\Gamma^2}\frac{k^2}{k^2+q^2}\right), \tag{15}$$

where s_{AD} and s_{IS} are an adiabatic and isothermic sound speed, respectively, and we have used a thermodynamic identity

$$s_{AD}^2 = \frac{p_T}{\varepsilon_T}\left(\frac{\varepsilon+p}{\rho} - \varepsilon_\rho\right) + p_\rho. \tag{16}$$

The isothermal sound speed is simply $s_{IS}^2 = p_\rho$. Instead of thermal conductivity κ, we have introduced thermal diffusivity χ:

$$\chi = \frac{\kappa}{\varepsilon_T}. \tag{17}$$

Thermal diffusivity has the dimension of length square divided by time, so that the quantity $1/\chi(k^2+q^2)$ defines some characteristic thermal diffusion time over the scale $\lambda\!\!\!^{-}$:

$$\tau = \frac{s_{AD}^2}{s_{IS}^2(k^2+q^2)\chi_0}. \tag{18}$$

Equations (15) and (18) generalize the analysis of Ryutov (2000) in two respects: first, they describe matter with arbitrary thermodynamic properties (not just a monatomic ideal gas) and, second, they are not limited by the assumption that $k \gg q$.

In the limit a negligible thermal conductivity ($\chi \to 0$), one recovers from (15) the dispersion relation for adiabatic perturbations:

$$\Gamma^2 = \Gamma_{AD}^2 \equiv \frac{gk^2}{k^2+q^2}\left(\frac{\rho'}{\rho} + \frac{g}{s_{AD}^2}\right). \tag{19}$$

Conversely, in the limit of very high thermal conductivity (short wavelengths), one finds an entirely different class of perturbations whose stability or instability is determined by the sign of the product $p_C C'$:

$$\Gamma^2 = \Gamma_{IS}^2 \equiv -\frac{gk^2}{k^2+q^2}\frac{p_C C'}{\rho s_{IS}^2}. \tag{20}$$

The quantities Γ_{AD} and Γ_{IS} represent growth rates in the adiabatic and isothermal regimes, respectively. If the system is stable in either of these regimes, the corresponding quantity becomes imaginary and represents the frequency of stable oscillations.

As we have mentioned in the paragraph preceding (3), we consider C as an atomic concentration of the heavier component (e.g., copper in copper-beryllium alloy). To be specific regarding the sign of the unstable gradient of concentration, we assume that p_C is negative, so that the isothermal instability develops if C grows in the upward direction, $C' > 0$.

For further analysis, it is convenient to normalize all the quantities of dimension of time to a "characteristic" e-folding time $\sqrt{L/g}$, where L is the vertical length scale. In particular, a normalized thermal conduction time becomes

$$\hat{\tau} \equiv \tau \sqrt{\frac{g}{L}}. \tag{21}$$

The condition $\hat{\tau} < 1$ determines the range of parameters where thermal conduction plays a dominant role. We denote by λ_0 the wavelength for which $\hat{\tau} = 1$. By noting that $k = 2\pi/\lambda$ and assuming $q = k$, we obtain from (18), (21) that

$$\lambda_0 = 2^{3/2}\pi \frac{s_{IS}}{s_{AD}} \left(\frac{L\chi^2}{g}\right)^{1/4}. \tag{22}$$

3 Characteristic numerical parameters

Although the previous analysis is quite general in terms of possible specific parameters of the system, we make numerical estimates for the set of parameters that are not too different from those of the previous successful experiments on the studies of a "conventional" laser-driven Rayleigh-Taylor instability (Remington et al. 1995; Budil et al. 1996; Glendinning et al. 1997). We consider the matter at about solid-state density and temperature ~1 eV. As "representative" values of the thermal diffusivity, kinematic viscosity, and sound speed we assume:

$$\chi \sim 1\ \text{cm}^2/\text{s}, \quad \nu \sim 10^{-2}\ \text{cm}^2/\text{s}, \quad s \sim 3\ \text{km/s}. \tag{23}$$

This would roughly correspond to parameters of beryllium or beryllium-copper alloy. As a rough estimate, we do not distinguish between s_{AD} and s_{IS}, although the first is universally higher than the second (for the ideal monatomic gas, the first is 30% higher). For scale-height L and acceleration g we assume:

$$L \sim 10^{-2}\ \text{cm}, \quad g \sim 2 \times 10^{13}\ \text{cm/s}^2. \tag{24}$$

As we have already mentioned, for a particular experiment, the parameters may differ, but equations of Sect. 2 provide a way of scaling to any set of input parameters.

With that, the characteristic growth rate will be $\sim\sqrt{g/L} \sim 4 \times 10^7\ \text{s}^{-1}$. So, for an acceleration phase of duration 150 ns, the number of e-foldings will be 6, enough to detect significant growth of the initial perturbation. The 150 ns duration of a continuous acceleration is larger than what has been achieved to date with high-intensity lasers. On the other hand, it may become feasible when new large lasers would be put into operation. It can also be achieved on the basis of pulsed-power technology. Conversely, one can sacrifice the number of e-foldings used to measure the growth-rate.

The thermal diffusivity becomes a dominant effect for the perturbation wave-lengths (22) $\lambda \sim \lambda_0 \sim 6$ μm. As we will see below, a strong effect of the thermal diffusivity will take place even at significantly higher wavelengths, up to $\lambda \sim$ 30–50 μm.

Note that the unperturbed temperature distribution does not significantly change for the whole duration of acceleration phase, be it as long as 150 ns. Indeed, for the thermal diffusivity as in (23), and L as in (24), the temperature relaxation at the global scale will take at least 2500 ns.

4 Analysis of the dispersion relation

By using (19) and (20), one can rewrite (15) in the form identical to that used in Ryutov (2000):

$$\Gamma^2 = \Gamma_{AD} + \frac{\Gamma_{IS} - \Gamma_{AD}}{1 + \Gamma\tau}. \tag{25}$$

Using results of Ryutov (2000), one can obtain the stability diagram for (25) shown in Fig. 2. In the heavily shaded domain 1, the system is unstable with respect to adiabatic perturbations, that is those corresponding to a low thermal conductivity or large wavelengths. In the lightly-shaded domain 2, perturbations dominated by thermal conduction become unstable. In domain 3, both adiabatic (19) and isothermal (20) perturbations are stable, but finite thermal diffusivity makes them overstable (instability of oscillatory growth). Only region 4 is totally stable.

These properties can be used to find the thermal diffusivity. Indeed, by putting the system into domain 2 by choosing the composition and temperature stratification so as to make adiabatic perturbations stable, and, at the same time, isothermal perturbations unstable, would allow one to evaluate the thermal diffusivity from the measurements of the growth rate: at zero thermal diffusivity, there is no instability; therefore, if the system is unstable, the growth rate is a direct measure of χ.

The dependence of the growth rate vs. the wavelength of a pre-imposed perturbation is shown in Fig. 3. These graphs shows that it is advisable to perform this experiment close to the stability boundary of adiabatic perturbations, as in this

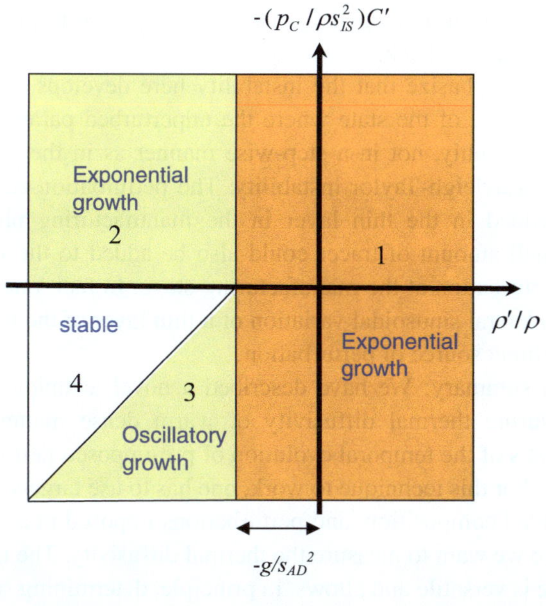

Fig. 2 Stability diagram for (25). The heavily shaded zone 1 is unstable at all wavelengths and at any thermal diffusivity. In principle, one can use this for our purpose by detecting deviation of the growth-rate from that predicted for $\chi = 0$. But this approach requires subtraction of two large numbers and will probably yield very poor accuracy. In zones 2 and 3, the existence of the instability requires finite χ, thereby providing a more robust tool for its measurement

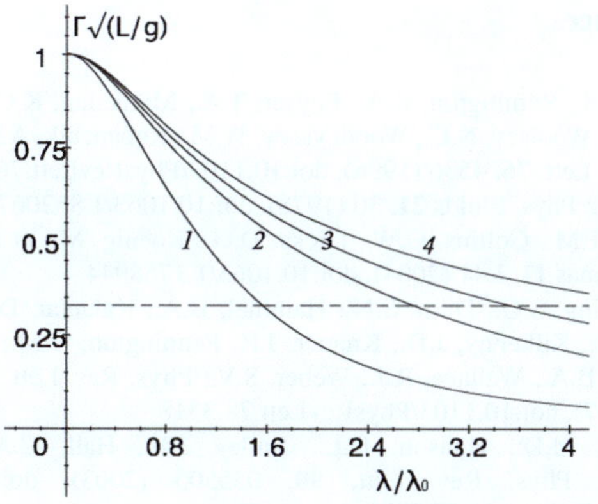

Fig. 3 The growth rate normalized to $\sqrt{g/L}$ vs the perturbation wavelength for $\Gamma_{IS}/\sqrt{g/L} = 1$. *Curve* 1 corresponds to $|\Gamma_{AD}|/\Gamma_{IS} = 1$, *curve* 2 to $|\Gamma_{AD}|/\Gamma_{IS} = 1/2$, *curve* 3 to $|\Gamma_{AD}|/\Gamma_{IS} = 1/3$, and *curve* 4 to $|\Gamma_{AD}|/\Gamma_{IS} = 1/10$. *Dashed horizontal line* is explained in the text

case the growth rate remains large even for quite long (easier to observe) wavelengths.

For the numerical example presented in Sect. 3, the maximum growth-rate shown in Fig. 3 is 4×10^7 s^{-1}. For the duration of a quasi-continuous drive of 150 ns, the number of e-folding times for the maximum growth rate is 6. A horizontal dashed line in Fig. 3 corresponds to the number of e-folding times of ~2, sufficient to see a significant growth of the initial perturbation. For curve 1, this would correspond to a wavelength of $\lambda = 1.6\lambda_0 \sim 10$ μm, for curve 2, $\lambda \sim 17$ μm, for curve 3, $\lambda \sim 21$ μm, and for curve 4, $\lambda \sim 30$ μm. These are wavelengths that are in the same range as those used in Remington et al. (1995), Budil et al. (1996), Glendinning et al. (1997). For a constant acceleration phase of duration ~80 ns, two e-folding times would be reached for $\lambda > 10$ μm for curves 2–4.

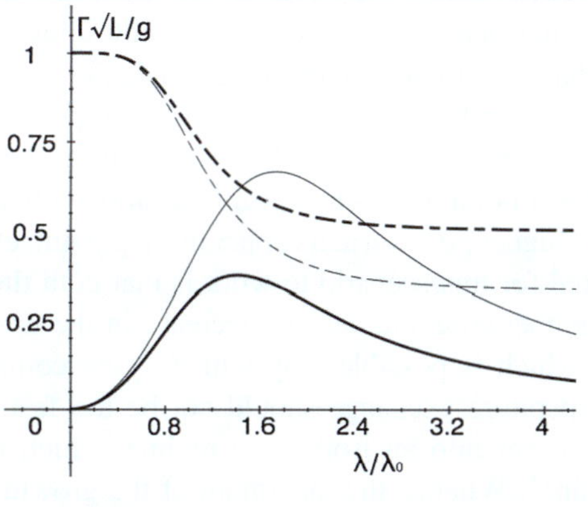

Fig. 4 The normalized growth rates (*solid lines*) and oscillation frequencies (*dashed lines*) for domain 3 in Fig. 2. *Bold lines* correspond to $|\Gamma_{AD}|/|\Gamma_{IS}| = 1/2$, *thin lines* correspond to $|\Gamma_{AD}|/\Gamma_{IS} = 1/3$

To formulate a qualitative explanation of the instability in domain #2 (Fig. 2), one can use the same arguments as those used in Ryutov (2000) for an ideal plasma. Imagine, that we take a parcel of fluid of scale smaller than $\lambda\!\!\!^{-}_0$, so that, in the course of its displacement in the upward direction the temperature inside it is promptly equilibrated with the ambient temperature. The composition of the parcel in question does not change because of negligible interspecies diffusivity. The pressure in the new position is equal to the ambient pressure. Then, the density inside the parcel becomes less than the ambient density if $p_C C' < 0$, and the parcel rises further (the perturbation grows)—exactly the condition for isothermal instability described by (20). At intermediate wavelengths, the temperature equilibration lags behind, and the growth rate decreases—as is clearly seen in Fig. 3.

An interesting possibility is associated with putting the system into domain #3, where perturbations experience the oscillatory growth,

$$\propto \exp(\mathrm{Re}\,\Gamma t)\cos(\mathrm{Im}\,\Gamma t + \psi). \qquad (26)$$

This domain is defined by the condition that both isothermal and adiabatic perturbations are stable, $\Gamma_{AD}^2 < 0$, $\Gamma_{IS}^2 < 0$ and, at the same time, $|\Gamma_{AD}| < |\Gamma_{IS}|$. The dependence of the growth-rate and frequency vs. λ for $\Gamma_{IS}^2 = -g/L$ is presented in Fig. 4. What is interesting here is that, for every shot we obtain two parameters, $\mathrm{Re}\,\Gamma$ and $\mathrm{Im}\,\Gamma$, and thereby obtain more information about the properties of the system than in domain #2 (see next section). The qualitative explanation of the instability in this domain is as follows. In the absence of thermal conductivity, any parcel of fluid will be oscillating between its lower and upper turning points with the frequency determined by (19). If we now allow for some heat exchange with the ambient medium, in the domain 3 the parcel will lose heat near the upper turning point, thereby becoming denser, and falling lower than at the beginning. At the lower point, it gets heat from the ambient matter, and reaches a higher position at its upper turning point, etc. What is required for this scenario to work is that both the unperturbed temperature and density decrease in the upward direction, which is possible only with varying composition. The temperature exchange should not be too fast, as otherwise we get into an isothermal regime, which is stable in domain 3. Whence the maximum of the growth rate vs. wavelength in Fig. 4.

5 Discussion

To determine the thermal diffusivity at a given z-position, one has to analyze the growth of pre-imposed perturbations created at this level during the manufacturing of the package. The accuracy of the measurement of the growth-rate is directly related to the accuracy of evaluation of the thermal diffusivity, via (25). If there exist information of the thermodynamic properties of the material, then all the coefficients in (25) are known, except for τ, which directly depends on the thermal diffusivity via (18).

If some of these parameters are not known, one can, in principle, derive them from making measurements at several wavelengths. The thermodynamic parameters that enter the dispersion relation (25) are: the adiabatic sound speed, and the ratio $p_C/\rho s_{IS}^2$ that enters expression (20) for the isothermal growth-rate. There is also a ratio s_{IS}^2/s_{AD}^2 in (18), but it enters (20) only in combination $\chi s_{IS}^2/s_{AD}^2$ with the unknown thermal diffusivity, χ. So, if thermodynamic parameters are unknown, even an ideal experiment, would give us only the combination $\chi s_{IS}^2/s_{AD}^2$, not just χ. In the absence of any a priori knowledge of thermodynamic parameters, one would have to make three measurements, which will allow one to determine s_{AD}, $p_C/\rho s_{IS}^2$, and $\chi s_{IS}^2/s_{AD}^2$. Any additional measurements would be used to check the consistency of (25).

An alternative way of assessing the problem would be to analyze directly (25), not assuming that Γ_{AD}^2 and Γ_{IS}^2 are known. What is important, is that they do not depend on the wavelength, whereas τ depends on the wavelength in a well-defined fashion (18). Again, the measurements at 3 wavelengths would yield Γ_{AD}^2, Γ_{IS}^2, and $\chi s_{IS}^2/s_{AD}^2$.

An interesting possibility is associated with running the experiment in the domain 3, where the instability is of oscillatory nature. Here, every shot would produce two numbers, the growth-rate and the reversal period. Accordingly, even measurements at two wavelengths would already allow one to determine Γ_{AD}^2, Γ_{IS}^2, and $\chi s_{IS}^2/s_{AD}^2$, and to provide a consistency check.

We emphasize that the instability here develops on the background of the state where the unperturbed parameters vary smoothly, not in a step-wise manner as in the "classical" Rayleigh-Taylor instability. The perturbation can be embedded in the thin layer in the manufacturing phase. A small amount of tracer could also be added to the zone of perturbation at the manufacturing stage. In fact, one can use a lateral sinusoidal variation of a thin layer of the tracer as a direct source of perturbation.

In summary: We have described a novel technique for measuring thermal diffusivity of warm dense matter by analysis of the temporal evolution of pre-imposed perturbations. For this technique to work, one has to use targets with a graded composition, and perturbations imposed in a layer where we want to measure the thermal diffusivity. The technique is versatile and allows, in principle, determining some thermodynamic parameters of the medium, alongside with the thermal diffusivity.

Acknowledgements Prepared by LLNL under contract DE-AC52-07NA27344.

References

Budil, K.S., Remington, B.A., Peyser, T.A., Mikaelian, K.O., Miller, P.L., Woolsey, N.C., Wood-Vasey, W.M., Rubenchik, A.M.: Phys. Rev. Lett. **76**, 4536 (1996). doi:10.1103/PhysRevLett.76.4536

Catto, P.J.: Phys. Fluids **21**, 30 (1978). doi:10.1063/1.862067

Celliers, P.M., Collins, G.W., Hicks, D.G., Koenig, M., et al.: Phys. Plasmas **11**, L41 (2004). doi:10.1063/1.1758944

Glendinning, S.G., Dixit, S.N., Hammel, B.A., Kalantar, D.H., Key, M.H., Kilkenny, J.D., Knauer, J.P., Pennington, D.M., Remington, B.A., Wallace, R.J., Weber, S.V.: Phys. Rev. Lett. **78**, 3318 (1997). doi:10.1103/PhysRevLett.78.3318

Knudson, M.D., Hanson, D.L., Bailey, J.E., Hall, C.A., Asay, J.R.: Phys. Rev. Lett. **90**, 035505 (2003). doi:10.1103/PhysRevLett.90.035505

Koenig, M., Benuzzi-Mounaix, A., Batani, D., Hall, T., Nazarov, W.: Phys. Plasmas **12**, 012706 (2005) doi:10.1063/1.1830488

Landau, L.D., Lifshitz, E.M.: Fluid Mechanics. Pergamon Press, New York (1987)

Lindl: J. Phys. Plasmas **2**, 3933 (1995). doi:10.1063/1.871025

Lorenz, K.T., Edwards, M.J., Glendinning, S.G., Jankowski, A.F., McNaney, J., Pollaine, S.M., Remington, B.A.: Phys. Plasmas **12**, 056309 (2005). doi:10.1063/1.1873812

Remington, B.A., Weber, S.V., Marinak, M.M., Haan, S.W., Kilkenny, J.D., Wallace, R.J., Dimonte, G.: Phys. Plasmas **2**, 241 (1995). doi:10.1063/1.871096

Remington, B.A., Drake, R.P., Ryutov, D.D.: Rev. Mod. Phys. **78**, 755 (2006). doi:10.1103/RevModPhys.78.755

Ryutov, D.D.: Phys. Plasmas **7**, 4797 (2000). doi:10.1063/1.1321316

Widmann, K., Ao, T., Foord, M.E., Price, D.F., Ellis, A.D., Springer, P.T., Ng, A.: Phys. Rev. Lett. **92**, 125002 (2004). doi:10.1103/PhysRevLett.92.125002

Astrophys Space Sci (2009) 322: 147–150
DOI 10.1007/s10509-008-9927-z

ORIGINAL ARTICLE

Radiation of electrons in Weibel-generated fields: a general case

Mikhail V. Medvedev

Received: 16 June 2008 / Accepted: 2 October 2008 / Published online: 31 October 2008

Abstract Weibel instability turns out to be the a ubiquitous phenomenon in High-Energy Density environments, ranging from astrophysical sources, e.g., gamma-ray bursts, to laboratory experiments involving laser-produced plasmas. Relativistic particles (electrons) radiate in the Weibel-produced magnetic fields in the Jitter regime. Conventionally, in this regime, the particle deflections are considered to be smaller than the relativistic beaming angle of $1/\gamma$ (γ being the Lorentz factor of an emitting particle) and the particle distribution is assumed to be isotropic. This is a relatively idealized situation as far as lab experiments are concerned. We relax the assumption of the isotropy of radiating particle distribution and present the extension of the jitter theory amenable for comparisons with experimental data.

Keywords Radiation production · Weibel instability · Laboratory astrophysics · High-energy-density physics · Gamma-ray bursts · Shock waves

1 Introduction

In general, high Mach number shocks, e.g., relativistic shocks of gamma-ray bursts, must be highly turbulent. It has been shown that the Weibel instability (Weibel 1959; Fried 1959) is responsible for the GRB shock formation (Medvedev and Loeb 1999) on the microscopic level. This instability is driven by the anisotropy of the particle distribution function (PDF) associated with a large number of particles reflected from the shock into the upstream region. This theoretical prediction has recently been confirmed in a number of state-of-the-art numerical simulations (Silva et al. 2003; Frederiksen et al. 2004; Nishikawa et al. 2003; Spitkovsky 2008). The Weibel instability is also observed in laser-plasma experiments. In particular, it is the goal of the Hercules experiment at the university of Michigan (Huntington et al. 2008; for technical details, see Maksimchuk et al. 2008) to create and diagnose the Weibel instability and turbulence in the laboratory high-energy density plasmas, as a part of the Laboratory Astrophysics and High-Energy Density Physics programs.

The state of the Weibel turbulence corresponds to the self-organized nonlinear regime of the Weibel instability, which is characterized by reorganization of the currents and magnetic fields via strong interaction and hierarchical merger process of current filaments. The Weibel-generated magnetic fields are very small-scale, of order several plasma skin depths, c/ω_p, which is much smaller than the typical Larmor radius of particles in such fields. Thus, if relativistic electrons are present in such a plasma, they will produce emission that is different from the standard synchrotron radiation. Such radiation, referred to as the "jitter radiation" has distinct spectral properties (Medvedev 2000), has been predicted to be emitted from the Weibel turbulence. It has also been predicted that jitter radiation can explain various observational data from GRBs. At last but not least, jitter radiation has been suggested as an interesting diagnostic of the Weibel turbulence in laser plasma experiments. Here we elaborate more on jitter radiation theory in application to the experiments.

M.V. Medvedev (✉)
Department of Physics and Astronomy, University of Kansas, Lawrence, KS 66045, USA
e-mail: medvedev@ku.edu

M.V. Medvedev
Russian Research Center "Kurchatov Institute", Moscow, Russia

2 Weibel turbulence

The instability under consideration was first predicted by Weibel (1959) for a non-relativistic plasma with an anisotropic distribution function, and the physical interpretation was provided later by Fried (1959), who considered the extreme case of anisotropy—two counter-streaming particle (plasma) beams. In essence, the two electron-proton plasma streams experience deflections in seed magnetic fields due to the Lorentz force, $e(\mathbf{v} \times \mathbf{B})/c$, so that protons (and electrons) moving in opposite directions concentrate in spatially separated current filaments. The magnetic field of these filaments appears to increase the initial magnetic field fluctuation. The growth rate and the wavenumber of the fastest growing mode (which, in fact, sets the spatial correlation scale of the produced field) are of order the plasma frequency, ω_p, and the plasma skin depth, c/ω_p, respectively. The current in the filaments and the associated magnetic fields increase until the energy density in the fields reaches about $\epsilon_B \sim 10\%$ of the kinetic energy density of the streaming particles, which is enough to rapidly isotropize the particle distribution; hence the instability quenches.

At longer times, the plasma with the Weibel-generated currents and fields enters the turbulent state, referred to as the "Weibel turbulence", in which current filaments begin to interact with each other, forcing like currents to approach each other and merge. The filament coalescence is a hierarchical and self-similar process (Medvedev et al. 2005). For filaments with the initial separation $\sim D_0$ the magnetic field correlation length in the non-relativistic and relativistic filaments regimes are

$$\lambda_B(t) = D_0 2^{t/(2\tau_{0,NR})}, \quad \lambda_B(t) \simeq ct. \tag{1}$$

Here, the typical non-relativistic time-scale is determined by (1). The coalescence time may be written as

$$\tau_{0,NR} \sim= \frac{(c/v)}{\sqrt{\epsilon_B}\,\omega_p} \sim 10^4\,\omega_p^{-1}, \tag{2}$$

Here v is the beam velocity and we assumed the typical value: $\epsilon_B \sim 10^{-1}$. Numerical PIC simulations showed (Medvedev et al. 2005) that both a non-power-law non-relativistic regime and a power-law regime are clearly present in the dynamics. The power-law fits yield $\lambda_B(t) \propto t^\alpha$ with $\alpha \approx 0.8$. It should be noted that the field scale growth is somewhat analogous to the inverse cascade in two-dimensional magnetohydrodynamic (MHD) turbulence, with the crucial difference that the former is an entirely *kinetic* process since at such small scales $\sim c/\omega_p$ the MHD approximation is completely inapplicable.

3 Jitter radiation

The jitter regime realizes when the deflection angle, α, of a particle in the magnetic field is smaller than the relativistic radiation beaming angle $\sim 1/\gamma$, that is $\alpha \ll \Delta\theta$. In this case, the velocity $\mathbf{v}$ of a particle is almost constant whereas its acceleration $\mathbf{w} \equiv \dot{\mathbf{v}}$ varies with time. Calculating the Fourier component of the electric field using the Liénard-Wiechart (retarded) potentials, one arrives at the following expression for the total energy emitted per unit solid angle $d\Omega$ per unit frequency $d\omega$:

$$dW = \frac{e^2}{2\pi c^3}\left(\frac{\omega}{\omega'}\right)^4 \left|\mathbf{n} \times \left[\left(\mathbf{n} - \frac{\mathbf{v}}{c}\right) \times \mathbf{w}_{\omega'}\right]\right|^2 d\Omega \frac{d\omega}{2\pi}, \tag{3}$$

where $\mathbf{w}_{\omega'} = \int \mathbf{w} e^{i\omega' t}\, dt$ is the Fourier component of the particle's acceleration, $\omega' = \omega(1 - \mathbf{n}\cdot\mathbf{v}/c)$, and $\mathbf{n}$ is the unit vector pointing towards the observer. We need to express the temporal Fourier component of the Lorentz acceleration, $\mathbf{w} = (e/\gamma mc)\mathbf{v} \times \mathbf{B}$, taken along the particle trajectory in terms of the Fourier component of the field in the spatial and temporal domains Medvedev (2006). In the static case, i.e., when the magnetic field is independent of time, the ensemble-averaged acceleration spectrum reads:

$$\langle|\mathbf{w}_{\omega'}|^2\rangle = (2\pi V)^{-1} \int |\mathbf{w}_\mathbf{k}|^2 \delta(\omega' + \mathbf{k}\cdot\mathbf{v})\, d\mathbf{k}, \tag{4}$$

$$|\mathbf{w}_\mathbf{k}|^2 = (ev/\gamma mc)^2 (\delta_{\alpha\beta} - v^{-2} v_\alpha v_\beta)\, V K_{\alpha\beta}(\mathbf{k}), \tag{5}$$

$K_{\alpha\beta}(\mathbf{r}, t) = T^{-1} V^{-1} \int B_\alpha(\mathbf{r}', t') B_\beta(\mathbf{r}' + \mathbf{r}, t' + t)\, d\mathbf{r}' dt$ is the second-order correlation tensor of the magnetic field.

We adopt the following geometry: the Weibel current filaments are aligned with the z direction and their magnetic fields lie predominantly in the x–y plane. For a GRB, this geometry corresponds to a shock which is located in the x–y-plane and is propagating along z-direction. As the shock is propagating through a medium, the produced field is transported downstream (in the shock frame) whereas new field is continuously generated at the shock front and the foreshock region. Thus, the field is also random in the parallel (z) direction. Similar structure of the magnetic fields is expected in the laboratory experiments with the field being random in the plane perpendicular to the beam propagation direction. Thus, Weibel turbulence shall be *highly anisotropic*. Both the theoretical considerations and realistic 3D simulations of relativistic shocks indicate that the dynamics of the Weibel magnetic fields in the shock plane and along the normal to it are decoupled. Hence, the Fourier spectra of the field in the x–y plane and in z direction are independent. Thus, for the Weibel fields at shocks, the correlation tensor has the form

$$K_{\alpha\beta}(\mathbf{k}) = C(\delta_{\alpha\beta} - s_\alpha s_\beta) f_z(k_\parallel) f_{xy}(k_\perp), \tag{6}$$

where $\mathbf{s}$ is the unit vector along the filaments (and normal to the shock front in the GRB case), C is the normalization constant proportional to $\langle B^2\rangle$, f_z and f_{xy} are the magnetic field spectra along **s** and in the perpendicular plane, respectively, $k_\perp = (k_x^2 + k_y^2)^{1/2}$ and $k_\| = k_z$, and finally, the tensor $(\delta_{\alpha\beta} - s_\alpha s_\beta)$ is symmetric and its product with **s** is zero, implying orthogonality of **s** and **B**.

Numerical simulations (Frederiksen et al. 2004) also indicate that the field transverse spectrum, f_{xy}, is well described by a broken power-law with the break scale comparable to the skin depth. We expect that the spectrum f_z, has similar properties. Therefore, we use the following models:

$$f_z(k_\|) = \frac{k_\|^{2\alpha_1}}{(\kappa_\|^2 + k_\|^2)^{\beta_1}}, \quad f_{xy}(k_\perp) = \frac{k_\perp^{2\alpha_2}}{(\kappa_\perp^2 + k_\perp^2)^{\beta_2}}, \tag{7}$$

where $\kappa_\|$ and $\kappa_\perp$ are parameters determining the location of the peaks in the spectra, α_1, α_2, β_1, β_2 are power-law exponents below and above a spectral peak ($\beta_1 > \alpha_2 + 1/2$ and $\beta_2 > \alpha_2 + 1$, for convergence at high-k).

We now evaluate (4), (5). The scalar product of the two tensors is

$$\begin{aligned}(\delta_{\alpha\beta} &- v_\alpha v_\beta/v^2)(\delta_{\alpha\beta} - s_\alpha s_\beta)\\ &= 1 + (s_\alpha v_\alpha)^2/v^2 = 1 + \cos^2\Theta,\end{aligned} \tag{8}$$

where we used that $\delta_{\alpha\alpha} = 3$. Here Θ is the angle between the filament direction (and normal to the shock for GRBs) and the particle velocity (in an observer's frame), which is approximately the direction toward an observer, that is $\mathbf{v}\|\mathbf{k}$ for an ultra-relativistic particle (because of relativistic beaming, the emitted radiation is localized within a narrow cone of angle $\sim 1/\gamma$). Equation (4) becomes

$$\begin{aligned}\langle|\mathbf{w}_{\omega'}|^2\rangle &= \frac{C}{2\pi}(1 + \cos^2\Theta)\int f_z(k_\|) f_{xy}(k_\perp)\\ &\times \delta(\omega' + \mathbf{k}\cdot\mathbf{v})\, dk_\| d^2k_\perp.\end{aligned} \tag{9}$$

Equations (3), (9) fully determine the spectrum of jitter radiation from relativistic electrons propagating through the Weibel turbulence.

3.1 Isotropic electron distribution: the GRB case

First, we can simplify the vector expression in (3). Indeed, in the ultrarelativistic case, the longitudinal component of the acceleration is small compared to the transverse component, $w_\|/w_\perp \sim 1/\gamma^2 \ll 1$. Therefore **v** and **w** are approximately perpendicular to each other. Second, the dominant contribution to the integral over $d\Omega$ comes from small angles $\theta \sim 1/\gamma$ with respect to the particle's velocity. Therefore, we approximately write $\omega' \simeq \omega(1 - v/c + \theta^2/2) \simeq \frac{1}{2}\omega(1 - v^2/c^2 + \theta^2) = \frac{1}{2}\omega(\theta^2 + \gamma^{-2})$. We now can replace integration over the solid angle $d\Omega \simeq \theta\, d\theta\, d\phi$ with integration over $d\phi\, d\omega'/\omega$ and integrate equation (3) over the azimuthal angle, ϕ, from 0 to 2π. The angle-averaged spectral power emitted by a relativistic particle moving through

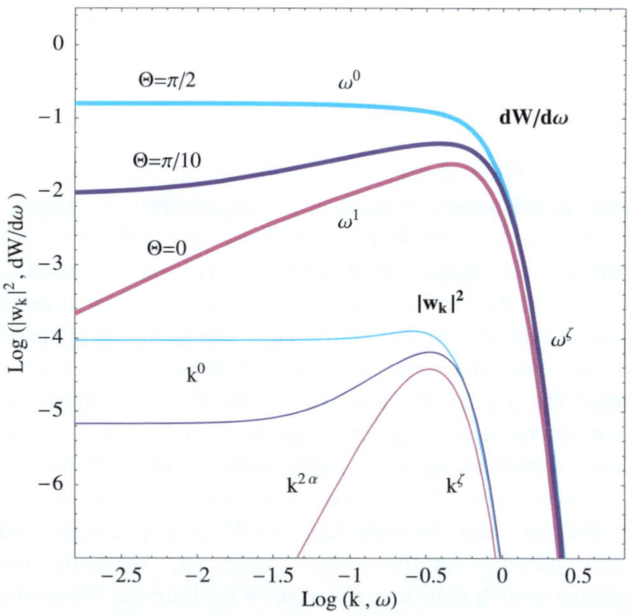

Fig. 1 The log–log plots of $|\mathbf{w_k}|^2$ vs k (*thin lines*) and of $dW/d\omega$ vs ω (*thick lines*), for three viewing angles $\Theta = 0,\ \pi/10,\ \pi/2$. The axes units are arbitrary. In this calculation we used $f_z = f_{xy}$ with $\alpha = 2$, $\beta = 20$, $\kappa = 10$, $v = 1$. The exponent $\zeta = \zeta(\alpha, \beta)$ is model dependent. We also chose $\gamma = 1$ in order to align the peaks of $|\mathbf{w_k}|^2$ and $dW/d\omega$. Note that the actual peaks are at values k, ω lower than 10 by a factor two or three. Note also that the spectrum $dW/d\omega$ levels off at oblique angles at frequencies much smaller than $\kappa v \gamma^2 \sin\Theta$, whereas $|\mathbf{w_k}|^2$ indeed starts to flatten at $k \sim \kappa v \sin\Theta$

small-scale random magnetic fields, under the assumption that the deflection angle is negligible and the particle trajectory is a straight line, has been derived elsewhere (Landau and Lifshitz 1971; Medvedev 2000):

$$\begin{aligned}\frac{dW}{d\omega} &= \frac{e^2\omega}{2\pi c^3}\int_{\omega/2\gamma^2}^{\infty} \frac{|\mathbf{w}_{\omega'}|^2}{\omega'^2}\\ &\times \left(1 - \frac{\omega}{\omega'\gamma^2} + \frac{\omega^2}{2\omega'^2\gamma^4}\right) d\omega'.\end{aligned} \tag{10}$$

A spectrum from a shock viewed at an arbitrary angle, $0 \le \Theta \le \pi/2$, is illustrated in Fig. 1, which represents full numerical solutions of (10), (7), (9) for three different viewing angles. In calculation of $dW/d\omega$, the emitting electrons were assumed monoenergetic, for simplicity. An important fact to note is that the jitter radiation spectrum varies with the viewing angle. When filaments are (and a shock velocity is, in the GRB case) along the line of sight, the low-energy spectrum is hard $F_\nu \propto \nu^1$, harder than the "synchrotron line of death" ($F_\nu \propto \nu^{1/3}$). As the viewing angle increases, the spectrum softens, and when the filaments are orthogonal to the line of sight, it becomes $F_\nu \propto \nu^0$. Another interesting feature is that at oblique angles, the spectrum does not soften simultaneously at all frequencies. Instead, there appears a smooth spectral break, which position depends on Θ. The spectrum approaches $\sim \nu^0$ below the break and is harder above it.

3.2 Beam electron distribution: the lab case

Unlike a GRB shock, the distribution of radiating electrons is anisotropic in most of lab experiments. In particular, to diagnose the Weibel turbulence in the Hercules experiment (Huntington et al. 2008), it has been suggested to launch a probe, nearly monoenergetic electron beam through plasma with the Weibel fields. Thus, the geometrical shape and the electron energy-momentum distribution of the probe beam are important. Here we assume the electrons to be monoenergetic and neglect the geometrical divergence of the beam, for simplicity (in the experiment, it is likely somewhat smaller than the relativistic beaming cone of $1/\gamma$ anyway).

We again start from the Liénard-Wiechart potentials and the expression for the emitted power (3). We adopt the geometry such that the unit vector $\mathbf{s}$ is along the filaments, $\mathbf{v}$ is the particle's velocity and the unit vector $\mathbf{n}$ is toward an observed. We also define the unit vector $\hat{\mathbf{v}} = \mathbf{v}/v$ and $\beta = v/c$. Unlike the isotropic case, we do not neglect terms proportional to $|\mathbf{n} \cdot \mathbf{w}_{\omega'}|^2$ although they are small compared to those proportional to $|\mathbf{w}_{\omega'}|^2$. Using a similar approach, we arrive at the following expression for the emitted power per frequency, per solid angle, per electron:

$$\begin{aligned}\frac{dW_\omega}{d\omega d\Omega} &= \frac{e^2}{(2\pi)^2 c^3} \frac{1}{(1-\beta(\mathbf{n}\cdot\hat{\mathbf{v}}))^4} \\ &\quad \times \left(\frac{e\beta}{\gamma m}\right)^2 \beta^2 \frac{C}{2\pi} \int f_z(k_\parallel) f_{xy}(k_\perp) \\ &\quad \times \delta(\omega(1-\beta(\mathbf{n}\cdot\hat{\mathbf{v}})) - \mathbf{k}\cdot\mathbf{v})\, dk_\parallel d^2k_\perp \\ &\quad \times \left\{(\mathbf{n}\cdot\hat{\mathbf{v}})^2(1+(\mathbf{s}\cdot\hat{\mathbf{v}})^2) + \left[(\mathbf{s}\cdot\mathbf{n})^2 + (\mathbf{s}\cdot\hat{\mathbf{v}})^2 \right.\right. \\ &\quad \left.\left. - 2(\mathbf{s}\cdot\mathbf{n})(\mathbf{s}\cdot\hat{\mathbf{v}})(\mathbf{n}\cdot\hat{\mathbf{v}})\right]\right\}. \end{aligned} \tag{11}$$

One can readily see that the emitted power along the electron probe beam, i.e., when $\mathbf{n} \cdot \hat{\mathbf{v}} = 1$, is proportional to the electron acceleration spectrum:

$$\frac{dW_\omega}{d\omega d\Omega} \propto |\mathbf{w}_{\omega'}|^2, \tag{12}$$

which is represented by the integral over $\mathbf{k}$, up to a constant. Thus, the electron beam can directly probe and diagnose the structure of the magnetic field distributions: f_z if the beam is aligned with the Weibel filaments and f_{xy} if the filaments are probed "edge-on"; see Fig. 1 for the $\mathbf{w}$-spectrum. The angular shape of the radiation pattern (i.e., the angular distribution of intensity) can also be readily calculated by integrating over the frequency:

$$\frac{dW}{d\Omega} \propto \left(1-\beta(\mathbf{n}\cdot\hat{\mathbf{v}})\right)^{-5} \propto \left(1+(\gamma\vartheta)^2\right)^{-5}. \tag{13}$$

The latter expression is valid for small angles ϑ between the probe beam direction and the line of sight.

4 Conclusions

The jitter spectra can deliver much information on the structure of the Weibel magnetic fields. As one can see, the spectrum depends on the spatial spectra of the magnetic fields modeled by (7). In particular, when viewing angles are 0 and 90 degrees, the contributions of the parallel and transverse magnetic field spectra are decoupled. For instance, for $\theta = 0$, the peak of the jitter radiation spectrum and its high-energy asymptotic slope are uniquely determined by the parallel correlation length $\kappa_\parallel$ and the large-k magnetic field spectrum slope k^η with $\eta = 2\alpha_1 - 2\beta_1$. Similarly, the transverse jitter spectrum (at $\theta = \pi/2$) allows one to deduce these parameters for the transverse magnetic field spectrum, f_{xy}. At intermediate angles, one can determine the relative orientation of the current (and magnetic) filaments in the target and the radiation detector.

It seems feasible to obtain jitter radiation in a laser-plasma experiment, such as Hercules. The Weibel turbulence to be studied will have much in common with the upstream region of a gamma-ray burst collisionless relativistic shock. It may be so even up to and at the main shock compression, where the Weibel filaments are present. In the downstream region the filaments are destroyed and the fields are significantly isotropized, as follows from simulations (Spitkovsky 2008). In such a turbulence state, jitter radiation can still be present, but it will produce more isotropic, relatively soft spectra resembling those at $\theta \sim \pi/2$.

Acknowledgements This work has been supported by NASA grants NNX07AJ50G, NNX08AL39G, NSF grant AST-0708213, and DOE grant DE-FG02-07ER54940.

References

Frederiksen, J.T., Hededal, C.B., Haugbølle, T., Nordlund, Å.: Astrophys. J. **608**, L13 (2004)
Fried, B.D.: Phys. Fluids **2**, 337 (1959)
Huntington, C., et al.: Bull. Am. Astron. Soc. **40**, 192 (2008)
Landau, L.D., Lifshitz, E.M.: The Classical Theory of Fields. Pergamon Press, Oxford (1971)
Maksimchuk, A., et al.: Phys. Plasmas. **15**, 056703 (2008)
Medvedev, M.V.: Astrophys. J. **540**, 704 (2000)
Medvedev, M.V.: Astrophys. J. **637**, 869 (2006)
Medvedev, M.V., Loeb, A.: Astrophys. J. **526**, 697 (1999)
Medvedev, M.V., Fiore, M., Fonseca, R.A., Silva, L.O., Mori, W.B.: Astrophys. J. **618**, L75 (2005)
Nishikawa, K.-I., Hardee, P., Richardson, G., Preece, R., Sol, H., Fishman, G.J.: Astrophys. J. **595**, 555 (2003)
Spitkovsky, A.: Astrophys. J. **673**, L39 (2008)
Silva, L.O., Fonseca, R.A., Tonge, J.W., Dawson, J.M., Mori, W.B., Medvedev, M.V.: Astrophys. J. **596**, L121 (2003)
Weibel, E.S.: Phys. Rev. Lett. **2**, 83 (1959)

Astrophys Space Sci (2009) 322: 151–154
DOI 10.1007/s10509-009-0002-1

ORIGINAL ARTICLE

Laser-plasma simulations of artificial magnetosphere formed by giant coronal mass ejections

Yuri P. Zakharov · Arnold G. Ponomarenko · Konstantin V. Vchivkov · Wendell Horton · Parrish Brady

Received: 14 May 2008 / Accepted: 27 January 2009 / Published online: 24 February 2009

Abstract We employed the laboratory (Laser-Produced Plasmas, LPP) and numerical (3D/PIC-code) simulations to study the resulting state of very strong compression of magnetopause (MP) by CME with effective energy $E_0 \geq 10^{34}$ ergs directed to the Earth. During probable formation of an Artificial Magnetosphere (AM, *in a flow of CME' plasma around the Earth*) with the MP stand-off at R_{mp} up to (2–3)R_E, many catastrophic phenomena could occur in a space and ground networks due to very high curl electric fields induced by world-wide magnetic field's changes with a SC-rate >50 nT/s. The laboratory models of AM (with $R_{mp} \sim 0.1$–30 cm) were formed around high-field, 1D and 3D magnetic obstacles, overflowing by LPP-blobs with E_0 up to kJ and magnetized ions. The shape and internal structure of a large-scale AM were studied at KI-1 facility of the Russian team using a set of B-dot magnetic probes, while the main goal of UT's small-AM experiment was to explore a possible shock's generation and relevant electron acceleration. Preliminary results of KI-1 experiments show that the both R_m-size and SC (E_0) of AM could be described by modified Chapman-Ferraro Scaling, while the whole SC-distribution (in front "one-half" of equatorial plane)—by well-known "Image Dipole" model of the Earth's magnetopause field.

Y.P. Zakharov (✉) · A.G. Ponomarenko · K.V. Vchivkov
Institute of Laser Physics (ILP), Russian Academy of Sciences, Novosibirsk, 630090, Russia
e-mail: zakharov@plasma.nsk.ru

W. Horton · P. Brady
University of Texas (UT) at Austin, Austin, TX 78712, USA

Keywords Magnetopause super-compression · Coronal mass ejection · Laboratory and PIC-simulations · Chapman-Ferraro current · Flute-like instability · Magnetic dipole · Laser-produced plasma

1 Introduction

More than 3-fold compression of the Earth's magnetosphere could be caused by giant Coronal Mass Ejections (CME), driven by recently discovered Mega Solar Flares with the total released energy $\geq 10^{34}$ ergs. Its potential after-affects should be so hazardous that advanced and detailed investigations, including modern laboratory simulations are required (Horton and Chiu 2004; Horton et al. 2007a; Ponomarenko et al. 2005, 2007b). Physical conditions of such rare and enormous phenomena could be understood based on the concept of a non-stationary "Artificial Magnetosphere"—AM (Nikitin and Ponomarenko 1996; Zakharov et al. 1996; Ponomarenko et al. 2007b) which is formed when an exploding plasma flows around a magnetic dipole μ. According to the MHD-model (Nikitin and Ponomarenko 1996; Zakharov et al. 1996; Ponomarenko et al. 2007b), the stand-off distance R_{mp} of AM magnetopause is determined by the main energetic parameter of the problem $K = 3E_0R_0^3/\mu^2$ (for effective plasma energy E_0 and distance R_0 between dipole center and plasma release's point). For the explored case of $K \gg 1$ it could be expressed (Ponomarenko et al. 2007b) approximately as $R_{mp} = R_m^* \approx 0.75R_0/K^{1/6}$. This type of scaling was confirmed for the fist time in the Laser-Produced Plasma (LPP) experiment of ILP (Zakharov et al. 1996) at KI-1 target chamber ⌀120 cm and recently in our PIC-simulations (Ponomarenko et al. 2007a, 2007b) by 3D/Hybrid-code of Kyushu University (KU, Japan). Our first experiment on the simulation of AM-formation during anti-asteroid explosions (Zakharov et al. 1996) was characterized by unmagnetized ions as their directed Larmor radius $R_L = mcV_0/zeB_d^*$ at magnetopause (for $B_d^* = \mu/R_m^{*3}$) was larger than R_m^*, i.e. criterion $\varepsilon_m = R_L/R_m^* \sim 3(>1)$, similarly to pioneer work of Bostick et al. (1966) with plasma gun.

To correctly simulate AM-formation around the Earth we carried out Artificial Magnetosphere EXperiment (AMEX) in the required MHD range of dimensionless criteria $K \gg 1$ and $\varepsilon_m < 1$. So far the problem was explored in this range only in 2D/3D PIC—simulations, but only for stationary overflowing, while for the non-stationary and CME related phenomena, a 3D/MHD-code (Ridley et al. 2006) appears recently. To analyse our laboratory data and apply them to extreme geophysical phenomena we have used a 3D/PIC hybrid code of KU, which was verified in turn by AMEX data. Our laboratory simulation is based on the *principle of limited simulation* (Podgorny and Sagdeev 1970; Zakharov 2003) and one of its first goal was to reproduce the structure and dynamics of super-compressed magnetic field in the sunward part of highly-disturbed Earth's magnetosphere.

2 Large scale experiment AMEX and relevant 3D/PIC-simulation

Figure 1 shows a principal scheme of AMEX in the main cross-section of KI-1 target chamber (Zakharov 2003) with the length 5 m and high-vacuum up to 10^{-7} Torr. This stainless steel chamber is supplied by the ms-source of high-density plasma background's flow and external coil of steady uniform magnetic field B_0, while the whole KI-1 facility consists of a set of CO_2-laser generator and amplifier with the output up to 400 J in 100 ns-pulse. We used a flat or convex Nylon6 plastic target with a few cm^2 laser spot to produce a LPP with moderate front velocity $V_0 \approx 180$–200 km/s (maximal along to target normal, where $\theta = 0$), consisting of a H^+ and C^{+3}/C^{+4} ions with the average $\langle m/z \rangle \approx 2.5$ a.m.u. The initial shape of LPP appears as a sphere touched to target and consequently it had angular distribution like $V_0 \propto \cos\theta$, for which LPP of uniform density should have a total kinetic energy $E_k = E_0/16$, where $E_0 = (dE/d\Omega)_0 \times 4\pi$ is effective LPP energy for maximal $(dE/d\Omega)_0$ along to target normal. So for the real energy of LPP up to $E_k \approx 25$ J we have its effective value (into 4π) $E_0 \approx 400$ J.

This approach allows us to compare directly the results of our experiments with theoretical models of the interaction between expanding spherical plasma clouds (of energy E_0) and various magnetized media. It was based on the data of a KI-1 test experiment (Ponomarenko et al. 2007b) with an LPP-blob decelerated by uniform field $B_0 = 600$ G. As

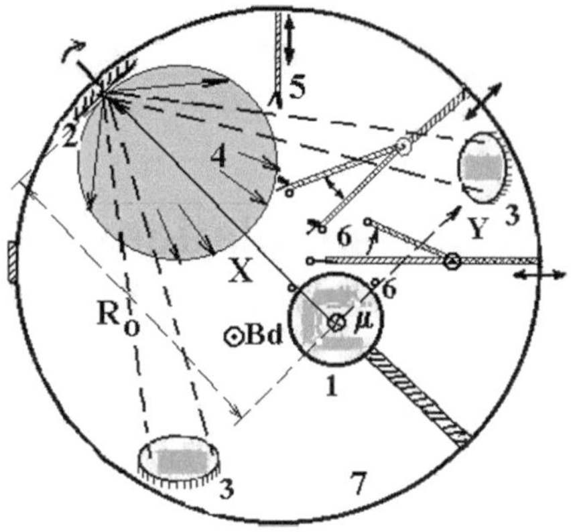

Fig. 1 General scheme of AMEX experiment at KI-1 facility of ILP: **1**—dipole; **2**—laser target; **3**—mirrors for beams of CO_2-laser; **4**—Laser-Produced Plasma (LPP); **5**—set of Langmuir probes; **6**—set of magnetic (B-dot) probes; **7**—chamber ⌀120 cm; B_d—dipole field (in the shown equatorial plane). Axis of Gated Optical Imager (GOI) was directed into the paper

we recently reported (Zakharov et al. 2008) such LPP can be used as the model of giant CME in the laboratory simulation with LPP—dipole interaction, to test a new magnetopause scale R_m^* of AM, very important for the problem of probable catastrophic changes of the Earth's magnetosphere. Earlier we verified of R_m^*—scaling in the PIC-runs with a spherical plasma clouds (Ponomarenko et al. 2007a) and LPP-blobs (Zakharov et al. 2008), as presented below. A 3D/PIC-model of KU with our simple MHD-model are briefly described by Ponomarenko et al. (2007b) and we used up to 3 000 000 ions in the grid with a minimal size up to 4–5 mm for the PIC-runs. The most runs were conducted under the same conditions as for experiments AMEX, with the initial distance $R_0 = 75$ cm (mainly GOI photos for LPP with $E_0 \approx 300$–400 J) and its final value $R_0 = 61.5$ cm (for LPP with $E_0 \approx 200$ J), which was more appropriate for the detailed probe measurements along to main X-axis (Fig. 1). They were done for two main regimes (I and II) of dipole field with the moments $\mu = 1.1 \times 10^7$ and 2×10^6 G·cm^3. Such 3D-fields were generated by many-turns coil (⌀16 cm) covered by thin-wall metallic ended cylinder of ⌀20 cm and supplied by ms-pulsed capacitor scheme for multi-kA current.

3 Results and discussion

Figure 2 shows a typical examples of AM-shape in AMEX with apparent non-MHD ($\lambda \leq R_L$), flute-type instability, developing (at left) from the early stage ($t = 1.5$–2 μs) of LPP deceleration by field (Zakharov et al., 2006, 2008). It could be a reason of plasma penetration inside of MP, up to the dipole's surface at small $\mu \leq 10^6$ G·cm^3, since the gap ≈10 cm (between it and MP, here) is comparable with the directed ion Larmor radius $R_L \geq 10$ cm (at $B_d^* \approx 300$ G).

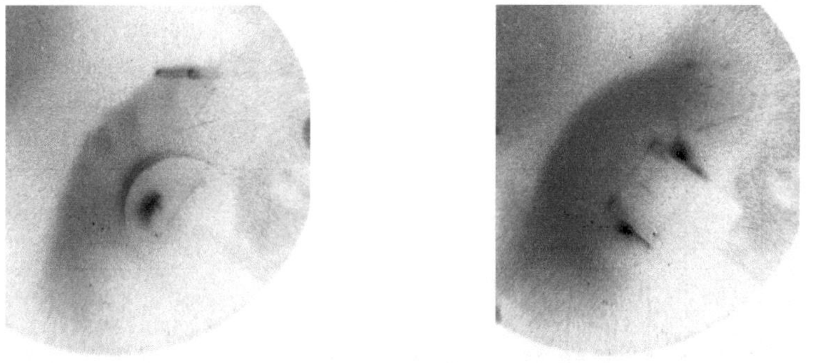

Fig. 2 Initial GOI data on the shape of magnetopause around dipole (∅20 cm) with the moment: $\mu = 1.1 \times 10^6$ G · cm^3 (#35 at the left, $t = 4.2$ μs for $K \approx 4000$) and 2×10^6 G · cm^3 (#41 at right, $t = 4.4$ μs for $K \approx 1000$, where dipole was rotated to show "meridional plane"). In both cases $R_0 = 75$ cm, $E_0 = 300$–400 J and H_2-gas added at 0.1 mTorr (to enhance luminosity of MP)

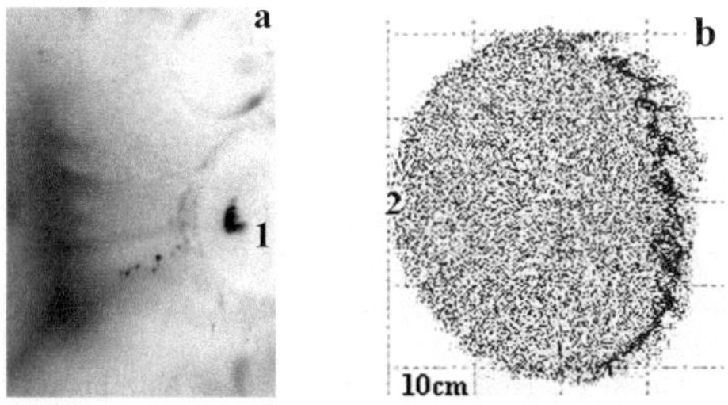

Fig. 3 Comparison at the same scale of AMEX #224 (**a**) with added $B_0 = 115$ G and data of its PIC-modeling (**b**). Here $K \sim 10$ (for $E_0 = 200$ J, $R_0 = 61.5$ cm and $\mu = 1.1 \times 10^7$ G · cm^3) and such LPP's stop ($t = 2.2$–2.3 μs) at $R_{mp} \approx 26$ cm ($< R_m^*$) is better described by MHD-model; **1** is front part of dipole (with a spot of plasma penetrated along to B_d) and **2**—point of plasma release

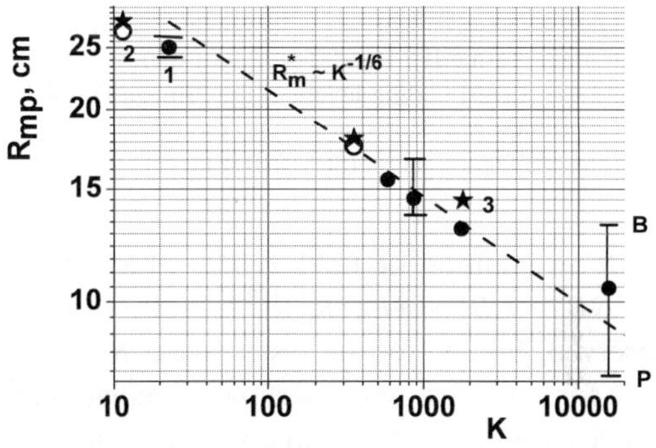

Fig. 4 Combined data on magnetopause stand-off R_{mp} in a flow of exploding plasma, ejected at $R_0 = 61.5$ cm: **1**—PIC-modeling for plasma energy $E_0 = 400$ J; **2**—PIC for $E_0 = 200$ J; **3**—AMEX laboratory simulation for $E_0 = 200$ J (and μ up to 1.1×10^7 G · cm^3, at the left point); **P** and **B**—plasma boundary and its diamagnetic boundary (i.e. MP), correspondingly; R_m^*—is estimation of stand-off via scaling $R_m^* \approx 0.75 R_0 / K^{1/6}$ for $E_0 = 400$ J

Figure 3(b) shows the result of PIC-run with the energy $E_0 = 200$ J and the same LPP-geometry, V_0 and $\langle m/z \rangle$ as in AMEX. The dipole-like field "was created" by one turn of radius 8 cm like in AMEX coil. PIC-modeling data demonstrate a similar LPP deceleration along to X-axis or development of flute instability and reveal that up to the moment of LPP's stop it excludes all magnetic field in accordance with AMEX experimental results (Zakharov et al. 2008). In result, a general relation $R_{mp}(K)$ for the case $R_0 = 61.5$ cm was obtained (Fig. 4) and important verification of suggested approximation $R_m^*(K)$ was done in a wide range of criterion K.

Beside expected and pronounced deviation from $R_m^*(K)$ at finite $K \sim 10$, a substantial difference between a plasma' (**P**) and magnetic field' (**B**) boundary at large K in the given case (of fixed E_0, V_0 and R_0) is apparent effect of low ion magnetization. To the definition of $R_L \propto V_0/B_d^*$ and also due to fixed value of field $B_d^* \propto (E_0/R_0^3)^{1/2}$ at MP (as R_m^*), the value of ion Larmor also should be fixed, while the own MP's size essentially decrease to high K numbers (and small μ).

Therefore ion magnetization criterion $\varepsilon_m = R_L/R_m^*$ became ≥ 1 (at $K > 10^4$), instead of minimal $\varepsilon_m \sim 0.3$ at $K \sim 10$–20, where the difference between positions of plasma and field boundary is very small (Fig. 4). As result we had chosen for AMEX the intermediate regime of $K \sim 1000$, in which ε_m is enough small for MHD-like structure of MP, while its whole boundary could be enough bended around dipole (Fig. 2), similar to the case of Earth's MP.

So, for the main regime II of AMEX at $K \approx 350$ (with $E_0 = 200$ J, $R_0 = 61.5$ cm and $\mu = 2 \times 10^6$ G · cm^3) it was shown, that the 1D-distribution of magnetic field' compressional disturbances $\Delta B(X)$ inside of AM could be described (Zakharov et al. 2008) by well-known model of "Image Dipole" (Shabansky 1968). It allows us to determine the magnitude of world-wide ΔB-effect for the given size $R_m^*(E_0)$ and shape of AM, if our simulative AMEX measurements out of X-axis also correspond to this model and the data of Fig. 5 do confirm this.

Another important magnetic disturbances were measured directly at the dipole' conductive surface (Figs. 1, 2), where an essential local amplification by factor 1.5 could be possible due to its "Telluric" currents and for its spherical shape (Shabansky 1968). We did not see such effect, most probably because of a shape of used dipole in a form of closed cylinder of finite height 14 cm $\sim$ ∅. Finally, according to "Image Dipole" model, for enough "bended" MP-shape (around the dipole), a global compression effect of dipole field could be evaluated in magnitude as a

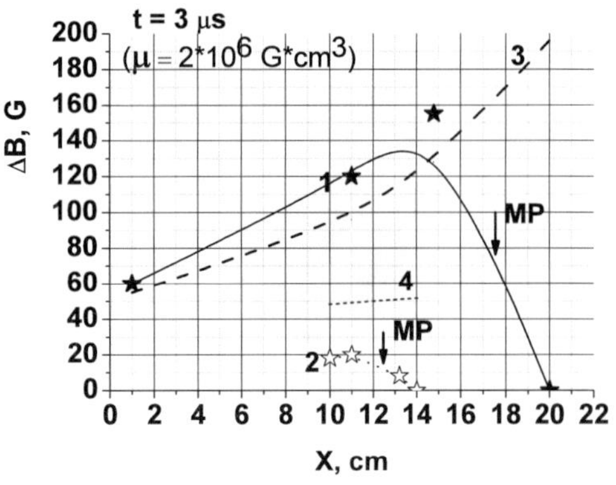

Fig. 5 Compression of magnetic field inside of Artificial Magnetosphere (in main regime II): **1** and **2**—AMEX data at Y = 19 cm and 34 cm; **3** and **4**—corresponding levels of the "Image Dipole" field for its $\mu_i = 7.7 \times 10^6$ G $\cdot$ cm^3 suited at X = 49 cm (Y = 0); **MP**—*arrows* show locations of magnetopause (at Y = 0 it lies at 19.5 cm)

quasi-uniform addition of field ΔB^+, which we can express as $\Delta B^+ \approx B_d^* \approx (E_0/R_0^3)^{1/2} \sim 100$ G (comparable with observed, Fig. 5) for the tested scale R_m^* of our MP. This scale for explosive-type overflowing replaces the usual stand-off $R_m = (\mu^2/4\pi P_d)^{1/6}$ in a stationary flow with dynamic pressure P_d (of Background Plasma—BP) and therefore instead of traditional Chapman-Ferraro Scaling (CFS) for $\Delta B \propto P_d^{1/2}$ we should use in the CME's case a founded modified CFS as $\Delta B^+ \propto E_0^{1/2}$ that was tested for the first time by AMEX recently (Zakharov et al. 2008).

4 Conclusions

Thus, we validated the novel approach developed in recent laser experiments (Horton et al. 2007b; Zakharov et al. 2008) for simulation of various magnetospheric phenomena, opening new opportunities in this field. Moreover, the experimental capabilities of KI-1 facility (Zakharov 2003), supplied by source of Background Plasma (BP) allow us to plan a combined type of simulative experiment, when so called Terrella (model of stationary magnetosphere) would be compressed by BP-shock (from LPP) and the additional "erosion" of MP could occur due to reconnection between the dipole field and magnetic field frozen into shocked BP.

Acknowledgements This work was supported by Grant # RUP2-2683-NO-05 of the USA Civilian Research and Development Foundation (CRDF). Russian Team would like to acknowledge the KI-1 Team of ILP: Vladimir Antonov, Eduard Boyarintsev, Alexandr Melekhov, Vitaliy Posukh and Ildar Shaikhislamov for assistance in carrying out of AMEX experiment as well as Prof. Hideki Nakashima from the Kyushu University (Japan) for the opportunity to use 3D/PIC-code and useful discussions.

References

Bostick, B.H., Byfield, H., Jemarkin, A.: Phys. Fluids **9**, 2287 (1966)
Horton, W., Chiu, C.: Phys. Plasmas **11**, 1645 (2004)
Horton, W., Chiu, C., Ditmire, T., et al.: Adv. Space Res. **39**, 358 (2007a)
Horton, W., Mays, L., Brady, P., Ditmire, T., Ponomarenko, A., Zakharov, Yu.: Invited paper #93. In: 9 Int. Workshop "Interrelationship between Plasma Experiments in Laboratory and Space", Cairns, 5–10 August 2007 (2007b)
Nikitin, S.A., Ponomarenko, A.G.: J. Appl. Mech. Tech. Phys. **36**, 483 (1996)
Podgorny, I.M., Sagdeev, R.Z.: Sov. Phys. Uspekhi **98**, 445 (1970)
Ponomarenko, A.G., Zakharov, Yu.P., Antonov, V.M. et al.: In: Abstract Book of 4th Int. Conf. on Inertial Fusion Science and Applications, Biarritz, 4–9 Sept. 2005, p. 287
Ponomarenko, A.G., Zakharov, Yu.P., Antonov, V.M., et al.: In: Bothmer, V., Hady, A.A. (eds.) Solar Activity and Its Origin, IAU Symp., No. 233. Cambridge University Press, Cambridge (2007a)
Ponomarenko, A.G., Zakharov, Yu.P., Antonov, V.M., et al.: IEEE Trans. Plasma Sci. **35**, 813 (2007b)
Ridley, A.J., DeZeeuw, D.L., Manchester, W.B., et al.: Adv. Space Res. **38**, 263 (2006)
Shabansky, V.P.: Space Sci. Rev. **8**, 366 (1968)
Zakharov, Yu.P.: IEEE Trans. Plasma Sci. **31**, 1243 (2003)
Zakharov, Yu.P., Nikitin, S.A., Ponomarenko, A.G.: Comput. Technol. **1**, 36 (1996) (in Russian)
Zakharov, Yu.P., Antonov, V.M., Boyarintsev, E.L., et al.: Plasma Phys. Rep. **32**, 183 (2006)
Zakharov, Yu.P., Antonov, V.M., Boyarintsev, E.L., et al.: J. Phys. Conf. Ser. **112**, #042011 (2008)

Astrophys Space Sci (2009) 322: 155–159
DOI 10.1007/s10509-009-0012-z

ORIGINAL PAPER

Collisionless interaction of an energetic laser produced plasma with a large magnetoplasma

C. Constantin · W. Gekelman · P. Pribyl · E. Everson · D. Schaeffer · N. Kugland · R. Presura · S. Neff · C. Plechaty · S. Vincena · A. Collette · S. Tripathi · M. Villagran Muniz · C. Niemann

Received: 11 June 2008 / Accepted: 17 February 2009 / Published online: 5 March 2009

Abstract We have commissioned a high-energy glass-laser at the Large Plasma Device (LAPD) to study the interaction of a dense laser-produced plasma with a large (17 m) magnetized plasma. First experiments with an energy of the laser blow-off an order of magnitude higher than previous work (Gekelman et al. in J. Geophys. Res. 108(A7):1281, 2003) produced large amplitude Alfvén waves ($\delta B_{\perp}/B_0 \approx 15\%$). We will discuss the potential of this facility for collisionless laboratory astrophysics experiments.

Keywords Alfvén waves · Collisionless shocks

1 Introduction

High-power laser produced plasmas can model certain aspects of astrophysical phenomena in a controlled laboratory setting, by doing well-scaled physics despite orders-of-magnitude discrepancies in spatial and temporal scales (Ryutov et al. 2001; Remington et al. 2006). In combination with an ambient magnetized plasma, novel laser experiments can be devised (Drake 2000) that are relevant to collisionless processes, such as shocks (Sagdeev 1966), magnetic turbulence, magnetic reconnection, or wave-particle interactions. These laboratory experiments are crucial to benchmarking the codes developed to model space physics and astrophysical phenomena of interest, such as planetary bow-shocks, coronal mass ejections, and supernova remnants.

C. Constantin (✉) · W. Gekelman · P. Pribyl · E. Everson · D. Schaeffer · N. Kugland · S. Vincena · A. Collette · S. Tripathi · C. Niemann
University of California Los Angeles, Los Angeles, CA 90095, USA
e-mail: constant@physics.ucla.edu

R. Presura · S. Neff · C. Plechaty
University of Nevada Reno, Reno, NV 89506, USA

M.V. Muniz
Universidad Nacional Autónoma de México, Mexico, Mexico

2 Experimental arrangement and diagnostics

The experiments described here are performed at the Large Plasma Device (LAPD) at UCLA (Gekelman et al. 1991). The LAPD creates highly magnetized plasmas of unsurpassed dimensions (60 cm diameter, 17 m length). These experiments differ thus from previous work in that they allow studies of laser-plasma coupling to the ambient plasma over much larger length and time-scales. In addition, the surrounding plasma is large enough to support Alfvén waves. The long linear vacuum chamber is surrounded by a large solenoidal magnet array that creates a uniform axial dc-magnetic field between 300 and 2000 G. The discharge is initiated once per second in a variety of gases (He to Xe) using a pulsed cathode. The plasma is uniform, quiescent, current-free, and reproducible, with a typical density of 2×10^{12} cm^{-3}, an electron temperature of 5 eV, and an ion temperature of $\sim$1 eV. For most ions the ionization stage is $Z = 1$ and there are about 30% neutral atoms. The discharge reaches a steady regime of constant density after a few ms and lasts for 10 ms.

The unconverted beam from a flashlamp-pumped Nd:glass laser (20 J, 5 ns, 1064 nm, circularly polarized) entered the LAPD chamber (Fig. 1) from a side port, at the time when the LAPD plasma was operating in a steady regime. An $f/22$ lens created an elliptical laser focal spot of 200×500 μm^2 onto a thick graphite target. The resulting peak intensity was 3×10^{12} W/cm^2. The target was mounted

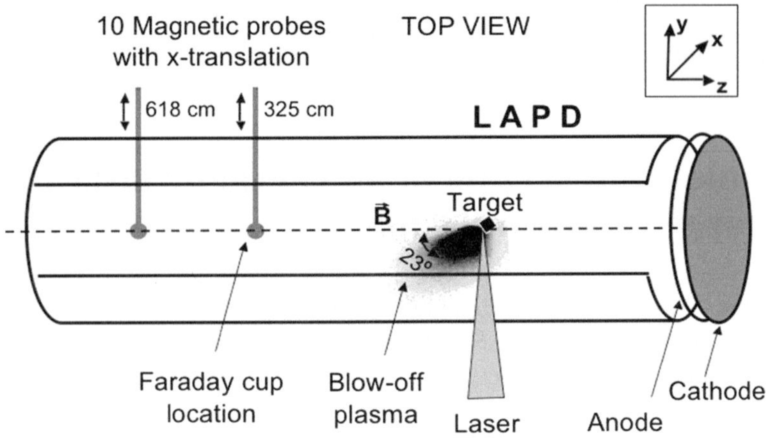

Fig. 1 Experimental setup showing the blow-off laser plasma emerging from a graphite target embedded in the LAPD plasma, and some of the diagnostics

from the top, 5 m away from the cathode, and the interaction region was >10 m in the opposite direction. The target normal was oriented at an angle of 23° relative to $\vec{B}$ (quasi-parallel), while the laser beam was incident perpendicular to $\vec{B}$. The target surface was refreshed after each shot. A fast shutter camera was used to image the temporal evolution of the laser produced plasma from below, with several filters for CI to CIII emission. These images, together with time-of-flight measurements taken with a Faraday cup 3.3 m away from the target location provided information on the blow-off expansion velocity. Faraday cup measurements were taken during shots without background plasma. An array of 10 magnetic pickup coils (sensitive to $d\vec{B}/dt$ in each direction, x, y, and z) was distributed along the machine at distances Δz between 30 cm and 1000 cm to the target to measure the magnetohydrodynamic response of the background plasma to the laser blow-off. These probes were differentially wound on a 3 mm cube core (10 turn per axis) and were calibrated to frequencies up to 100 MHz. The probe signals were recorded for different transverse positions x using custom-built 200 MHz differential amplifiers coupled to 14 bit digitizers. The signals were numerically integrated using an algorithm that applies a frequency-dependent calibration for both amplitude and phase in the frequency domain.

3 Scaled laboratory experiments and first results

Creating collisionless shocks in the laboratory is a challenge (Drake 2000) because it requires facilities that can produce highly magnetized, high-β plasmas at densities sufficiently low to minimize collisions (where β is the ratio between the plasma pressure and the magnetic field pressure). Previous laboratory experiments used magnetic-pinch plasmas (e.g. Paul et al. 1967; Stamper and DeSilva 1968; Keilhacker et al. 1969; Martone and Segre 1970; Mourenas et al. 2003) or laser-produced plasmas (Tsuchimory et al. 1968; Borovsky et al. 1984; Ripin et al. 1987; Dimonte and Wiley 1991; Ditmire et al. 2000), creating the first test beds where observations and models could be compared. Although limited in size and duration at scales that allow for a detailed study of shock formation and particle acceleration, these pioneering experiments made a valuable progress in understanding various aspects of collisionless shock physics (see reviews in Drake 2000 and Zakharov 2003).

Past laser experiments at the LAPD (VanZeeland et al. 2001; Gekelman et al. 2003) with low laser energies (~1 J) and intensities (~5×10^{10} W/cm^2) discovered that shear Alfvén waves are radiated from the laser-plasma. These experiments also studied the evolution of the laser plasma expansion in magnetized environments and the consequent formation of a diamagnetic cavity (VanZeeland and Gekelman 2004). Recently, interactions between colliding plasmas were also studied (Gekelman et al. 2007), but were limited to laser-plasma kinetic energies of $E_0 < 1$ J and blow-off velocities of $v_{lp}/v_A \leq 0.3$ (where v_A is the Alfvén speed). As a result, a small conversion efficiency of laser energy into Alfvén waves $E_A/E_0 \leq 1\%$ was observed, in agreement with numerical simulations.

Some of the first results that we obtained with increased laser energy and intensity, in a He plasma with a density of 2×10^{12} cm^{-3}, are shown in Fig. 2. The velocity distribution of the blow-off plasma ions was measured by the 5 mm diameter Faraday cup, biased at −90 V. The velocity of C ions peaks at around 250 km/s, in agreement with fast shutter photographs obtained with filters for wavelength ranges around $\lambda = 460$ nm, specific to CIII emission. This is consistent with the on-target intensity and is below the Alfvén velocity in He ($v_A = 460$ km/s). Different velocities are attributed to different charge states of the blow-off ions. We observe Alfvén waves with amplitudes as large as 100 G ($\delta B_{\perp}/B_0 \approx 15\%$) emerging from the laser-produced plasma in the He-discharge at $B_0 = 600$ G (Fig. 2a and 2c). Van Campernolle et al. (2006) has shown that these waves are created through Cherenkov radiation by fast electrons.

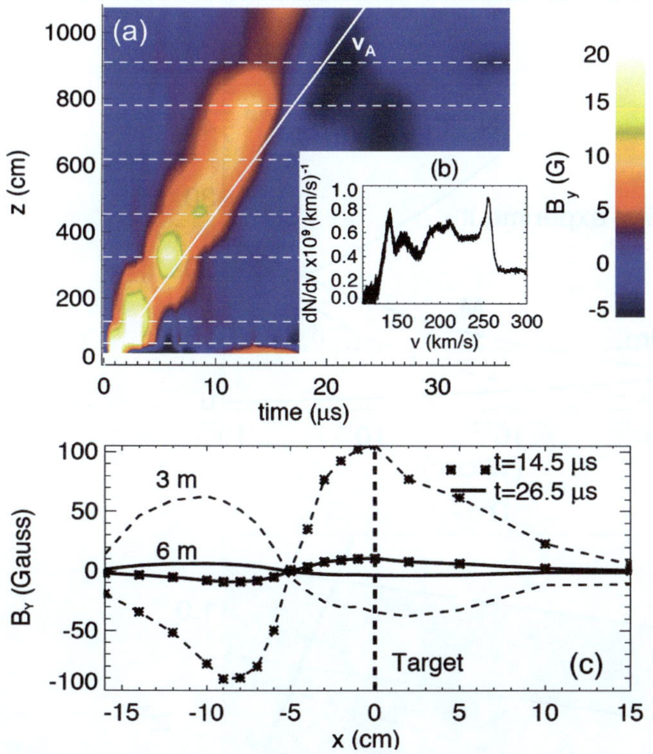

Fig. 2 (**a**) Evolution of a magnetic perturbation caused by a large amplitude Alfvén wave along the LAPD, at position $x = 2$ cm. (**b**) Velocity distribution of the carbon ions ejected by the laser plasma, as inferred from the Faraday cup measurements, 3 m away from the target. (**c**) Magnetic field signals measured 3 m (*dashed line*) and 6 m (*solid line*) away from the target in a He plasma, at $B = 600$ G. Two different times are shown

A typical Alfvén wave behavior, causing B_y field oscillation along the transverse direction x was measured with a Bdot-probe at a distance of $\Delta z = 6$ m from the target, as shown in Fig. 2c. The probe was moved to a new x position after 3 laser shots and the experiment was repeated with a laser energy stability better than 10%. The measured Alfvén frequency is 160 kHz (70% f_{ci}, where f_{ci} is the ion gyrofrequency), which is consistent with previous work (Gekelman et al. 2003), but an order of magnitude larger in amplitude. With the high-power laser now available, experiments could in principle drive shocks in plasmas that approach the collisionless regime. The laser-produced blow-off plasma acts as a piston upon the ambient plasma. Drake (2000) shows that a large ambient plasma size L is crucial to drive collisionless shocks in a laboratory experiment. A shock velocity v_s in excess of the Alfvén speed $v_A \sim B/\sqrt{mn}$ is required to ensure that the upstream magnetic turbulence does not escape the shock (Alfvénic Mach number $M_A = v_s/v_A \geq 1$). Here B is the external magnetic field, while m and n are the mass and the density of the ambient plasma-ions, respectively. The plasma $\beta \sim nT/B^2$ must be as large as possible so that the shock can affect the magnetic field, and yet the plasma must be highly magnetized (i.e. magnetization $\Gamma_m = d/r_{Li} \gg 1$, where d is the relevant background plasma size scale, and $r_{Li} \sim \sqrt{mT}/B$ is the ion Larmor radius). The parameter d will be determined in principle by the size of the expanding diamagnetic cavity formed by the laser plasma (see details in Bashurin et al. 1984). In the perpendicular case the cavity radius is given by the equal-charge radius: $R' = (3Z_iN_i/(4\pi Zn))^{1/3} \equiv (mZ_i/(m_iZ))^{1/3}R_*$, where Z, m, and Z_i, m_i are the ion charge state and the mass of laser-plasma and background plasma, respectively, $R_* = (3N_im_i/(4\pi nm))^{1/3}$ is the equal-mass radius, with N_i the number of ions in the laser-plasma, and n the density of ions in the background plasma. In the parallel case, the laser cloud extends over larger distances, as indicated by our measurements (more than 3 m).

Another key parameter to be considered in the shock formation is the Larmor-coupling efficiency (Bashurin et al. 1984; Golubev et al. 1978; Winske and Gary 2007) that depends strongly on the magnetic laminar mechanism (MLM) criterion: $\delta = R'^2/(R_L \cdot R_{L*})$ must be greater than 1, where R_L is the Larmor radius of the laser-plasma ions, which is proportional to the laser-plasma front speed v_{lp}, and R_{L*} is the directed Larmor radius of the background ions, also proportional to v_{lp} (Zakharov 2003).

In addition, the experiment must last long enough for the shock to form. In the quasi-perpendicular case ($\vec{v}_s \perp \vec{B}$ within $\pm 45°$), the shock-formation takes a time of the order of the ion gyroperiod $T_{ci} \sim m/B$. In the quasi-parallel case ($\vec{v}_s || \vec{B}$ within $\pm$ 45°), the shock formation time is believed to be much longer, of the order of the growth-rate of the firehose instability ($\sim 10T_{ci}$) (Draine 1993). This means that the number of gyroperiods per shock transit time $N_T = (L/v_s)/T_{ci}$ must be larger than 1 in the quasi-perpendicular case and larger than 10 in the quasi-parallel case. Therefore, even though the size is larger in the axial direction ($L_\parallel \geq 1000$ cm) compared to $L_\perp \sim 40$ cm in the transverse direction, creating collisionless shocks can prove more challenging due to the increased growth time of the shock formation and requires more laser energy.

Figure 3a shows that a high-power laser focused to moderate intensities around 10^{13} W/cm^2 could create conditions in the LAPD where M_A and Γ_m are above unity. We note that Fig. 3a shows a lower bound of β. The actual plasma pressure will be higher close to the laser target, since the blow-off density is initially orders of magnitude larger than the density of the ambient plasma. Here we use an empirical scaling law for the blow-off velocity $v_{lp} \equiv 3/4 v_s = 2.5 \times 10^7\ I_{12}^{0.2}$ for 1064 nm laser light, where I_{12} is the intensity in units of 10^{12} W/cm^2 (Key et al. 1983; Meyer and Thiell 1984). While this regime is interesting in itself one can see that these conditions are not maintained long enough for a shock to form (i.e. $N_{T,\perp} < 1$ and $N_{T,\parallel} < 10$). A simple scaling analysis shows that a small ion mass is required for shock formation. Using the standard relation for the temperature of the shocked ions $T \sim mv_s^2$ we

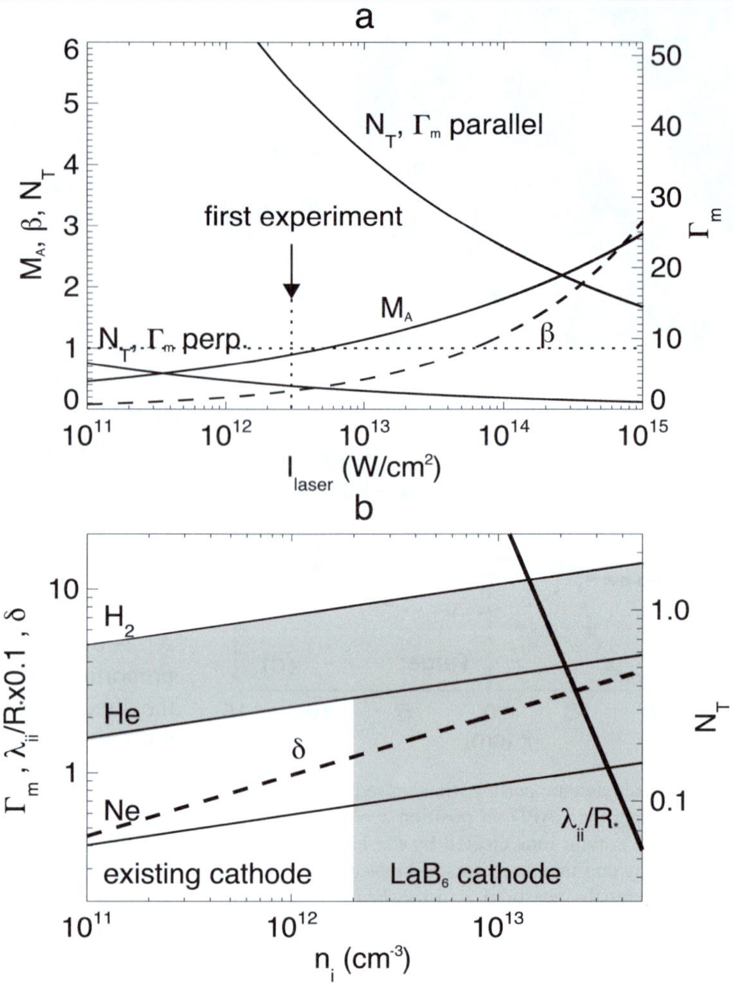

Fig. 3 (**a**) Scaling of M_A, β, N_T, and Γ_m with laser intensity for a He-discharge in 600 G, with a density of 2×10^{12} cm^{-3}, over distances $L_\parallel = 1000$ cm, and $L_\perp \sim 40$ cm. (**b**) Γ_m and N_T as a function of n_i for different gases at $M_A = 1$, in a perpendicular geometry. The bold line indicates the proton mean free path $\lambda_{ii} \sim 1/n_i$ in units of plasma size (scaled down by a factor of 10 in this plot), and the *dashed bold line* indicates the MLM parameter δ, calculated for a H background plasma, and blow-off velocities of 170 km/s, in a 600 G magnetic field. The *shaded areas* indicate the parameters accessible with the existing and with the new cathode

find that for a constant M_A shock ($v_s \sim B/\sqrt{mn}$) neither the magnetization Γ_m, nor N_T and the plasma β vary with the external magnetic field. The reduced gyroperiod at higher B is compensated by the shorter shock transit time, while the temperature of the shocked ions must increase with B keeping the gyro-radius constant. Furthermore, Γ_m and N_T have the same scaling dependence, and increase with plasma size and ion-density, but decrease with ion-mass. Thus, maximizing density while minimizing mass will create the best experimental conditions for shock formation. At the densities accessible by the LAPD the mean free path of the shocked ambient ions $\lambda_{ii} \sim v_{lp}^4/n$ is typically much larger than the ambient plasma size such that collisionless processes dominate (Fig. 3b). From the laser-driver point of view a lower B will require the smallest blow-off speeds and smallest laser intensities, maximizing the number of particles that can be energized by the limited laser energy absorbed.

4 Conclusion and outlook

A high-energy laser coupled to the LAPD provides a unique experimental capability to study the interaction between a laser plasma and an ambient magnetized plasma at scale-lengths, time-scales, and energies previously inaccessible. First experiments with 20 J have created blow-off plasma speeds below the Alfvén velocity in He ($0.5v_A$), creating Alfvén waves with amplitudes an order of magnitude higher than in previous work. Experiments in the near future will take advantage of a new lanthanum-hexaboride cathode (LaB_6) that is chemically compatible with hydrogen and will also support higher densities (up to 5×10^{13} cm^{-3}). In particular, scaling shows that collisionless shocks are possible in H-plasmas at these higher densities and magnetic fields of 300 G (as low as LAPD can support). $M_A \sim 1$ shocks with $N_T \geq 30$ in the quasi-parallel case, and $N_T \sim 2$ in the quasi-perpendicular case could then be produced. Here, the perpendicular case benefits from the compressed field ($B \approx 2B_0$ for $M_A = 1$). Gases heavier than He are not suitable for the formation of collisionless shocks in the LAPD. At laser intensities around 10^{11} W/cm^2 we expect to ablate around 10^{17} particles from the laser-target (Meyer and Thiell 1984). An energy of 25 J (more than 4 times the total energy in the ambient plasma) would be sufficient to maintain the required ion-temperatures for a $M_A = 1$

shock. The ions in the laser plasma are well magnetized in these conditions, since the Larmor radius decreases and the $R_* \sim 40$ cm increases, at blow off velocities of 170 km/s. The MLM interaction parameter δ is then around 1. These parameters show that conditions favorable for studying the Larmor coupling mechanism, the collisionless shock formation and propagation, as well as its interaction with the ambient magnetic field are possible and can be of great value for testing the existing models. Ultimately, strong ($M_A \geq 4$) shocks and studies of the associated particle acceleration thought to be caused by them can be then attempted at facilities providing laser energies at kilojoule levels, as discussed in Drake's and Zakharov's reviews.

Acknowledgements We thank A. Ng (UBC) for the donation of the laser system, as well as D. Hargrove and S. Glenzer for providing crucial experimental components. We thank G. Morales, J. Maggs, D. Larson, and T. Carter for fruitful discussions. This work is supported by the NSF/DOE partnership in Basic Plasma Science Facility, and by the DOE/NNSA grant DE-FC52-06NA27616.

References

Bashurin, V., Golubev, A.I., Terekhin, V.A.: J. Appl. Mech. Tech. Phys. **24**, 614 (1984)
Borovsky, J.E., Pongrats, M.B., Roussel-Dupre, E.A., Tan, T.H.: Astrophys. J. **280**, 802 (1984)
Dimonte, G., Wiley, L.G.: Phys. Rev. Lett. **67**, 1755 (1991)
Ditmire, T., Rubenchik, A., Mirnov, V.V., Ucer, D.: J. Suppl. Ser. **127**, 293 (2000)
Draine, B.T.: Annu. Rev. Astron. Astrophys. **31**, 373 (1993)
Drake, R.P.: Phys. Plasmas **7**(11), 4690 (2000)
Gekelman, W., Pfister, H., Lucky, Z., Bamber, J., Leneman, D., Maggs, J.: Rev. Sci. Instrum. **62**, 2875 (1991)
Gekelman, W., Van Zeeland, M., Vincena, S., Pribyl, P.: J. Geophys. Res. **108**(A7), 1281 (2003)
Gekelman, W., Collete, A., Vincena, S.: Phys. Plasmas **14**, 062109 (2007)
Golubev, A.I., Solov'ev, A.A., Terekhin, V.A.: J. Appl. Mech. Tech. Phys. **19**, 6020 (1978)
Keilhacker, M., Kornherr, M., Steuer, K.H.: Z. Phys. **223**, 385 (1969)
Key, M.H., Toner, W.T., Goldsack, T.J., Kilkenny, J.D., Veats, S.A., Cunningham, P.F., Lewis, C.L.S.: Phys. Fluids **26**(7), 2011 (1983)
Martone, M., Segre, S.E.: Plasma Phys. **12**, 205 (1970)
Meyer, B., Thiell, G.: Phys. Fluids **27**(1), 302 (1984)
Mourenas, D., Vierne, J., Simonet, F., Krauz, V.I., Nikulin, S., Mialton, V.V., Karakin, M.A.: Phys. Plasmas **10**, 605 (2003)
Paul, J.W.M., Goldenbaum, G.C., Iioshi, A., Holmes, L.S., Hardcastle, R.A.: Nature **216**, 363 (1967)
Remington, B.A., Drake, R.P., Ryutov, D.D.: Rev. Mod. Phys. **78**, 755 (2006)
Ripin, B.H., MacLean, E.A., Manka, C.K., Pawley, C., Stamper, J.A., Peyser, T.A., Mostovych, A.N., Grun, J., Hassam, A.B., Huba, J.: Phys. Rev. Lett. **59**, 2299 (1987)
Ryutov, D.D., Remington, B.A., Robey, H.F.: Phys. Plasmas **8**, 1804 (2001)
Sagdeev, R.Z.: Rev. Plasma Phys. **4**, 23 (1966)
Stamper, J.A., DeSilva, A.W.: Phys. Fluids **12**, 1435 (1968)
Tsuchimory, N., Yamanaka, T., Yamanaka, C.: Jpn. J. Appl. Phys. **7**, 84 (1968)
Van Campernolle, B., Gekelman, W., Pribyl, P.: Phys. Plasmas **13**, 092112 (2006)
VanZeeland, M., Gekelman, W., Vincena, S., Dimonte, G.: Phys. Rev. Lett. **87**(10), 105001 (2001)
VanZeeland, M., Gekelman, W.: Phys. Plasmas **11**(1), 320 (2004)
Zakharov, Y.P.: IEEE Trans. Plasma Sci. **31**(6), 1243 (2003)
Winske, D., Gary, S.P.: J. Geophys. Res. **112**, A10303 (2007)

Astrophys Space Sci (2009) 322: 161–165
DOI 10.1007/s10509-008-9950-0

ORIGINAL ARTICLE

Propagation of relativistic electrons in low density foam targets

B. Ramakrishna · P.A. Wilson · K. Quinn · L. Romagnani · M. Borghesi · A. Pipahl · O. Willi · L. Lancia · J. Fuchs · R.J. Clarke · M. Notley · W. Nazarov

Received: 19 June 2008 / Accepted: 14 November 2008 / Published online: 25 November 2008

Abstract The propagation of laser-driven, relativistic electron beams in plasmas is a phenomenon of relevance to applications such as the Fast Ignition of fusion targets. In this paper, we report on experiments where the transport of hot electron currents through foam targets has been studied using the proton imaging technique. Strong filamentation has been observed, possibly due to electromagnetic instabilities of the Weibel type. The experiments were performed at the RAL Petawatt laser facility.

Keywords Laser-plasma interaction · Electron transport · Filamentation

B. Ramakrishna (✉) · P.A. Wilson · K. Quinn · L. Romagnani · M. Borghesi
Centre for Plasma Physics, School of Mathematics and Physics, The Queen's University of Belfast, Belfast, UK
e-mail: bramakrishna01@qub.ac.uk

A. Pipahl · O. Willi
Institut fur Laser-Und Plasma Physik, Heinrich Heine-Universitat, Dusseldorf, Germany

L. Lancia · J. Fuchs
Laboratoires pour L'Utilisation des Laser Intenses, Ecole Polytechnique, Palaiseau, France

R.J. Clarke · M. Notley
Central Laser Facility, Rutherford Appleton Laboratory, Chilton, Oxfordshire, England

W. Nazarov
School of Chemistry, University of St Andrews, St Andrews, KY16 9ST, UK

1 Introduction

High intensity laser pulses focused on solid targets produce very large currents of relativistic electrons. The transport of these currents through dense matter may be affected by the large electric and magnetic fields they generate. This is a crucial issue for the Fast Ignitor (FI) approach to Inertial Confinement Fusion (Tabak et al. 1994), where an impulsive, laser-driven current of relativistic electrons is employed to trigger the ignition of a pre-compressed target. Experiments with innovative targets employing a re-entrant cone to allow direct laser interaction close to the compressed core have demonstrated effective electron heating of the core plasma (Kodama et al. 2001, 2002).

Scaling laws for the fast electrons produced during ultrahigh intensity interactions give electron temperatures $K_B T_{hot} \sim U_{pond} \sim 1\ \text{MeV} \times (I^2/10^{19}\ \text{W cm}^2\ \mu\text{m}^2)^{0.5}$, with up to 50% of the laser energy converted into these relativistic electrons, where U_{pond} is the ponderomotive energy of laser and the wavelength is 1.05 μm (Hatchett et al. 2000). Acceleration of these electrons in the steep plasma gradients present in solid targets is mainly due to processes such as resonance absorption, $j \times B$ heating, or possibly direct laser acceleration if an extended preplasma is present (Wilks et al. 1992). It is estimated that the magnitude of the hot electron current density could be as large as 10^{12} A/cm^2 (Wilks et al. 1992). These currents exceed the critical Alfven limit given by $J_A = 17.1\beta\gamma$ kA, where $\beta = v/c$ and γ is the relativistic Lorentz factor (Alfven 1939). Depending on the resistivity of the target a return current balancing the current of fast electrons is set up to maintain quasi-neutrality (i.e. $J_{hot} + J_{return} \sim 0$) and allow propagation of hot electrons (Bell et al. 1997). Instabilities in plasma arise due to the presence of these two counterstreaming currents (Pukhov 1999) which can affect the spatial distributions of the hot electron current. An example of electromagnetic instability is given by the Weibel instability which grows with wavelengths of the order of the collisionless skin depth (Forslund et al. 1985). Besides the development of instabilities, a reduction of the propagation range from the purely collisional range may arise due to the effect of space-charge and inductive fields associated to the hot electron propagation. Relativistic electron propagation in this regime is mainly been modelled by using hybrid Particle-in-Cell-fluid codes such as LSP (Evans 2006).

A number of experiments investigating the propagation and filamentation of laser produced relativistic electron beams have been performed using metal and plastic foils or glass slabs (Teng et al. 2003; Fuchs et al. 2003; Santos et al. 2002; Borghesi et al. 1999). The propagation through low/medium density foams has also been investigated in some experiments (Zheng et al. 2004; Jung et al. 2005). Low density foam targets offer a different approach for the study of electron-beam transport through dense plasmas over long distances. In terms of target resistivity this approach is possibly more relevant to the FI scheme (Zheng et al. 2004). Being much less collisional than solid target plasmas, they are also ideal test beds for simulations (Evans 2006).

Experimental observations obtained using foam targets of different densities are reported in this paper. The electrons are launched into the target forming filaments which are probed by multi-MeV proton beams. This diagnostic allows to obtain maps of the electric and magnetic fields arising from the electron beam propagation, with high spatial and temporal resolution.

2 Diagnostic approach

The unique properties of protons from high intensity laser-matter interactions, particularly in terms of spatial quality and temporal duration, have opened up a totally new area of application of proton probing/proton radiography (Borghesi et al. 2006). This technique has been applied successfully to the detection of highly transient fields in a number of experiments (Romagnani et al. 2005).

The proton beams emitted from a laser-irradiated foil are highly laminar (Cowan et al. 2004) and, for projection purposes, can be described as emitted from a virtual, point-like source located in front of the target (Borghesi et al. 2004; Brambrink et al. 2006). A point-projection imaging scheme is therefore automatically achieved. Electric or magnetic fields in the sampled region can be revealed via the proton deflection and the associated modifications in the proton density pattern.

In a typical set-up, a high-intensity laser pulse is used to accelerate a beam of MeV protons from the rear surface of a thin metal foils, and this beam is used as a transverse particle

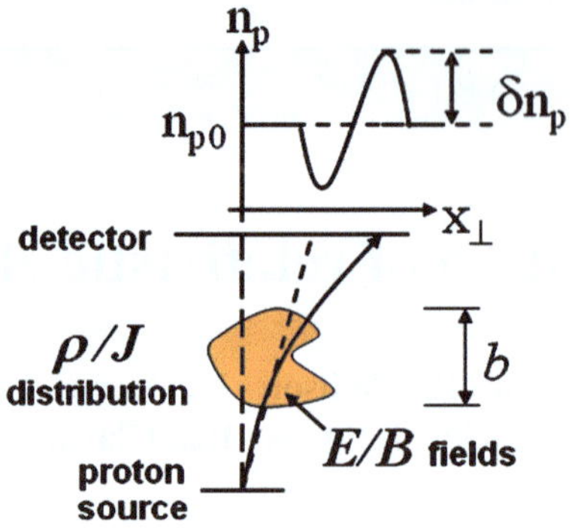

Fig. 1 Proton imaging technique

probe of the interaction of a second short pulse with a separate target. Generally, protons originate from the impurities present on the target surface. A general analysis method consists of using particle tracing codes to follow the propagation of the protons through a given three-dimensional field structure, which can be modified iteratively until the computational proton profile reproduces the experimental ones. State-of-the-art tracers allow realistic simulations including experimental proton spectrum and emission geometry, as well as detector response.

As the proton beam originates (or appears to) from a small source, when a thin object is placed between the source and a detector, there will be a one-to-one correspondence between the points of the object plane and of the detector plane (we assume purely geometrical propagation for the protons). Small deflections due to E/B fields present in the plasma will cause density variations in the proton probe cross section, from which one can obtain information about the field distribution in the object plane (Borghesi et al. 2002) (see Fig. 1). The broad spectral content of the beam is an advantage when probing evolving field distributions due to the different propagation velocity of the various energy components. The protons will be spread temporally when they reach the object plane. As the detector performs spectral selection, each layer will convey information pertaining to different stages of the target evolution. Therefore the technique is intrinsically multiframe. In principle, probing with sufficiently high-energy protons provides the unique possibility of accessing field distribution, and therefore diagnosing electron transport, directly in the interior of laser-irradiated targets. However collisional scattering may affect the probe protons, intrinsically reducing the spatial resolution of the diagnostic (Kar et al. 2008) limiting the areal density which can be probed efficiently with protons of a given energy.

3 Experiment

The experiments were carried out at the VULCAN Petawatt (PW) laser facility at Rutherford Appleton laboratory. The

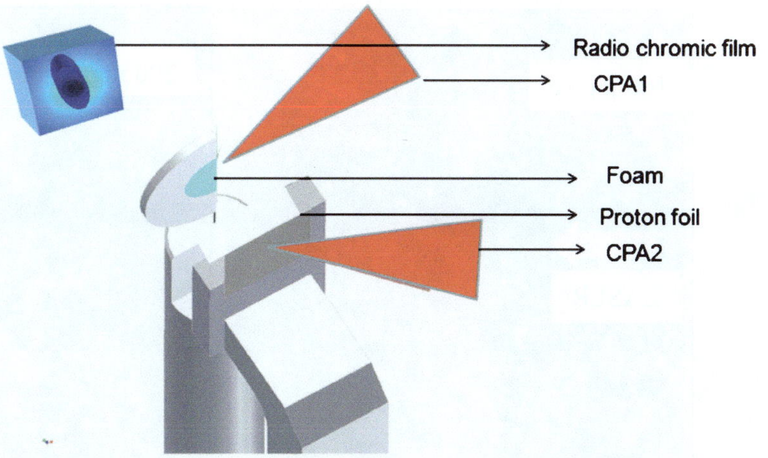

Fig. 2 Experimental arrangement

proton probing technique was used throughout the experiment. The set up for the experiment is shown in Fig. 2.

A dual CPA configuration was employed. CPA1 was used to produce high energetic proton beams and CPA2 was used to interact with the target. A 500 J laser pulse (CPA1) with a wavelength of 1053 nm and 700 fs in duration was focused with an $f/3.2$ off-axis-parabola, reaching intensities of the order $\sim 5 \times 10^{20}$ W/cm^2. This intense pulse was focused on to 20 μm thin gold foil to produce energetic proton beam of around 30 MeV. The Second CPA (CPA2) beam which is extracted from the main PW arm using a pick up mirror assembly, which gives a 20–30 TW beam, is used as an interaction beam which is focused on the foam target at an intensity $\sim 10^{19}$ W/cm^2.

The foam targets were enclosed in an aluminium half washer of internal diameter 1.5 mm and thickness 300 μm with its sides closed by an 8 μm plastic foil. The foams had a cell size of 1 μm and densities ranging from 50 to 200 mg/cm^3. The front side of the foams was coated with 75 nm of gold layer. This was done with the aim to generate a radiation wave produced by the leading edge of the laser pulse for partial preionization of the foam, and to have a constant electron source for all the different types of foams employed. The thickness used for these targets was a compromise between the need of limiting scattering degradation of the beam's spatial quality and ease of efficient target irradiation. The density was also kept relatively low to minimize scattering and allow a high-resolution view of the electron propagation region.

A stack of radio chromic film (RCF) of the type Gafchromic HD 810[1] was used as detector. The RCFs are self developing films sensitive to different kinds of ionizing radiations like electrons, ions, X-rays and gamma rays etc. The basic components of this film are a radiation sensitive micro crystalline monomer belonging to the family of diacetylene.

[1]http://www.ispcorp.com/products/dosimetry/index.html (2008).

4 Experimental results

Typical RCF images of foam targets are shown in Fig. 3. The thickness was 250 μm.

The densities were 100 and 200 mg/cc respectively. The images clearly show filamentary structures inside the interior of the target immediately after the interaction, arising due to the transport of the hot electrons inside the target. These are more pronounced in the 100 mg/cc foam. From the data it can be inferred that there is a variation in the cone angle aperture, decreasing with increase in target density (see Fig. 3). This is possibly an indication for the confinement of electrons by magnetic field due to Weibel like resistive instabilities (Jung et al. 2005). The data also indicates the growth of filamentary structures within the electron cone, which also shows a variation with the foam density, with more pronounced filaments observed at lower foam densities. Indeed, varying the foam density corresponds to varying the ratio of beam electron density to background electron density, with consequent effects on the growth of electromagnetic instabilities (Evans 2006). Particle tracing simulations (see Fig. 4) are currently being employed to reproduce the proton deflection maps and obtain information on the field distributions in the target. Although the analysis of the data is still in progress, the general trends and features observed in the experiment are in qualitative agreement with published computational results for similar interaction conditions (Evans 2006).

5 Conclusion

To summarize, relativistic electron beam propagation through dense plasmas using foam targets have been studied. A proton probing technique is implemented to obtain spatially and temporally resolved information on the electric

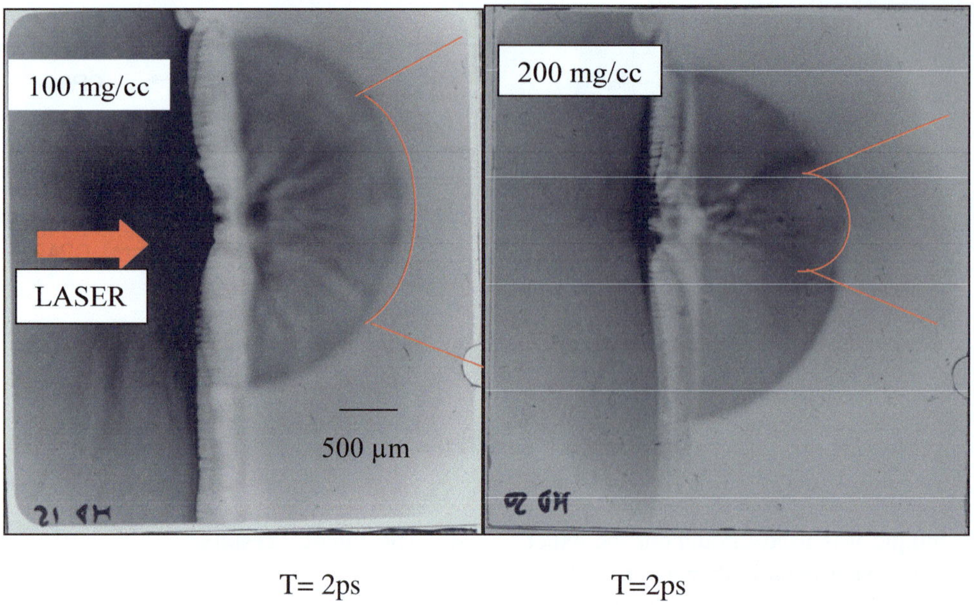

Fig. 3 RCF images of foam targets following CPA interaction. The density of the foam is indicated in figure. Timing relative to the interaction is indicated below the figures. The hemi-circular shadow on the right side of the images is due to the half-washer enclosing the foam (see Fig. 2), while the *vertical white band* at the centre of the images corresponds to the foam-vacuum interface. Beside plasma expansion from the interface into the vacuum, the images clearly show filamentary structures appearing inside the bulk of the target. The cone containing the filaments expanding from the interaction point is indicated by the *red lines* on the right of the figure

Fig. 4 Comparison between the experimental data and preliminary particle trace simulation. (**a**) RCF image shows the evolution of the filamentation inside the foam, ∼5 ps after the interaction. The density of the foam target is 100 mg/cc. (**b**) A particle tracing simulation attempting to reproduce the features observed in the experimental data. These simulations use peak electric field (radial inside the filaments), $\sim 10^9$ V m^{-1} and peak magnetic field (azimuthal around the filaments) $\sim 10^3$ T

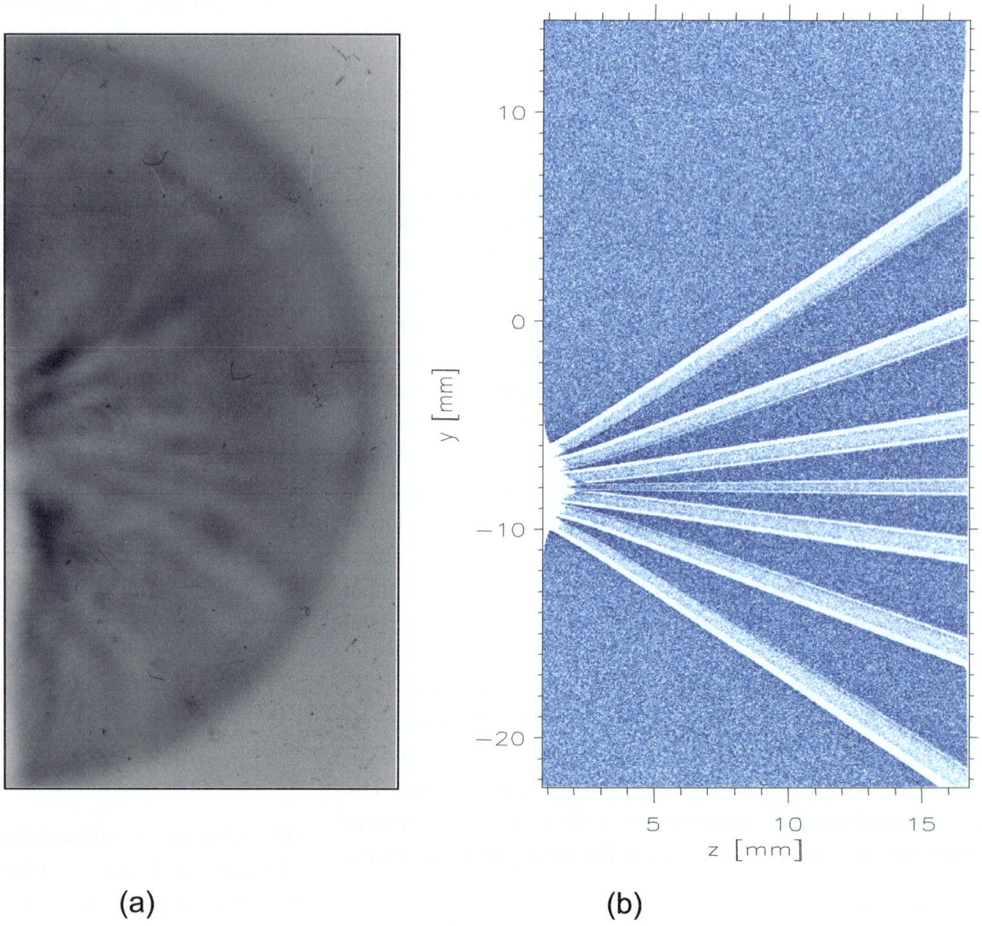

and magnetic fields arising in the interior of the target following the propagation of the electrons injected by the interaction taking place at the front surface. RCF images clearly indicate the appearance of filamentary structures in the interior of the target. The observations are in broad agreement with published Hybrid simulation results.

Acknowledgements This work was supported by RAL Facility Access scheme, by EPSRC grants EP/P501210/1 and EP/C003586/1 and by a British Council/Alliance grant. We acknowledge discussions with Prof. R.G. Evans (Imperial College).

References

Alfven, H.: Phys. Rev. **55**, 425 (1939)
Bell, A.R., et al.: Plasma Phys. Contr. Fus. **39**, 653 (1997)
Borghesi, M., et al.: Phys. Rev. Lett. **83**, 4309 (1999)
Borghesi, M., et al.: Phys. Plasmas **9**, 2214–2218 (2002)
Borghesi, M., et al.: Phys. Rev. Lett. **92**, 055003 (2004)
Borghesi, M., et al.: Fusion Sci. Techn. **49**, 412–439 (2006)
Brambrink, E., et al.: Laser Part. Beams **2**, 163–168 (2006)
Cowan, T.E., et al.: Phys. Rev. Lett. **92**, 204801 (2004)
Evans, R.G.: High Energy Density Phys. **2**, 35 (2006)
Forslund, D.W., et al.: Phys. Rev. Lett. **54**, 558 (1985)
Fuchs, J., et al.: Phys. Rev. Lett. **91**, 255002 (2003)
Hatchett, S., et al.: Phys. Plasmas **7**, 2076 (2000)
Jung, R., et al.: Phys. Rev. Lett. **94**, 195001 (2005)
Kar, S., et al.: High Energy Density Phys. **4**, 26 (2008)
Kodama, R., et al.: Nature (Lond.) **412**, 798 (2001)
Kodama, R., et al.: Nature (Lond.) **418**, 933 (2002)
Pukhov, A.: Phys. Plasmas **6**, 2847 (1999)
Romagnani, L., et al.: Phys. Rev. Lett. **95**, 195001 (2005)
Santos, J., et al.: Phys. Rev. Lett. **89**, 025001 (2002)
Tabak, M., et al.: Phys. Plasmas **1**, 1626 (1994)
Teng, H., et al.: Phys. Rev. E **67**, 026408 (2003)
Wilks, S.C., et al.: Phys. Rev. Lett. **69**, 1383 (1992)
Zheng, J., et al.: Phys. Rev. Lett. **92**, 165001 (2004)

Astrophys Space Sci (2009) 322: 167–177
DOI 10.1007/s10509-009-0001-2

ORIGINAL ARTICLE

High energy density physics generated by intense heavy ion beams

D.H.H. Hoffmann · V.E. Fortov · M. Kuster · V. Mintsev · B.Y. Sharkov · N.A. Tahir · S. Udrea · D. Varentsov · K. Weyrich

Received: 31 May 2008 / Accepted: 27 January 2009 / Published online: 14 February 2009

Abstract Intense ion beams from accelerators are now available to generate high energy density matter and to study astrophysical phenomena in the laboratory under controlled and reproducible conditions. A detailed understanding of interaction phenomena of intense ion- and laser radiation with matter is important for a large number of applications in different fields of science, extending from basic research of plasma properties to application in energy science and the investigation of processes occurring in stellar atmospheres or even in the interior of stars and planets. Energy loss processes of heavy ions in plasma and cold matter are important for the generation of high energy density states in general and especially in the hot dense plasma of an inertial fusion target. Of special interest are phase transitions and the associated time scales when matter passes the warm dense matter regime of the phase diagram at high density but relatively low temperature. We present an overview on recent results and developments of beam plasma, and beam matter interaction processes studied with heavy ion beams and laser beams combined with accelerator and nuclear physics technology. A natural example of hot dense plasma is provided by our neighbouring star the sun, and allows a deep insight into the physics of fusion, the properties of matter at high energy density, and is moreover an excellent laboratory for astroparticle physics. As such the sun's interior plasma can even be used to probe the existence of novel particles and dark matter candidates with a combination of equipment and methods from accelerator technology and high resolution plasma spectroscopy.

Keywords High energy density physics · Dense plasma · Energy loss · Inertial fusion · Warm dense matter · Interior of planets and stars

D.H.H. Hoffmann (✉) · M. Kuster · S. Udrea
Technische Universität Darmstadt, Institut für Kernphysik, Schlossgartenstr. 9, 64289 Darmstadt, Germany
e-mail: hoffmann@physik.tu-darmstadt.de

V.E. Fortov
Institute of Thermophysics of Extreme States, Joint Institute for High Temperatures, Russian Academy of Sciences, Moscow, Russian Federation

V. Mintsev
Institute of Problems of Chemical Physics, Chernogolovka, Russian Federation

B.Y. Sharkov
A.I. Alikhanov Institute of Theoretical and Experimental Physics, Moscow, Russian Federation

N.A. Tahir · D. Varentsov · K. Weyrich
Gesellschaft für Schwerionenforschung, GSI-Darmstadt, Darmstadt, Germany

1 Introduction

In the environment that we are used to, matter occurs predominantly as baryonic matter in the solid, liquid or gaseous phase. This, however, does not represent the general situation in the universe at large, where most of the matter is in the form of dark matter (Zioutas et al. 2004) or dark energy and we know only very little about the nature of this species of matter. Only 4% of the total mass of the universe is made up of baryons that we are familiar with. If we turn to this small fraction of matter in the universe, then this matter is mainly found to be in the plasma state, in the interior of stars, in stellar atmospheres, in interplanetary space or even in the interior of planets where matter exists in the form of dense strongly coupled plasma. High energy density matter often consists of a mixture of electrically charged particles such

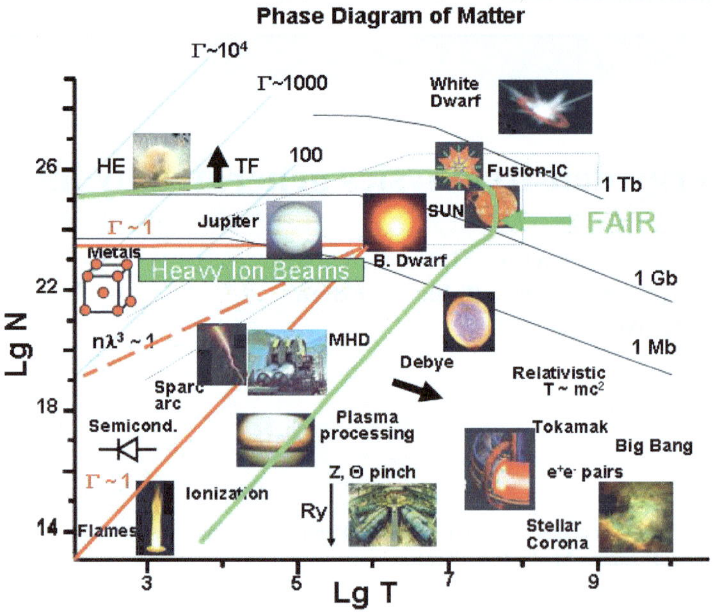

Fig. 1 Phase diagram of matter with density as function of temperature. The area encircled by the *green line* is the area of matter phase space that is available for investigation with intense heavy ion beams with intense beams from the future FAIR facility at GSI

as ions and electrons and neutral particles as well. In this situation, the long-range Coulomb force is the factor that determines the statistical properties of the sample.

Figure 1 displays part of the phase diagram of matter and the location of familiar phenomena like flames lightning, the interior of the sun and other plasmas are indicated. Different areas of this diagram can be investigated by different experimental methods. Examples for the Tokamak, Z-pinch and high explosives are indicated and it can be seen that a large part of the parameter space may be addressed by intense heavy ion beams as shown by the area inside the green line.

Investigation of astrophysical phenomena in the laboratory requires the generation of high energy density matter under controlled and reproducible conditions. Lasers, pulsed power machines, high explosives and intense particle beams from accelerators are commonly used experimental tools to generate high energy density states in matter (Fortov et al. 2007; Hora 2007; Tahir et al. 2007b; Varentsov et al. 2007a; Zou et al. 2006; Zvorykin et al. 2007). Among these techniques heavy ion beams are relatively new (Varentsov et al. 2007b). However, due to their unique way to interact with matter they open new pathways to generate and to diagnose high energy density matter.

The Virtual National Laboratory on Heavy Ion Fusion (Berkeley) with a series of medium-scale experiments to study aspects of intense-beam physics (Logan et al. 2007b), the Institute of Theoretical and Experimental (Moscow) with the TeraWatt Accumulator (Sharkov 2007) and the Gesellschaft für Schwerionenforschung (GSI-Darmstadt) with their heavy ion accelerators UNILAC and SIS 18 (Hoffmann et al. 2005b) are leading laboratories in this field (see Fig. 2). The CERN large hadron collider (LHC) and the facility for antiproton and ion research (FAIR) presently under construction at GSI in Darmstadt (Henning 2007) will add to the prospects of using intense particle beams from accelerators to study high energy density matter (Tahir et al. 2005a, 2005b). In recent years accelerator technology has improved sufficiently to provide intense heavy ion beams to heat macroscopic samples of matter and reach the regime of warm dense matter (Hoffmann et al. 2005a, 2007; Logan et al. 2007a).

2 Stopping power of ionized matter

Heating of matter with intense ion beams requires a basic understanding of energy deposition processes of ions in matter. During the heating process the material changes from solid to gas and eventually to dense ionized matter. In general the energy deposition process of heavy ions in matter is a topic of scientific interest since many decades, and there still are unchartered regimes, even in interaction processes with solid state and gaseous matter and this is why many research groups are attracted to this field. However, when matter undergoes the transition to plasma or even to a dense strongly coupled plasma there are only few experimental data available to benchmark theoretical predictions.

Since the interaction is mediated by the well known Coulomb force, it is due to collective phenomena of beam target coupling which still leads to surprising results, for example the unexpectedly high energy loss of ions in dense ionized matter.

In general the ion energy deposition is extremely high, since a single atom can well be excited up to and above 20 eV. The electromagnetic pulse experienced by the target atom is extremely short like 10^{-17} s to 10^{-13} s, depending

Fig. 2 Heavy Ion accelerator laboratories: GSI Darmstadt. ITEP Moscow, HIFS-VNL Berkeley

on the energy of the ion. Only ultra-short laser pulses are comparable with this process. The relaxation process then leads to heating of a macroscopic sample of matter and the timescale involved here is 10^{-13}–10^{-11} s and is often associated with particle emission. Moreover ions penetrate deep into the bulk of the target material, and the penetration depth can be finely tuned with the particle energy. The phenomena get even more exciting when relativistic heavy ions are involved. If an atomic electron is considered, while a completely ionized Uranium ion passes by the atom at an impact parameter of 1 nm, then the interaction time may be as short as 0.1 as at a relativistic energy corresponding to $\gamma = 5$ (4 GeV/u), and the intensity of the electromagnetic pulse experienced by the electron is 10^{21} W/cm^2. This corresponds to the highest intensity of modern high power lasers, and it can still be increased easily at higher ion energies which will be available at high energy heavy ion accelerators.

Another property associated with heavy ion matter interaction are the static electromagnetic fields of highly charged ions, which are experienced by the few electrons that are present in highly charged ionic systems. The electron in the hydrogen atom experiences an electric field of 10^{10} V/cm at a binding energy of 13 eV. Already at a nuclear charge of 10 the 1 s binding energy is of the order of 3×10^{12} V/cm which corresponds to an intense laser field of 10^{22} W/cm^2. In hydrogen like uranium the field is 1.8×10^{16} V/cm and this is far above the limit that is accessible with current laser fields. Thus it is the realm of quantum electrodynamics (QED),

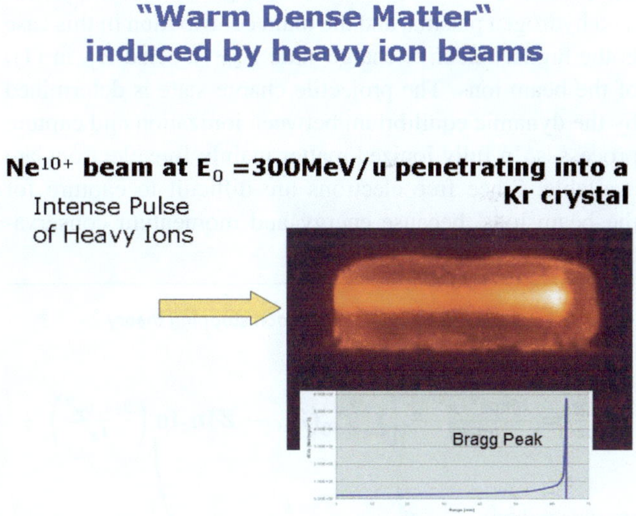

Fig. 3 Intense heavy ion beam penetrating into matter. Below the open shutter photograph of a neon beam penetrating into cryogenic krypton the calculated energy loss according to SRIM 2007 is given

where perturbation theory is no longer applicable since the expansion parameter αZ is of the order of unity.

In ordinary cold matter, which is the situation when the beam first enters into a sample, the stopping power is well described by the SRIM calculation (Ziegler 2004, 2006). This situation is shown in Fig. 3 where a neon beam penetrates into a Kr crystal. Cryogenic targets of rare gases as well as hydrogen and deuterium have been developed at GSI for experiments addressing laboratory planetary science issues as well as part of a program to develop targets for ion driven fusion (Funk et al. 1998; Meyertervehn et al. 1990; Ulrich et al. 1987). The range can actually be measured with a yardstick and agrees very well with simulations. Contrary to common belief there is no "Bragg peak" at low projectile energies below the maximum of the stopping power, which is below 10 MeV/u for uranium ions. We have verified this experimentally by measuring the projectile velocity inside the target with Doppler spectroscopy of the projectile emission (Rosmej et al. 2005) (see Fig. 4).

Pioneering ion-beam-plasma interaction experiments were initially planned and implemented at Orsay (Deutsch 1986; Deutsch et al. 1989; Gardes et al. 1989) at GSI (Dietrich et al. 1992; Hoffmann et al. 1990, 1994, 2000; Jacoby et al. 1995; Weyrich et al. 1989) to compare the stopping power of hydrogen gas and fully ionized hydrogen plasma. Plasmas in linear discharge geometry and a dense z-pinch plasma have been employed to cover the density regime from $n_e = 10^{16}$ cm^{-3} to 10^{19} cm^{-3}. These experiments clearly demonstrated the increased stopping power of fully ionized plasmas, due to the interaction of beam ions with free electrons in plasma.

The increase in stopping power that was measured is close to a factor of 40 for low energy ions in fully ionized hydrogen plasma, and the main contribution in this case is the higher effective charge state Z_{eff} (labeled Z_p in (1)) of the beam ions. The projectile charge state is determined by the dynamic equilibrium between ionization and capture processes. In fully ionized matter mainly free electrons are available. Since free electrons are difficult to capture for the beam ions, because energy and momentum conservation requirements are difficult to be fulfilled at the same time with free electrons higher projectile charge states are observed. With bound electrons, however, momentum conservation is generally no problem due du the strong binding energy. Charge state analysis of beam ions penetrating the fully ionized hydrogen plasma clearly showed the effect of an enhanced charge state.

Laser heated plasmas provide higher temperatures and densities than gas discharges. Consequently, a powerful Nd:YAG/Nd:glass laser system (nhelix: nanosecond high energy laser for ion experiments) at the Z6 experimental area of GSI has been installed (Schaumann et al. 2005). It is now possible to investigate plasmas of the critical density 10^{21} cm^{-3} corresponding to the laser wavelength (1064 nm). Currently experiments using the nhelix and PHELIX (Petawatt High Energy Laser for Ion Experiments (Neumayer et al. 2005)) laser system at GSI for beam plasma interaction studies are underway. Simulations show that the temperature of laser generated plasma of carbon foils may well exceed 100 eV. In this kind of interaction experiments the laser focal spot is kept larger than the ion beam radius, which is of the order of 1 mm, in order to achieve low gradients in density and temperature. This constraint led to the necessity of high energy pulse laser systems. The nhelix laser currently provides 120 J on target and the PHELIX laser achieved 240 J in 2 ns.

The main subjects under investigation in beam plasma interaction experiments are the stopping power of the plasma and the charge states of the ions. These quantities directly enter the Bethe-Bohr-Bloch formula.

Bethe - Bohr - Bloch stopping theory

$$-\frac{dE}{dz} = \frac{16\pi a_0^2 I_H^2 Z_p^2}{m_e v_p^2}\left[\sum_{Z=0}^{Z_K}(Z_K - Z)\, n_z \ln\left(\frac{2m_e v_p^2}{I_Z}\right) + n_e \ln\left(\frac{2m_e v_p^2}{\hbar\omega_P}\right)\right] \qquad (1)$$

effective projectle charge

bound electron contribution

free electron contribution

Future interaction experiments will benefit from planned improvement of the laser performance at the experimental area. There will be an immediate further development of the "nhelix" laser facility to provide a more homogeneous spot irradiation with total pulse energy up to 100 J. Currently the kjoule version of "PHELIX" is being commissioned and first experiments with ion and laser beams are underway. Fully ionized plasmas with atomic numbers up to and above 6 (carbon) will then be available. New target materials and designs give access to a more precise spectroscopic determination of the plasma parameters and new fields of investigation. Thus the development of thin cryogenic hydrogen films for laser plasma-ion beam interaction experiments will extend the stopping power studies for hydrogen plasmas to densities of almost solid state density. An improvement of spectroscopic techniques and new models

Fig. 4 The *upper part* shows the calculated energy loss for Uranium ions as function of range at low energy (SRIM2008). At this energy there is no Bragg peak. The *lower part* shows the measured velocity decrease of an Ar-ion beam at 11.4 MeV/u penetrating into an aerogel target

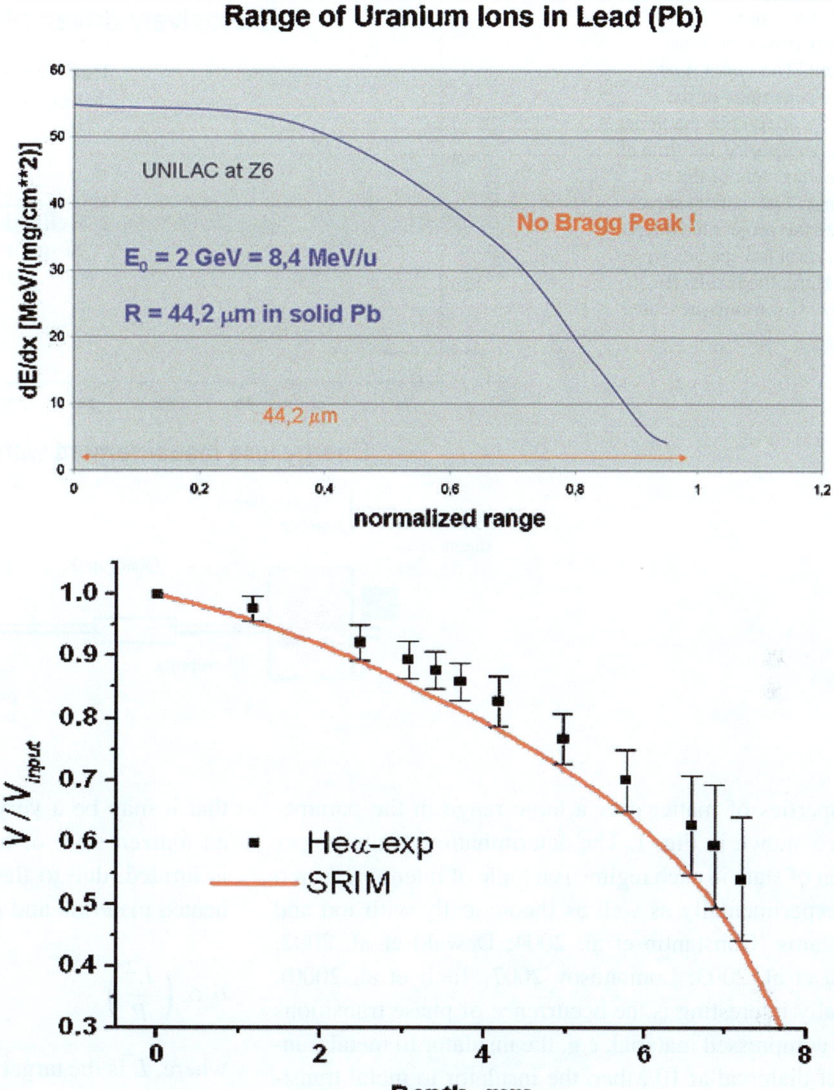

of calculation will enable us to find more precise descriptions of the interaction of heavy ion beams with high density plasmas, answering the questions that have been raised in the recent experiments.

Laser plasmas, due to their high temperature are usually nearly ideal even in the high density regime, and are characterized by a plasma parameter Γ much smaller than 1. The regime of dense, strongly correlated plasmas where $\Gamma = (e^2/4\pi\varepsilon_0\, r_D)/kT$, the ratio of potential over kinetic energy in the plasma, exceeds 1, is addressed in shock wave experiments, using high explosives to compress the target material at low plasma temperatures (see Fig. 5). In these experiments the Γ parameter was varied between 0.5 and 1.5 (Weyrich et al. 2006). However, in this range of non-ideality the influence on beam-plasma interaction processes is very small and we don't see a deviation from the Z^2 dependence of the stopping power within the precision of our experiment.

3 High energy density physics

Intense high-energetic heavy-ion beams are a tool well suited to generate dense plasmas. The advantage of this scheme is the homogeneous coupling of beam energy into matter over a range of several millimeters. This opens up the possibility of heating rather large volumes of matter at solid-state density, typically of the order of a few cubic millimeters. Experiments in fundamental research on the equation of state, phase transitions, and opacities of matter under high pressure become possible. Knowledge gained from such experiments is of considerable interest in the fields of astrophysics, geophysics, and inertial confinement fusion.

The properties of matter under conditions of high density and pressure are often described by an equation of state (EOS) that relates the pressure or energy density to the matter density of the sample. There is no such equation available that which is derived from first principles and describes

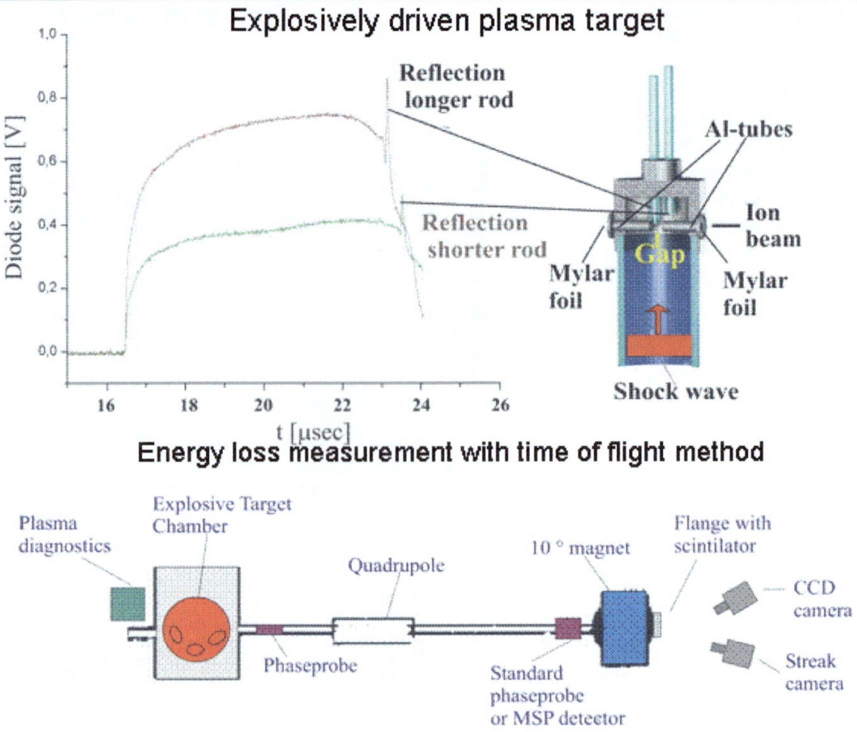

Fig. 5 Set-up for the explosively driven plasma experiment. The *upper part* shows the schematic of the explosively driven plasma target and the *lower part* is the time of flight measurement of the ion energy loss. The magnet serves to analyze the projectile charge states. Several ion species were studied. Here, the results for 5.9 MeV/u C ions are presented

the properties of matter over a large range in the parameter space shown in Fig. 1. The determination of the proper equation of state in each regime is a topic of intense research effort experimentally as well as theoretically with ion and laser beams (Constantin et al. 2004; Dewald et al. 2002; Godwal et al. 2003; Lomonosov 2007; Tahir et al. 2000). Especially interesting is the occurrence of phase transitions in cold compressed material, e.g. the insulator to metal transition of diamond at 10 Mbar, the insulator to metal transition of solid hydrogen above 5 Mbar, or the plasma phase transitions at temperatures of about 1 eV (Ray et al. 2006; Tahir et al. 2001a).

In laser matter interaction experiments it is the intensity I measured in W/cm^2 and the total power which are the relevant experimental parameters. Due to the nature of ion matter interaction, where ions penetrate deep into the target volume, the important parameter is the Energy E_s, deposited per gram of matter [J/g] and the deposition power P_s measured in W/g. Hence the physics of beam induced high energy density matter is governed by three equations (2–4):

$$E_s = (1.6 \times 10^{-19}) \cdot \frac{dE/dx \cdot N}{\pi \cdot r^2} \text{[J/g]} \tag{2}$$

where dE/dx is the stopping power of the material, N is the number of beam ions delivered by the accelerator πr^2 is the focal spot area. In order to achieve high deposition energy the accelerator has to provide the maximum beam intensity and the experiment has to care for an effective focusing. The stopping power is given by nature and we will show later that it may be a very different situation if the matter is in an ionized state or not. The time t_h to deliver this energy is limited, due to the hydrodynamic response of the beam heated material, and is approximately given by:

$$t_h \propto \left(\frac{L^2}{P}\right)^{1/3} \tag{3}$$

where, L is the target dimension. For a cylindrical target this is the target radius and t_H is essentially given by the time a rarefaction wave needs to travel over the distance L.

Combining these two equations yields the total deposition power P_s:

$$P_s = E_s/t_h \tag{4}$$

Thus it is obvious, that high energy density induced by heavy ion beams requires an intense bunch of short time duration (τ_B) which is focused into a sample with a small focal spot size (Tahir et al. 2001b). However, during the time the beam heats the sample, hydrodynamic response resulting in an expansion of the target material sets in. The time scale is set by the sound speed c_s. By the time the rarefaction of the target material has reached the outer edge of the focal spot the effective heating of the target material comes to an end. Thus the effective heating t_h time is given as the upper limit of the integral (5)

$$r_0 = \int_0^{t_h} c_s(T(P_s))dt \tag{5}$$

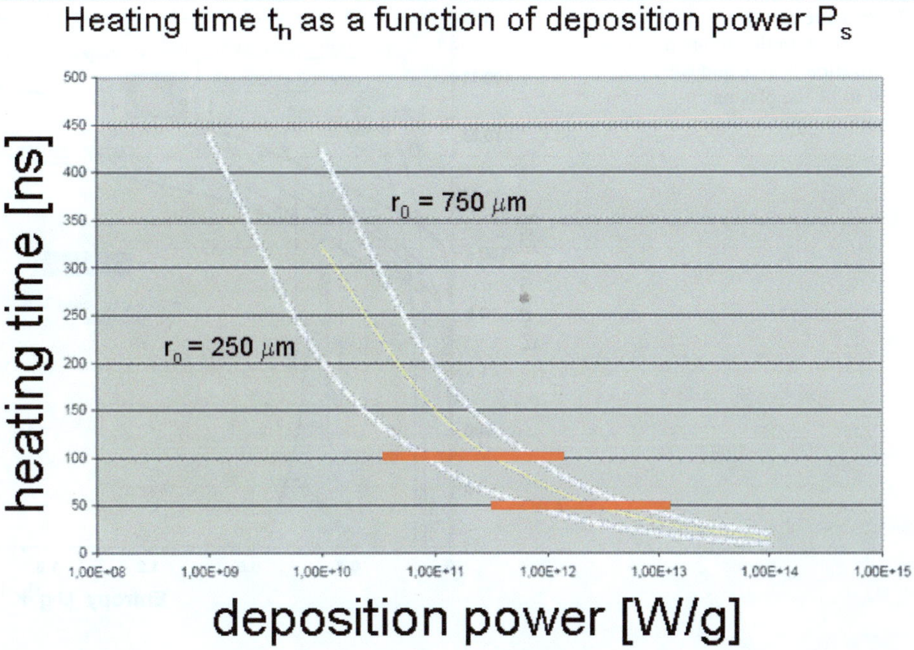

Fig. 6 Heating time as function of deposition power and focal radius

The effective heating time therefore depends on the proper equation of state. In the limit of a mono-atomic ideal gas with the adiabaticity coefficient $\kappa = 5/3$, and taking into account that $E_s = P_s t_h$ (from (4)) is defined as the energy deposited per mass, this can be written as:

$$r_0 = \int_0^{t_h} \sqrt{\kappa \cdot E_s} \cdot dt \tag{6}$$

The focal spot radius r_0, the specific energy deposition $E_s = P_s \tau_B$, and the beam bunch length τ_B are therefore interrelated by

$$r_0 = \sqrt{\kappa \cdot P_s \cdot \tau_B} \cdot \frac{2}{3} \cdot t_h^{3/2} \tag{7}$$

In Fig. 6 we plot the available time to heat a target material (t_h) as function of beam deposition power (E_ρ/τ_B). Typical values for our experimental conditions are $P_s = (E_s/\tau_B) \approx 10^{10}$ W/g. Therefore a focus radius of 250 μm is sufficient for a deposition power up to 0.1 TW/g. This limit will be reached after the intensity upgrade of the current GSI accelerator system within the coming years.

The presently available beam from the SIS 18 is focused by a fine focus system, and we make use of the electron cooling of the heavy ion beam in the synchrotron. Before this low emittance beam was available, a plasma lens was in operation (Boggasch et al. 1992b; Neuner et al. 2000b) to achieve a beam spot radius of 250 μm.

Many experiments to study high energy density matter are based on shock wave techniques starting from ambient pressure and solid density, or porous target material. With this technique high energy density states on the principal Hugoniot can be reached as can be seen from Fig. 7.

Figure 7 also indicates the areas of the phase diagram that can be addressed with different experimental techniques, such as shock wave compression by lasers, high explosives and others. The HIHEX quasi isochoric heating curve can be reached by intense heavy ion beams. Subsequent isentropic expansion allows reaching until now unexplored areas of the phase diagram. The area labelled SIS 100 will be accessible with heavy ion beams, when the new FAIR facility at GSI is in operation. Currently we are operating close to the critical point (CP). Lead, which is one of few materials where experimental data exist already, serves as a benchmark material. Once we have an assessment of the validity of the heavy ion heating and expansion method (HIHEX) we will explore unchartered areas of the phase diagram for many materials and make full use of the properties of heavy ion heated systems with their intrinsic large time and length scales offer a promising alternative to explore these areas of the phase diagram and phase transitions in precision experiments.

The experimental technique is based on the fact that penetrating into the binodal region or the critical point region, basic thermodynamic properties of matter change rapidly, causing an observable change in the hydrodynamic behaviour of the target material (Fortov et al. 1992; Novikov 1962).

Our experiments on heavy-ion-beam generated high-energy-density (HED) at GSI, make use of the intense uranium beam with up to 4.4×10^9 particles per bunch in approximately 100–125 ns. In these experiments, we investigated thermophysical properties and hydrodynamic response of various materials including Pb samples. High energy density states in HED states in the two-phase liquid-vapor region near binodal and the critical point have been

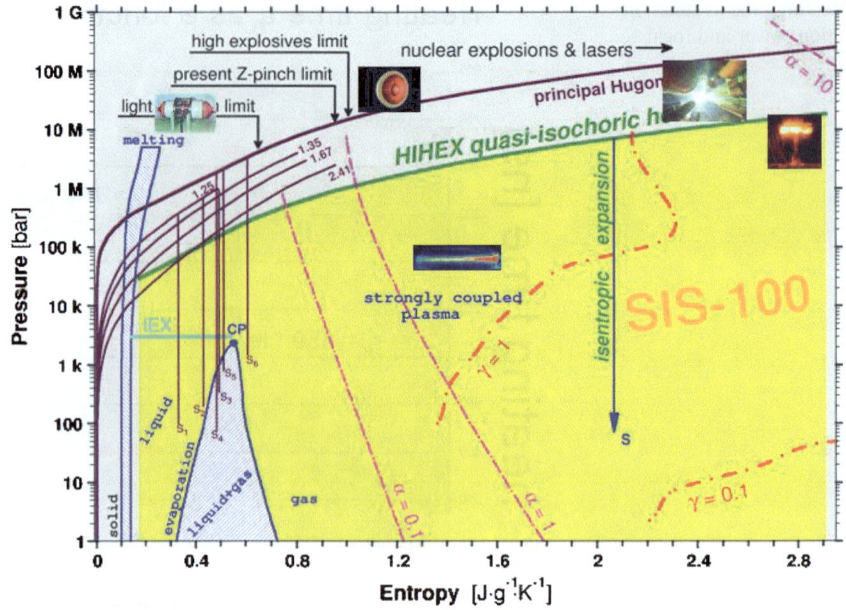

Fig. 7 Pressure entropy diagram of lead. In this figure α is the ionization degree and we choose γ to be the plasma parameter

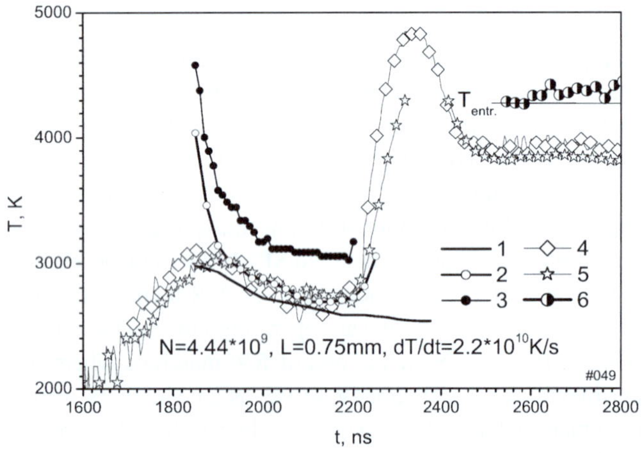

Fig. 8 Evaporating lead temperature along release isentropes. Shown are brightness temperatures on shock-loading release isentropes: *1*—151 GPa; *2*—180 GPa; *3*—223 GPa at 700 nm and in heavy-ion-beam driven experiment (beam intensity $N = 4 : 44\_109$ ions/pulse, expansion length $L = 0 : 75$ mm) : *4*—at 650 nm; *5*—at 750 nm; *6*—grey-body approximation of physical temperature

obtained. A fast multi-channel radiation pyrometer has been used as a key diagnostics in these experiments (Ni et al. 2006). This allows for a quantitative comparison between the target expansion dynamics after ion-beam heating and the shock-wave loaded release isentrope data. Brightness temperature profiles on release isentropes measured for lead samples initially shocked up to 151; 180 and 223 GPa pressures are shown in Fig. 8 (solid lines labelled 1, 2, and 3) along with the corresponding results of ion-beam driven experiments. For the ion-beam experiments, the physical temperature was approximated by fitting all the data from the six-channel pyrometer to the grey-body model. It is seen from Fig. 8 that the temperature profile during the expansion phase observed in the ion-beam driven experiment is close to the 180 GPa release isentrope of shock-loaded lead (line "2" in Fig. 8). According to the EOS model for lead (Bushman and Fortov 1983a), this isentrope crosses the bimodal curve on the phase diagram at the point $P = 50$ MPa; $T = 4243$ K and $E = 0 : 727$ kJ/g: As it has been shown in shock-wave experiments, expansion of lead along this isentrope to the final pressures below 10 MPa is accompanied by the surface evaporation and generation of the boiling wave phenomena. In the corresponding beam-driven experiment, total duration of the heating pulse was about one microsecond with FWHM of 125 ns. During the heating phase, the sample foil is melted and starts to evaporate at the free surface. While expanding, the lead vapor is rapidly cooling down and at the moment when its temperature becomes sufficient for homogeneous nucleation (near the vapor spinodal line), liquid droplets appear inside the vapor and the boiling wave is formed. After this moment, the expansion dynamics is governed by the surface evaporation processes. These issues have to be taken into account in the hydrodynamic simulations for HEDP experiments. When the expanding lead which consists of a mixture of vapor and liquid metal droplets impacts on the sapphire window, it is compressed and heated further (see Fig. 8, $t \approx 2300$ ns). According to the EOS model (Bushman and Fortov 1983b), such weak (quasi-isentropic) shock compression will transform the material into pure vapor state causing the droplets disappear. On the other hand, a certain time is needed for such transformation. Therefore, if the heating rate is high enough, the liquid droplets can be overheated up to spinodal temperatures. To realize such overheated states for pressure below the critical pressure, a heating rate dT/dt which is at least higher than 10^9 K/s is necessary (Bennett 1971). The conditions in our experimental set-up the expansion length is

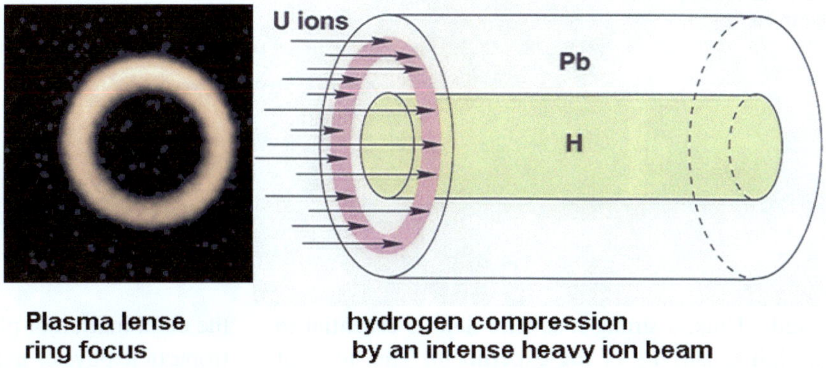

Fig. 9 An annular beam focus, achieved by plasma lens focusing is used to compress a hydrogen sample

short, and at the moment of the impact on sapphire the heating rate was about $dT/dt = 2.2 \times 10^{10}$ K/s and the overheating is clearly observed. The maximum temperature of about 4860 K measured in this experiment is close to that of the liquid spinodal at 50 MPa pressure.

The energy deposition characteristics of high-energy heavy ion beams in dense matter favor a cylindrical geometry. A simple experimental scenario is therefore a quasi-cylindrical plasma volume created by focusing an ion beam. A more complicated scenario with higher compression yields can be achieved with special beam focus geometry, e.g. a hollow cylindrical beam focus at the target position, as sketched in Fig. 9. This geometry is the basis for the LAPLAS experiment at the new FAIR facility (Funk et al. 2000; Tahir et al. 2004, 2007a, 2007c). This is a very challenging problem for the beam transport and focusing system (Temporal et al. 2003, 2005a). However, a ring focus was already demonstrated experimentally, which was achieved by carefully tuning nonlinear field gradients of the plasma (Boggasch et al. 1992a; Neuner et al. 2000a). Thus hollow cylindrical implosions become possible, and the simulations show that in such case the initial pressure, which is generated by the direct heating of target material, can be enhanced by more than a factor of 10. Alternative schemes to achieve a hollow beam are based on high frequency rotation of a beam spot on target. A careful analysis has shown that the symmetry in the cylindrical implosion can be maintained if the time scale of the rotation frequency is approximately ten times shorter than the heating time. However this strongly depends on the intensity distribution of the beam focus (Piriz et al. 2003; Temporal et al. 2005b).

4 Where astroparticle physics meets high energy density plasma physics

High energy accelerators and accelerator technology has become a useful tool in high energy density physics experiments. As mentioned earlier one of our proposals calls for the intense particle beam of LHC at CERN to be used to irradiate matter and generate high energy density states. The final part of our paper now addresses a topic where accelerator equipment and plasma spectroscopy techniques are used as a very novel tool for plasma diagnostic and astroparticle physics. We all are aware of the fact that the interior of the sun consists of dense hot plasma with interesting properties. There are, however very few tools to diagnose the temperature of this plasma in a direct way. For many reasons it is interesting to produce plasma with similar properties under laboratory conditions. Thus magnetic confinement fusion and inertial fusion are two different development paths to generate plasma conditions for energy production from fusion reactions. However, the sun is also an excellent laboratory for astroparticle physics. As such the sun can be used to probe the existence of novel particles and dark matter candidates like the axion.

The axion is a direct consequence (Peccei and Quinn 1977a; Weinberg 1978) of the theoretical solution of the CP problem in strong interactions proposed by (Peccei and Quinn 1977b), to explain the non-violation of CP (charge, parity) of the strong interaction. Thus the axion was introduced as a neutral pseudoscalar meson, similar to the neutral pion, however, with unknown mass. Astrophysical constraints leave open a small mass window around 1 eV/c^2 and the CAST is able to cover the mass range up to about 1 eV/c^2. Inside the core of the sun axions could be produced by coherent conversion of thermal photons interacting with the electromagnetic field of charged particles of the solar plasma (Primakoff effect). With the CAST experiment at CERN, we aim to detect such solar axions on earth by "converting" them back to X-ray photons inside a strong transversal magnetic field (inverse Primakoff effect, see Fig. 10).

The observed spectrum should very closely reflect the spectrum of a black body radiator at the temperature of the interior of the sun and this method would provide a direct way to measure the temperature of this interesting plasma.

The conversion probability of axions to photons is proportional to the square of the strength of the magnetic field

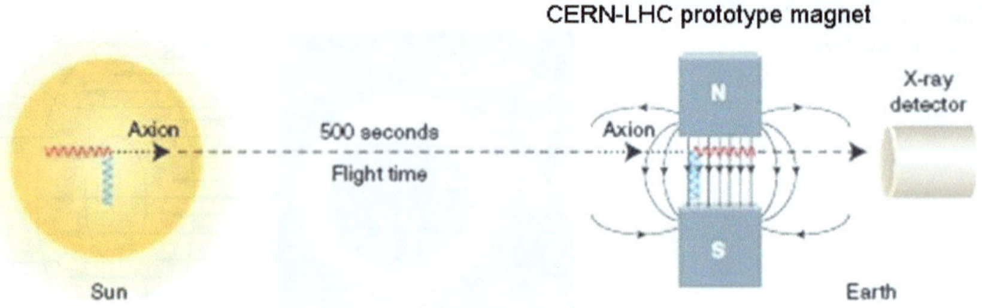

Fig. 10 Schematic set-up of the CAST experiment

and its length. Thus, a strong magnetic field is essential to achieve a high sensitivity of the experiment. The heart of CERN Axion Solar Telescope (CAST) is a prototype LHC superconducting magnet providing a dipole magnetic field of ≈ 9 T over a distance of 9.26 m. On both ends of the magnet X-ray detectors are looking for a potential axion signal as an excess signal over detector background. A TPC detector covers two magnet bores on one end looking for axions during sunset. On the opposite side of the magnet, a micro mesh gas detector and an X-ray telescope with a pn-CCD detector are looking for axions at sunrise. The magnet can be pointed towards the sun for about 1.5 h during sunrise and sunset. The remaining time is used for systematic background studies. The most sensitive detector system of CAST is the Wolter I type X-ray telescope which enhances the signal-to-background ratio by a factor of $\approx$100 by concentrating the potential signal flux on a small spot on the pn-CCD detector. The analysis of the data up to now reveals no significant excess signal over background and allows us to set a new upper limit on the axion to photon coupling (Zioutas et al. 2005). The preliminary corresponding upper limit for the axion to photon coupling constant g_a derived from the data is just below $<10^{-10}$ GeV^{-1} at 95% confidence level and for an axion mass below 0.02 eV/c^2.

The CAST experiment is now running in its third phase searching for axion masses up to 1 eV. The experiment is scheduled at CERN until 2010.

5 Conclusion

Investigation of interaction processes of heavy ion beams with ionized matter was originally motivated by inertial fusion physics, where intense heavy ion beams are proposed to ignite a fusion target. Since intense heavy ion beams are available today at different accelerator locations the emphasis has shifted to high energy density physics with applications such as laboratory astrophysics. Areas of specific interest are conditions that prevail in the interior of giant planets. Properties of matter under these conditions are hardly known and theoretical predictions call for an experimental verification. The detailed knowledge about interaction phenomena of intense fields with matter will certainly also influence the development towards inertial fusion. Moreover, the combination of plasma physics nuclear physics- and astroparticle physics using accelerator equipment and technology opened the gate to search for new particles and helps to address the dark matter problem.

Acknowledgements This work was supported by grants from various funding agencies in Germany and Russia including BMBF, GSI, and RFBR.

References

Bennett, F.D.: Physics of High Energy Density. Academic Press, New York and London (1971)
Boggasch, E., et al.: Appl. Phys. Lett. **60**, 2475 (1992a)
Boggasch, E., et al.: Appl. Phys. Lett. **60**, 2475 (1992b)
Bushman, A.V., Fortov, V.E.: Usp. Fiz. Nauk **140**, 177 (1983a)
Bushman, A.V., Fortov, V.E.: Usp. Fiz. Nauk **140**, 177 (1983b)
Constantin, C., et al.: Laser Part. Beams **22**, 59 (2004)
Deutsch, C.: Ann. De Phys. **11**, 1 (1986)
Deutsch, C., et al.: Nucl. Instr. Methods Phys. Res. Sect. a-Accel. Spectrom. Detect. Assoc. Equip. **278**, 38 (1989)
Dewald, E., et al.: Laser Part. Beams **20**, 399 (2002)
Dietrich, K.G., et al.: Phys. Rev. Lett. **69**, 3623 (1992)
Fortov, V., Lebedev, M., Ternovoi, V.: Rev. Gen. De Therm. **31**, 589 (1992)
Fortov, V.E., et al.: Phys. Rev. Lett. **99** (2007)
Funk, U.N., et al.: Nucl. Instr. Methods Phys. Res. Sect. a-Accel. Spectrom. Detect. Assoc. Equip. **415**, 68 (1998)
Funk, U.N., et al.: J. De Phys. IV **10**, 215 (2000)
Gardes, D., et al.: Radiat. Eff. Defects Solids **110**, 49 (1989)
Godwal, B.K., Rao, R.S., Verma, A.K., Shukla, M., Pant, H.C., Sikka, S.K.: Laser Part. Beams **21**, 523 (2003)
Henning, W.F.: J. Phys. G-Nucl. Part. Phys. **34**, S551 (2007)
Hoffmann, D.H.H., Weyrich, K., Wahl, H., Gardes, D., Bimbot, R., Fleurier, C.: Phys. Rev. A **42**, 2313 (1990)
Hoffmann, D.H.H., et al.: Nucl. Instr. Methods Phys. Res. Sect. B-Beam Interact. Mater. Atoms **90**, 1 (1994)
Hoffmann, D.H.H., et al.: Nucl. Instr. Methods Phys. Res. Sect. B-Beam Interact. Mater. Atoms **161**, 9 (2000)
Hoffmann, D.H.H., et al.: Laser Part. Beams **23**, 47 (2005a)
Hoffmann, D.H.H., et al.: Laser Part. Beams **23**, 47 (2005b)
Hoffmann, D.H.H., et al.: Eur. Phys. J. D **44**, 293 (2007)
Hora, H.: Laser Part. Beams **25**, 37 (2007)
Jacoby, J., et al.: Phys. Rev. Lett. **74**, 1550 (1995)
Logan, B.G., et al.: Nucl. Instr. Methods Phys. Res. Sect. a-Accel. Spectrom. Detect. Assoc. Equip. **577**, 1 (2007a)
Logan, B.G., et al.: Nucl. Instr. Methods Phys. Res. Sect. a-Accel. Spectrom. Detect. Assoc. Equip. **577**, 1 (2007b)
Lomonosov, I.V.: Laser Part. Beams **25**, 567 (2007)
Meyertervehn, J., et al.: Phys. Fluids B-Plasma Phys. **2**, 1313 (1990)
Neumayer, P., et al.: Laser Part. Beams **23**, 385 (2005)

Neuner, U., et al.: Phys. Rev. Lett. **85**, 4518 (2000a)
Neuner, U., et al.: Phys. Rev. Lett. **85**, 4518 (2000b)
Ni, P., et al.: J. De Phys. IV **133**, 977 (2006)
Novikov, N.N.: Zh. Prikl. Mekh. Tekh. Fiz. **3**, 8 (1962)
Peccei, R.D., Quinn, H.R.: Phys. Rev. Lett. **38**, 1440 (1977a)
Peccei, R.D., Quinn, H.R.: Phys. Rev. Lett. **38**, 1440 (1977b)
Piriz, A.R., Temporal, M., Cela, J.J.L., Tahir, N.A., Hoffmann, D.H.H.: Plasma Phys. Controll. Fusion **45**, 1733 (2003)
Ray, A., Srivastava, M.K., Kondayya, G., Menon, S.V.G.: Laser Part. Beams **24**, 437 (2006)
Rosmej, O.N., et al.: Laser Part. Beams **23**, 79 (2005)
Schaumann, G., et al.: Laser Part. Beams **23**, 503 (2005)
Sharkov, B.Y.: Nucl. Instr. Methods Phys. Res. Sect. a-Accel. Spectrom. Detect. Assoc. Equip. **577**, 14 (2007)
Tahir, N.A., et al.: Phys. Rev. E **62**, 1224 (2000)
Tahir, N.A., et al.: Phys. Rev. E 6302 (2001a)
Tahir, N.A., Kozyreva, A., Spiller, P., Hoffmann, D.H.H., Shutov, A.: Phys. Rev. E **6303**, 036407 (2001b)
Tahir, N.A., et al.: Laser Part. Beams **22**, 485 (2004)
Tahir, N.A., et al.: Phys. Rev. Lett. **95**, 035001 (2005a)
Tahir, N.A., et al.: Phys. Rev. Lett. **94**, 135004 (2005b)
Tahir, N.A., et al.: Contrib. Plasma Phys. **47**, 223 (2007a)
Tahir, N.A., et al.: Laser Part. Beams **25**, 639 (2007b)
Tahir, N.A., et al.: Nucl. Instr. Methods Phys. Res. Sect. a-Accel. Spectrom. Detect. Assoc. Equip. **577**, 238 (2007c)
Temporal, M., Piriz, A.R., Grandjouan, N., Tahir, N.A., Hoffmann, D.H.H.: Laser Part. Beams **21**, 609 (2003)
Temporal, M., Lopez-Cela, J.J., Piriz, A.R., Grandjouan, N., Tahir, N.A., Hoffmann, D.H.H.: Laser Part. Beams **23**, 137 (2005a)
Temporal, M., Lopez-Cela, J.J., Piriz, A.R., Grandjouan, N., Tahir, N.A., Hoffmann, D.H.H.: Laser Part. Beams **23**, 137 (2005b)
Ulrich, A., et al.: J. Appl. Phys. **62**, 357 (1987)
Varentsov, D., et al.: Nucl. Instr. Methods Phys. Res. Sect. a-Accel. Spectrom. Detect. Assoc. Equip. **577**, 262 (2007a)
Varentsov, D., et al.: Nucl. Instr. Methods Phys. Res. Sect. a-Accel. Spectrom. Detect. Assoc. Equip. **577**, 262 (2007b)
Weinberg, S.: Phys. Rev. Lett. **40**, 223 (1978)
Weyrich, K., et al.: Nucl. Instr. Methods Phys. Res. Sect. a-Accel. Spectrom. Detect. Assoc. Equip. **278**, 52 (1989)
Weyrich, K., et al.: J. Phys. a-Math. Gen. **39**, 4749 (2006)
Ziegler, J.F.: Nucl. Instr. Methods Phys. Res. Sect. B-Beam Interact. Mater. Atoms **219–20**, 1027 (2004)
Ziegler, J.F.: www.srim.org (2006)
Zioutas, K., et al.: Phys. Rev. Lett. **94**, 121301 (2005)
Zioutas, K., Hoffmann, D.H.H., Dennerl, K., Papaevangelou, T.: Science **306**, 1485 (2004)
Zou, X.B., Liu, R., Zeng, N.G., Han, M., Yuan, J.Q., Wang, X.X., Zhang, G.X.: Laser Part. Beams **24**, 503 (2006)
Zvorykin, V.D., et al.: Laser Part. Beams **25**, 435 (2007)

Astrophys Space Sci (2009) 322: 179–188
DOI 10.1007/s10509-008-9962-9

ORIGINAL ARTICLE

High Energy Density physics and Laboratory Planetary Science using intense heavy ion beams at FAIR facility at Darmstadt: the HEDgeHOB collaboration

N.A. Tahir · A.R. Piriz · G. Wouchuk · A. Shutov · I.V. Lomonosov · C. Deutsch · D.H.H. Hoffmann · V.E. Fortov

Received: 28 April 2008 / Accepted: 1 December 2008 / Published online: 22 January 2009

Abstract Field of High Energy Density (HED) physics is very important from the scientific as well as technological point of view and due to these reasons, it has been an area of very active research over the past many decades. Static as well as dynamic experimental configurations have been used to study this subject. Recently, a new approach that involves isochoric and uniform heating of matter by intense ion beams, has been proposed to generate HED sates in the laboratory. Extensive theoretical work has been carried out to propose various experiment designs employing this technique. In this paper we describe one of these proposed experiments that can be used to study planetary interiors in the laboratory and is named **LAPLAS** (**LA**boratory **PLA**netary **S**ciences). Detailed analysis of implosion and hydrodynamic stability of the target is presented.

N.A. Tahir (✉)
Gesellschaft für Schwerionenforschung, 64291 Darmstadt, Germany
e-mail: n.tahir@gsi.de

A.R. Piriz · G. Wouchuk
E.T.S.I. Industriales, Universidad de Castilla-La Mancha, 1307 Ciudad Real, Spain

A. Shutov · I.V. Lomonosov · V.E. Fortov
Institute of Problems of Chemical Physics, Chernogolovka, Russia

C. Deutsch
Laboratoire de Physique des Gaz et des Plasmas, Universite Paris-Sud, 91405 Orsay, France

D.H.H. Hoffmann
Institut für Kernphysik, TU Darmstadt, 64289 Darmstadt, Germany

Keywords High Energy Density physics · Warm dense matter · Planetary science · Intense ion beams

1 Introduction

States of matter corresponding to an energy of 10^{11} J/m^3 or equivalently pressures of 1 Mbar and above, are classified as High Energy Density (HED) states. The importance of this subject is underscored by the fact that it spans over very wide areas of basic and applied physics including, for example, astrophysics, planetary sciences, geophysics, plasma physics, fluid dynamics, radiation hydrodynamics, magnetohydrodynamics, material sciences, condensed matter physics, interaction of intense radiation with matter, atomic and molecular physics and many others. In addition to that, this field has great potential for numerous very important industrial applications. A study of the fundamental properties of HED matter is therefore of considerable scientific importance as well as of great technological interest. Due to these reasons, this subject has been a very active research area over the past few decades. Various dynamic as well as static experimental schemes have been employed to research this very important field of science.

In this paper we describe a novel scheme that involves generation of HED states by isochoric and uniform heating of matter by intense particle beams. These beams can either be generated by traditional accelerators (Henning 2004) or by ultra high intensity lasers (Snavely et al. 2002; Schollmeier et al. 2008), although the former type of beams have a number of advantages over the latter ones. In this paper we present numerical simulations that have been carried out to study potential of intense uranium beams that will be generated at the future **F**acility for **A**ntiprotons and **I**on **R**esearch (**FAIR**) (Henning 2004) to induce HED states in matter. It has been shown that an intense ion beam can be employed using two very different experimental schemes to study this subject. First scheme named **HIHEX**, that means **H**eavy **I**on **H**eating and **EX**pansion, is suitable for studying equation-of-state (EOS) properties of HED matter at sub-solid densities and is described in detail elsewhere (Hoffmann et al. 2002; Tahir et al. 2005a, 2005b, 2005c).

The second scheme allows achievement of a low-entropy compression of a material like hydrogen or ice that is contained inside a cylindrical shell of a heavy material like gold or lead. Such a scheme is very suitable to study the problem of hydrogen metallization (Wigner and Huntigton 1935; Mao and Hemley 1994; Ross 1983; Nellis et al. 1983; Celliers et al. 2000; Caubel et al. 2000; Tahir et al. 2001, 2003; Piriz et al. 2002) and for generating physical conditions that are expected to exist in the interiors of the giant planets (Tahir et al. 2005c). This experimental configuration has been named **LAPLAS** which means **LA**boratory **PLA**natory **S**ciences.

In this paper we present theoretical work describing the LAPLAS scheme, that will allow one to study interiors of giant planets in our solar system as well as the exoplanets. This work has been done within the framework of the **HEDgeHOB** collaboration which stands for **H**igh **E**nergy **D**ensity Matter **ge**nerated by **H**eavy **I**on **B**eams. This international collaboration which includes more than 200 scientists that belong to very different areas of research, has been formed to construct the infrastructure necessary to perform experiments on HED matter at the future **F**acility for **A**ntiprotons and **I**on **R**esearch (**FAIR**) (Henning 2004) at Darmstadt. This collaboration will also be responsible for proposing, designing and carrying out the experiments in this field. It is expected that first experiments to study this type of problems will be carried out around the year 2015. The FAIR facility will include construction of a new heavy ion synchrotron, SIS100 that will have a magnetic rigidity of 100 Tm.

In Sect. 2 we discuss the general features of the LAPLAS scheme using two different beam–target configurations while numerical simulations of target are presented in Sect. 3. A detailed analysis of the hydrodynamic stability including the Rayleigh–Taylor and Richtmyer–Meshkov instabilities that can occur in the target during various implosion stages, is presented in Sect. 4 while conclusions drawn from this work are noted in Sect. 5.

2 Description of the LAPLAS experimental scheme for planetary science

Two possible beam–target configurations have been proposed for the LAPLAS experimental set up as described below.

2.1 LAPLAS scheme with an annular shaped beam focal spot

The target consists of a cylinder of frozen hydrogen that is surrounded by a thick shell of a heavy material, typically gold or lead. One face of the target is irradiated with an intense heavy ion beam that has an annular (ring-shaped) focal spot, as shown in Fig. 1(a). The inner radius of the annulus is larger than the radius of the sample material that avoids direct heating of the sample by the ion beam. Moreover, the outer radius of the focal spot ring is smaller than the outer radius of the surrounding shell. It is seen from Fig. 1(a), that a layer of cold material from the outer shell known as "pusher" or "payload", is created between the sample material and the beam-heated region. The payload plays an important role in placing the compression on the desired adiabat. It is also seen, that a cold shell around the beam-heated zone remains as a tamper that confines the implosion for a longer time.

The target length is assumed to be less than the range of the driver ions so that the energy deposition in the longitudinal direction is uniform. The pressure in the beam heated region increases substantially that launches a shock wave inwards, along the radial direction. The shock wave

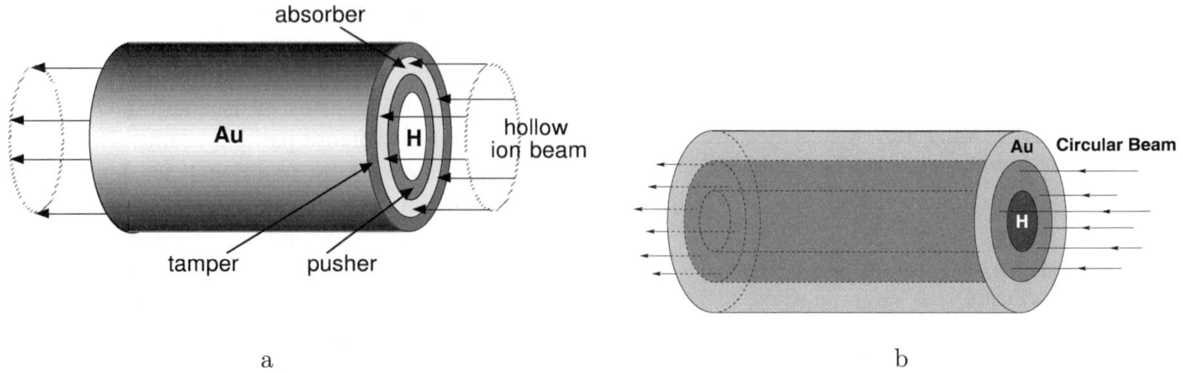

Fig. 1 Beam–target geometry using: (**a**) annular focal spot; (**b**) circular focal spot

enters the pusher and is subsequently transmitted into the hydrogen and is reflected at the cylinder axis. The reflected shock wave moves outwards along the radial direction and is re-reflected at the hydrogen–gold boundary. The boundary continues to move inwards, thereby compressing the hydrogen slowly. This scheme generates a low-entropy compression that leads to a very high material density with a relatively low temperature. Simulations have shown that using the parameters of the SIS100 beam at the FAIR, one would achieve a final hydrogen density of 1–3 g/cm^3, a pressure of 3–30 Mbar and a temperature of a few thousand K. This scheme is therefore more suited to study the problem of hydrogen metallization (Mao and Hemley 1994; Wigner and Huntigton 1935; Ross 1983; Nellis et al. 1983; Celliers et al. 2000; Caubel et al. 2000; Tahir et al. 2007, 2001, 2003; Piriz et al. 2002).

Generation of an annular focal spot is a challenging problem and one possible way of achieving this will be to employ an rf-wobbler that will rotate the beam with a very high frequency (of the order of a GHz) that will generate the required focal spot geometry. A detailed analysis of the symmetry issues of energy deposition in the target using such a system have been published elsewhere (Piriz et al. 2003a, 2003b). A wobbler will be constructed within the framework of the FAIR project. An alternative method for generating a hollow beam could be use of a plasma lense (Neuner et al. 2000), although this option is not under consideration at the moment.

2.2 LAPLAS scheme with a circular focal spot

Figure 1(b) shows the case where one uses a circular focal spot so that the hydrogen is also directly heated by the beam, but since the pressure in the surrounding shell is orders of magnitude higher than that in hydrogen, the hydrogen is still compressed to very high densities. However, the final temperature of hydrogen, in this case, is much higher (of the order of a few eV) than that in the previous one. This scheme is therefore suitable for studying the planet interiors.

3 Simulation results

In this section we present simulation results of implosion of the two targets shown in Figs. 1 and 2 respectively. These simulations have been carried out using a two-dimensional hydrodynamic computer code, BIG-2 (Fortov et al. 1996) that is based on an Arbitrary Lagrangian Eulerian (ALE) numerical scheme. The code can be used in plane and cylindrical geometry and can handle sophisticated multi-layered targets. Energy deposition by the projectile particles in the target is treated using a particle tracking scheme. Different phases of the target material are simulated using a sophisticated semi-empirical EOS model (Lomonosov 2007; Lomonosov and Tahir 2008).

3.1 Annular focal spot case

Two-dimensional hydrodynamic code, BIG2 (Fortov et al. 1996) described above, has been used to simulate the implosion of the LAPLAS beam–target geometry shown in Fig. 1. The length of the cylindrical target is assumed to be 1 cm, the radius of the frozen hydrogen region is 0.4 mm while the outer radius of the gold shell is considered to be 3 mm. The target is irradiated with a uranium beam having a particle

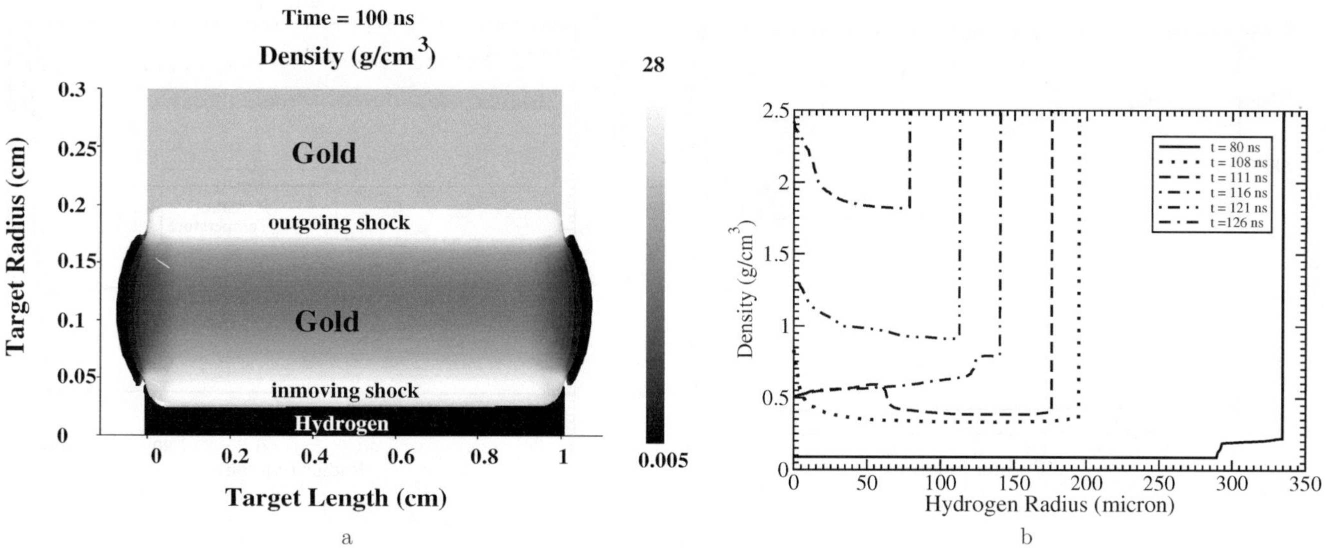

Fig. 2 (**a**) Target density on a length–radius plane at $t = 100$ ns, beam intensity, $N = 10^{12}$ ions, bunch length, $\tau = 50$ ns, particle energy = 2 GeV/u, annular focal spot, inner spot radius = 0.6 mm, outer spot radius = 1.6 mm. (**b**) Density vs. radius in hydrogen region at different times

energy of 2 GeV/u and an intensity of 10^{12} ions per bunch which is 50 ns long. The beam spot is ring shaped with an inner ring radius = 0.6 mm and an outer radius = 1.6 mm. These beam parameters deposit a specific energy of about 35 kJ/g in gold region that generates a high pressure of 3.5 Mbar. This high pressure launches strong shocks in the radial direction, inwards and outwards. This is seen from Fig. 2(a) where we plot the target density on length–radius plane at $t = 100$ ns.

The inmoving shock travels through the cold part of the gold that lies between the hydrogen–gold boundary and the inner radius of the focal spot ring. The gold layer is compressed and the shock is transmitted into hydrogen as is seen from Fig. 2(b), curve labeled with $t = 80$ ns. Vertical line shows the position of the hydrogen–gold boundary. It is seen that the shock arrives at the axis at $t = 108$ ns and is reflected. The position of the reflected shock is shown by curve labeled with $t = 111$ ns. The reflected shock is re-reflected from the boundary as shown in curve labeled with $t = 116$ ns and it arrives at the axis at $t = 121$ ns. The shock arrives at the axis third time at $t = 126$ ns. It is seen that every time the shock arrives at the axis, the hydrogen density is substantially increased. The hydrogen-gold boundary also slowly moves inwards, thereby compressing the sample to a final density of about 3 g/cm^3, pressure of the order of 30 Mbar and average temperature of about 7000 K, at $t = 135$ ns. Using a lower beam intensity of 5×10^{11}, one would achieve a density of about 2 g/cm^3, pressure of 15 Mbar and an average temperature of 4000 K. This scheme is suitable to study the problem of hydrogen metallization.

3.2 Circular focal spot case

We now present simulation results of compression of the target shown in Fig. 1(b) using a uranium beam with a particle energy of 2 GeV/u and an intensity, $N = 10^{12}$ ions that are delivered in a single bunch which is 50 ns. The FWHM of the transverse Gaussian intensity distribution is considered to be 1.5 mm.

The target length = 1 cm, radius of the hydrogen = 0.4 mm and the outer radius of the gold shell = 4 mm. In Fig. 3(a), we plot the temperature vs. radius in the hydrogen region at different times during irradiation. It is seen that at $t = 10$ ns, the hydrogen temperature has increased from an initial value of 10 K to 2000 K while at $t = 20$ ns the temperature becomes 5000 K. Also at $t = 30$ ns, the temperature in the inner 200 μm radius has increased to 8500 K while in the outer 100 μm, it is of the order of 10000 K. This is due to the shock that is transmitted from the high pressure gold region into the hydrogen. It is also seen that the hydrogen–gold boundary slowly moves inwards, thereby compressing the hydrogen to higher densities. The curve labeled with $t = 40$ ns shows high temperature at the cylinder axis which is due to the arrival of the first shock. This shock reverberates between the axis and the hydrogen–gold boundary, while the boundary slowly moves inwards, leading to hydrogen compression.

Figure 3(b) shows the density, temperature and pressure in the hydrogen vs. radius at $t = 80$ ns, which is the time of maximum compression. It is seen that a density of 1.2 g/cm^3, a pressure of 11 Mbar and a temperature of about 5 eV is achieved. These are the physical conditions predicted to exist in the interiors of the giant planets.

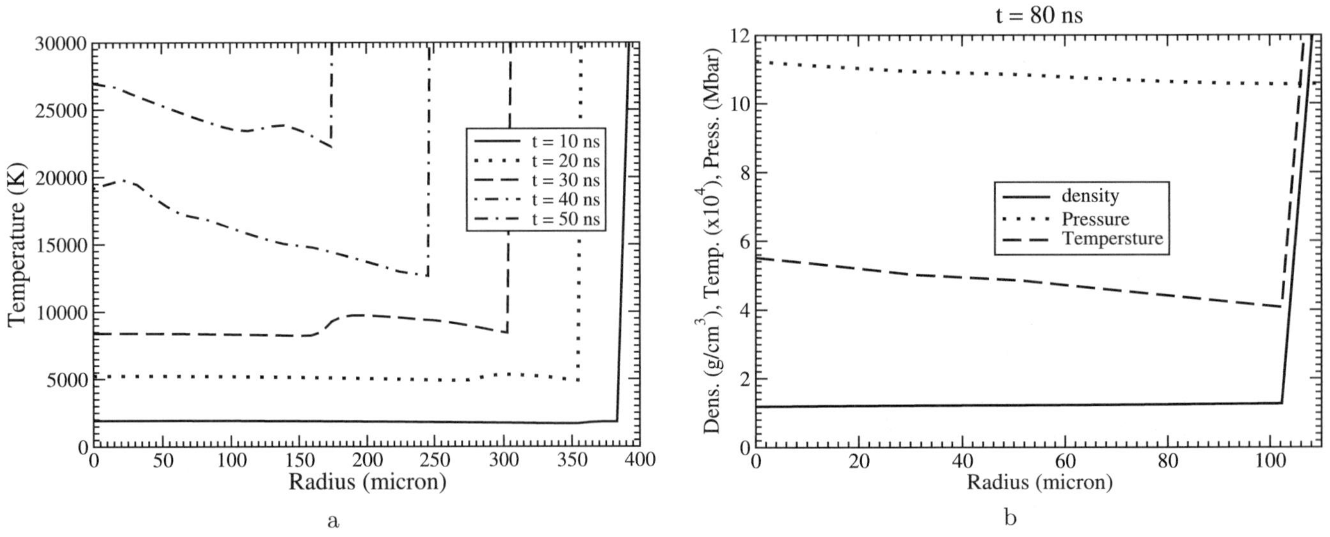

Fig. 3 (**a**) Temperature vs. radius in hydrogen at different times during irradiation, beam intensity, $N = 10^{12}$ ions, bunch length, $\tau = 50$ ns, particle energy = 2 GeV/u, circular focal spot, FWHM = 1.5 mm. (**b**) Density, temperature and pressure in hydrogen at maximum compression, $N = 10^{12}$ ions, bunch length, $\tau = 50$ ns, particle energy = 2 GeV/u, circular focal spot, FWHM = 1.5 mm

4 Hydrodynamic stability of the target

Two type of hydrodynamic instabilities can occur in the LAPLAS scheme. First, the Rayleigh–Taylor instability (Rayleigh 1954; Taylor 1950) that may develop during the implosion phase in the high-Z pusher region and second, the Richtmyer–Meshkov (Richtmeyer 1960; Meshkov 1969) instability that can occur if there is a corrugation at the hydrogen–gold interface. Both instabilities have the potential of destroying the implosion. A detailed analysis of these problems have been carried out and is presented below.

4.1 Rayleigh–Taylor instability in the pusher region

The well known Rayleigh–Taylor (RT) instability occurs when two fluids of different densities are placed in contact and are accelerated in a direction normal to their interface and directed towards the denser fluid.

The linear growth rate for this instability is given by

$$\gamma = \sqrt{A_t \cdot k \cdot a} \tag{1}$$

where

$$A_t = (\rho_2 - \rho_1)/(\rho_2 + \rho_1) \tag{2}$$

is the Atwood Number across the interface, k is the perturbation wave number and a is the acceleration.

The perturbation amplitude at time t is given by

$$\eta = \eta_0 e^{(\gamma \cdot t)} \tag{3}$$

4.1.1 Annular focal spot

The driver energy is deposited in part of the gold shell which expands, producing a low density region. The pusher part of the gold shell, on the other hand, is compressed to super solid densities due to shock compression and is accelerated by the high pressure in the beam heated region which may lead to the growth of RT instability.

In the case described in Sect. 3.1, pusher acceleration in the implosion phase starts at $t = 65$–70 ns when the shock emerges from the Au pusher and enters into the H. Since this moment, until the beginning of the deceleration phase that starts at $t = 120$ ns, the pusher acceleration is almost constant and is equal to $a = 6 \times 10^{12}$ cm s^{-2}. On the other hand, because of the absorber expansion, a density gradient is created at the interface pusher/absorber which has a minimum density scale length $L = \rho/(d\rho/dr)$ practically constant during the implosion and approximately equal to $L = 0.015$ cm (see Fig. 4(a)). These respective values of a and L allow us to estimate the maximum growth rate for the RT instabilities as:

$$\gamma_{max} = \sqrt{a/L} = 2 \times 10^7 \mathrm{s}^{-1}$$

During the implosion phase that lasts for a time interval, $\Delta t = 50$–70 ns the maximum number, n of e-foldings will be $n = \gamma_{max}\, \Delta t = 1$–$1.5$, and the perturbation amplitude η can not grow beyond a factor of 3 to 5 times its initial value η_0: $\eta_{max}/\eta_0 = \exp(\gamma_{max}\Delta t) = 2.7$–$4.5$. This means that if the initial perturbations are less than 1% and at the end on the implosion phase the interface has imploded to a half of its initial radius, the final symmetry level will be better than 5 to 10%.

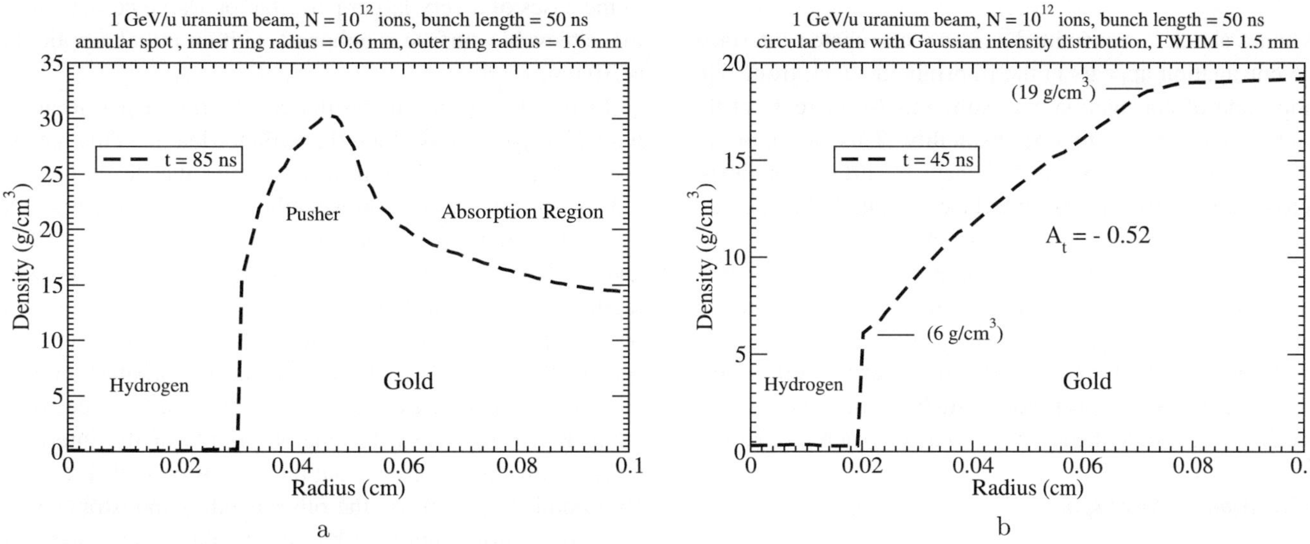

Fig. 4 (**a**) Density vs. cylinder radius at $t = 85$ ns using an annular focal spot, hydrogen radius = 0.4 mm, outer gold radius = 3 mm. (**b**) Density vs. cylinder radius at $t = 45$ ns using a circular focal spot, hydrogen radius = 0.4 mm, outer gold radius = 3 mm

It has been observed in the simulations that during the implosion phase the pusher remains mainly solid, although its internal layers are melted during the deceleration phase. This means that it will retain the elastic and plastic properties of the material. A detailed analytic study of the problem has shown that these effects will provide an additional stabilizing mechanism for the RT instability of the pusher during different phases of implosion (Piriz et al. 2005, 2006a).

Similar analysis has been carried out to investigate the Rayleigh–Taylor growth across the hydrogen–gold interface during the deceleration of the payload by the high pressure in the hydrogen. This analysis has shown that the target will remain stable during this final phase of implosion as well.

4.1.2 Circular focal spot

Since in the case described in Sect. 3.2, the focal spot is circular with a Gaussian transverse intensity distribution which has a FWHM = 1.5 mm, the material that lies within about 3 mm radius from the beam axis is heated. This includes the hydrogen and a part of the surrounding gold shell. The pressure in the heated gold is orders of magnitude higher than in hydrogen, a rarefaction wave therefore moves into the gold at the hydrogen–gold boundary. In Fig. 4(b) we plot the target density vs. radius at a longitudinal position, $L = 0.5$ cm at $t = 45$ ns. It is seen that the density profile in the gold region is such that denser part of gold with a density of 19 g/cm^3 accelerates material with a lower density of 6 g/cm^3. This configuration leads to a negative value of Atwood Number = −0.52, that results in a completely stable regime. The RT instability therefore does not pose any problems to this scheme.

4.2 Richtmyer–Meshkov instability at the H–Au interface

When a shock is transmitted across a corrugated interface, the corrugation appears in the transmitted shock front. The amplitude of corrugation may substantially increase at the contact surface due to RM instability. This problem can occur in the LAPLAS target if there is surface roughness across the hydrogen–gold interface in longitudinal or azimuthal direction. In case of a wobbler driven implosion, non-uniformities in energy deposition in azimuthal direction can also trigger this instability. It is also to be noted that this type of experiments can be used to study the problem of RM instability in cylindrical geometry in linear as well as nonlinear regime. Such experimental studies will be very useful because no analytic theory exists to model such problems.

4.2.1 Annular focal spot

In this case we consider an implosion driven by an annular focal spot with beam and target parameters described in Sect. 3.1. A corrugation on hydrogen–gold boundary in longitudinal direction is assumed. The cylinder length, $L = 4$ mm and we consider different perturbation wavelengths and amplitudes. For simplicity of numerical simulations, we assume that the facial boundaries of the target are fixed. The original target considered in LAPLAS calculations is 1 cm long and this approximation is therefore valid in the inner part of the target.

First we assume 4 wavelengths along the cylinder length which means $\lambda = 1$ mm that leads to a wavenumber, $k = 62.857$ cm^{-1}. Three different values for perturbation amplitude, A, namely, 0.1 μm, 1.0 μm and 10.0 μm, respectively, are considered. The respective values for the product kA become 6.28×10^{-4}, 6.28×10^{-3} and 6.28×10^{-2}. These numbers are very small which means that the instability will grow linearly throughout the implosion till the maximum compression is achieved at $t = 135$ ns. In Figs. 5(a) and 5(b) we plot radial position of the hydrogen–gold interface at $t = 0$, 80 ns and 135 ns, assuming $A = 0.1$ μm and 1 μm, respectively. It is seen that in both cases, the perturbation amplitude does not grow during implosion. Figure 5(c) represents the case with a larger initial perturbation amplitude, $A = 10$ μm. It is seen that at $t = 80$ ns, the amplitude of perturbation has been reduced substantially while at the time of maximum compression (at $t = 135$ ns) the amplitude has significantly increased, but has opposite phase compared to the initial perturbation. This is because when a shock propagates from a high-density (Au) medium to a low-density (H) medium, corrugated fronts in the expanding gold and hydrogen that is being compressed, deposit enough vorticity on the contact interface in such a way that it enhances the ripple growth in opposite direction from that at initial time (Meyer and Blewett 1972; Yang et al. 1994; Wouchuk 1994). However, the implosion is successful even in the cases of a very larger perturbation (large compared to the achievable surface roughness) with an initial amplitude of 10 μm.

In Fig. 6 we present results considering a perturbation with 10 wavelengths along the cylinder length. This means a perturbation wavelength, $\lambda = 0.4$ mm that leads to $k = 1.57 \times 10^2$ cm^{-1}. The corresponding values for the product kA for the above values of A are 1.57×10^{-3}, 1.57×10^{-2} and 1.57×10^{-1}. First two cases represent linear growth regime while the third one corresponds to nonlinear growth of the instability. The results are plotted in Fig. 6(a)–(c), respectively. Figure 6(a) shows that the perturbation amplitude does not increase and the compression is successful while Fig. 6(b) shows that the amplitude of the interface corrugation slightly grows, but the implosion is still reasonably good. Fig. 6(c), on the other hand, demonstrates that the perturbation amplitude has substantially increased that makes the compression to fail. It is interesting to note that the hydrogen is still compressed to very high densities but

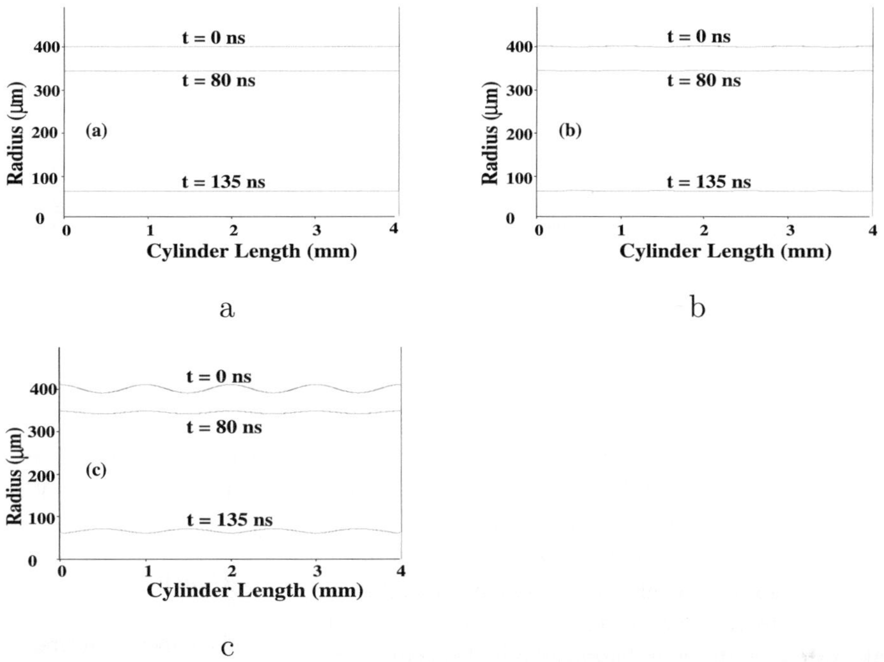

Fig. 5 Radial position of hydrogen–gold interface at three different times using $\lambda = 1$ mm; (**a**) $A = 0.1$ μm, (**b**) $A = 1.0$ μm, (**c**) $A = 10.0$ μm, imploded by annular focal spot

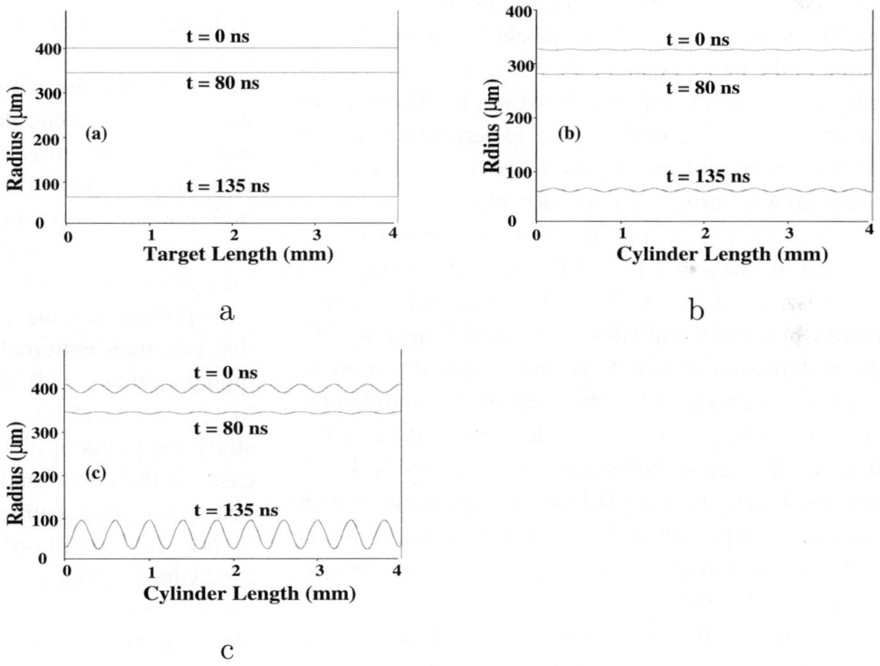

Fig. 6 Radial position of hydrogen–gold interface at three different times using $\lambda = 0.4$ mm; (**a**) $A = 0.1$ μm, (**b**) $A = 1.0$ μm, (**c**) $A = 10.0$ μm, imploded by annular focal spot

the sample cannot be diagnosed because of the gold fingers that enter the hydrogen.

The above study shows that the LAPLAS implosion will not be affected by R–M instability for small perturbation amplitudes, $a \leq 0.1$ μm for medium and large wavelengths compared to the target length. However, we note that in these analysis, we have considered an ideal fluid whereas in practice, one has non-ideal materials with constitutive properties like surface tension and viscosity for liquids, and shear modulus and yield strength for solids. The payload remains in a highly compressed solid state during the implosion and therefore retains its solid properties. Work on these difficult problems is in progress (Piriz et al. 2005, 2006a, 2006b, 2008) and we have strong indications that very short wavelengths will be stabilized due to these material properties.

4.2.2 *Circular focal spot*

In this section we use a circular focal spot to drive the implosion with beam and target parameters described in Sect. 3.2. In this case the maximum compression occurs

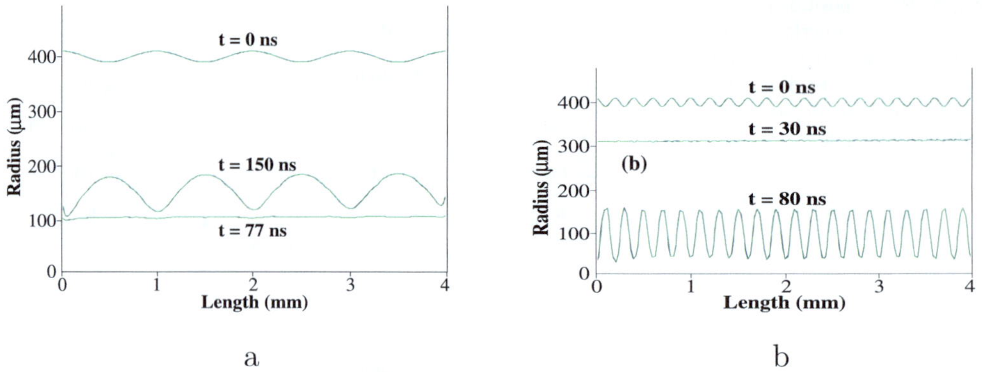

Fig. 7 Radial position of hydrogen–gold interface at three different times using $A = 10$ μm; (**a**) $\lambda = 1$ mm, (**b**) $\lambda = 0.2$ mm imploded by a circular focal spot

at $t = 80$ ns. In Fig. 7(a) we plot the radial position of hydrogen–gold boundary at three different times. A perturbation with $\lambda = 1$ mm and $A = 10$ μm is considered in longitudinal direction. The initial position of the interface is shown at $t = 0$ ns. As the interface moves inwards, it is seen that at $t = 77$ ns (at time of maximum compression), due to phase inversion, the perturbation amplitude becomes zero which means that in this particular case, the instability will be completely ameliorated. Curve labeled with $t = 150$ ns shows that as the target expands, the corrugation grows substantially with an inverted phase. However, by this time, all measurements should be made and the experiment would be over. Perturbations with smaller amplitude do not show any significant growth during the implosion phase.

In the results plotted in Fig. 7(b), we consider $\lambda = 0.2$ mm that means $k = 3.14 \times 10^3$ cm^{-1}. Assuming $A = 10$ μm the product $kA = 3.14$ which would lead to a non-linear growth of the perturbation. It is seen from Fig. 7(b) that the perturbation amplitude becomes zero at $t = 30$ ns due to phase inversion and at the time of maximum compression, at $t = 80$ ns, the amplitude grows substantially, which shows that the compression is not successful. However, for smaller amplitudes ≤ 0.1 μm, no significant growth of perturbation amplitude is observed and the implosion is stable. It is also seen that the phenomenon of phase inversion occurs faster for higher modes.

Figure 8 presents the cross sectional view of the target at three different times during the implosion. In this case we assume 10 wavelengths over the circumference of the interface (radius = 0.4 mm that leads to $k = 2.5 \times 10^2$ cm^{-2}) and $A = 1.0$ μm so that the product $kA = 2.5 \times 10^{-2}$ that represents a linear regime. However as a result of cylindrical convergence, kA increases essentially as a consequence of mass conservation [Bell–Plesset effect (Bell 1951; Plesset 1954)] and one finally enters a non-linear regime. Fig. 8 shows that at the time of maximum compression typical fingers of RM instability develop which destroy the compression. However, for $A \leq 0.1$ μm the instability does not show any significant development. Moreover, for smaller modes, larger values of A are tolerable. We also expect that the material properties of the solid payload will stabilize the very large wavenumbers.

5 Conclusions

This paper describes an experimental set up suitable to achieve a low-entropy compression of a sample material like hydrogen that is enclosed in a cylindrical shell of a high-Z material like gold or lead. The target can either be driven by a hollow beam with an annular focal spot or by a normal beam with a circular focal spot. In the former case, the hydrogen is not directly irradiated by the projectile particles but only part of the surrounding shell is heated. The high pressure in the heated region launches a shock wave into hydrogen that undergoes multiple reflections between the cylinder axis and the hydrogen–gold boundary. Such a compression scheme leads to a low-entropy compression that generates extremely high densities (1–3 g/cm^3), ultra-high pressures (3–30 Mbar) and relatively low temperatures (a few thousand K). This scheme is therefore suitable to study the problem of hydrogen metallization. In the latter case, on the other hand, hydrogen is directly heated by the beam, but the pressure in the heated part of the surrounding high-Z shell is still orders of magnitude higher than that in the hydrogen. One therefore achieves high compression ratio and very high pressures even in this case, but the temperature now is much higher (a few eV). These are the physical conditions that are expected to exist in the interiors of the giant planets. For this reason, this scheme is named LAPLAS which stands for **LA**boratory **PLA**netary **S**cience.

We have also carried out detailed analysis of hydrodynamic stability of the LAPLAS target. Our calculations have shown that the LAPLAS implosion is very stable against development of Rayleigh–Taylor instability in the high-Z shell for both cases of the focal spot geometry, during the acceleration as well as the deceleration phase. It has also been found that if there is a corrugation with an amplitude of ≤ 0.1 μm, the implosion is stable for medium and low wavenumber, whereas very high wave numbers may be stabilized due to the material properties of the solid payload.

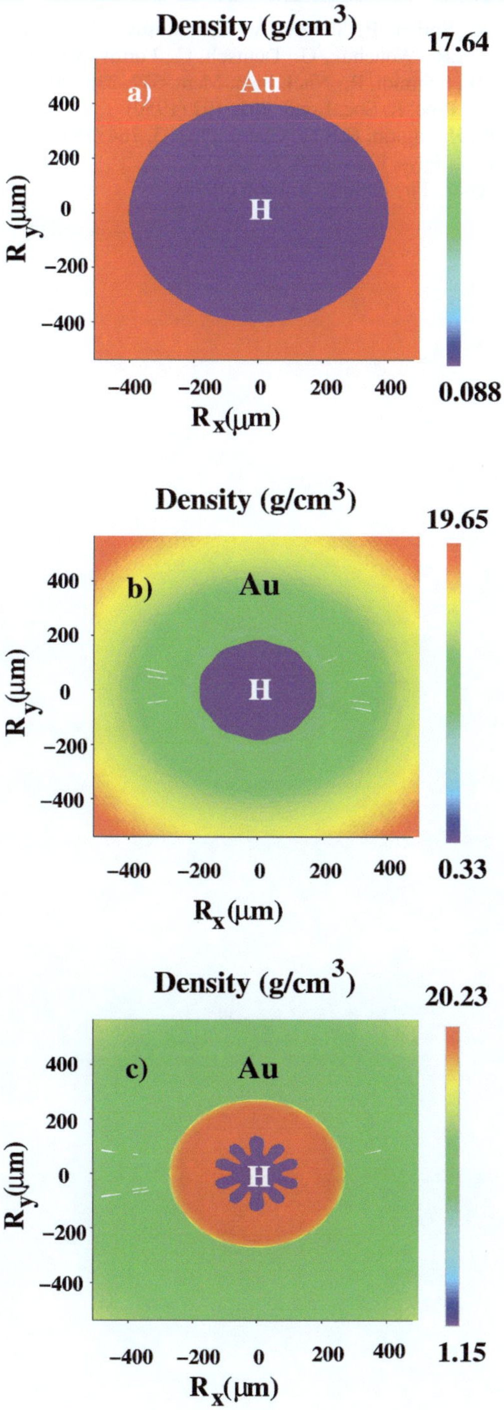

Fig. 8 Cross sectional view of the target, (**a**) at $t = 0$ ns, (**b**) $t = 50$ ns, (**c**) $t = 80$ ns imploded by circular focal spot; $\lambda = 2.5 \times 10^{-2}$ cm, $A = 1$ μm

An important outcome of this work is that in such experiments, one can study the interesting problem of RM instability in cylindrical geometry. Such studies will be very useful as no analytic models exist to treat such problems.

References

Bell, G.I.: Los Alamos National Laboratory Rep. La-1321 (1951)

Caubel, R., Celliers, P.M., Collins, G.W., Da Silva, L.B., Gold, D.M., Foord, M.E., Budil, K.S., Wallace, R.J.: Astrophys. J. Suppl. Ser. **127**, 267 (2000)

Celliers, P.M., Collins, G.W., Da Silva, L.B., Gold, D.M., Cauble, R., Wallace, R.J., Foord, M.E., Hammel, B.A.: Phys. Rev. Lett. **84**, 5564 (2000)

Fortov, V.E., Goel, B., Munz, C.-D., Ni, A.L., Shutov, A.V., Vorobiev, O.Y.: Nucl. Sci. Eng. **123**, 169 (1996)

Henning, W.F.: Nucl. Inst. Meth. B **214**, 211 (2004)

Hoffmann, D.H.H., Fortov, V.E., Lomonosov, I.V., Mintsev, V., Tahir, N.A., Varentsov, D., Wieser, J.: Phys. Plasmas **9**, 3651 (2002)

Lomonosov, I.V.: Laser Part. Beams **25**, 567 (2007)

Lomonosov, I.V., Tahir, N.A.: Appl. Phys. Lett. **92**, 101905 (2008)

Mao, H.K., Hemley, R.J.: Rev. Mod. Phys. **66**, 671 (1994)

Meshkov, E.E.: Fluid Dyn. **4**, 101 (1969)

Meyer, K.A., Blewett, P.J.: Phys. Fluids **15**, 753 (1972)

Nellis, W.J., Mitchell, A.C., van Theil, M., Devine, G.J., Trainor, R.J., Brown, N.: J. Chem. Phys. **79**, 1480 (1983)

Neuner, U., Bock, R., Roth, M., Spiller, P., Constantin, C., Funk, U.N., Geissel, M., Hakuli, S., Hoffmann, D.H.H., Jacoby, J., Kozereva, A., Tahir, N.A., Udrea, S., Varentsov, D., Tauschwitz, A.: Phys. Rev. Lett. **85**, 4518 (2000)

Piriz, A.R., Portugues, R.F., Tahir, N.A., Hoffmann, D.H.H.: Phys. Rev. E **66**, 056403 (2002)

Piriz, A.R., Temporal, M., Lopez Cela, J.J., Tahir, N.A., Hoffmann, D.H.H.: Plasma Phys. Control. Fusion **45**, 1733 (2003a)

Piriz, A.R., Temporal, M., Tahir, N.A., Hoffmann, D.H.H., Temporal, M.: Phys. Rev. E **67**, 017501 (2003b)

Piriz, A.R., Lopez Cela, J.J., Cortazar, O.D., Tahir, N.A., Hoffmann, D.H.H.: Phys. Rev. E **72**, 056313 (2005)

Piriz, A.R., Lopez Cela, J.J., Serna Moreno, M.C., Tahir, N.A., Hoffmann, D.H.H.: Laser Part. Beams **24**, 275 (2006a)

Piriz, A.R., Lopez Cela, J.J., Tahir, N.A., Hoffmann, D.H.H.: Phys. Rev. E **74**, 037301 (2006b)

Piriz, A.R., Lopez Cela, J.J., Tahir, N.A., Hoffmann, D.H.H.: Richtmeyer–Meshkov instability in elastic–plastic media. Phys. Rev. E (2008, submitted)

Plesset, M.S.: J. Appl. Phys. **25**, 96 (1954)

Rayleigh, J.W.: Scientific Press. Dover, New York (1954)

Richtmeyer, R.D.: Commun. Pure Appl. Math. **13**, 297 (1960)

Ross, M.: Phys. Rev. B **54**, R9589 (1983)

Schollmeier, M., Harres, K., Nürnberg, F., Blazevich, A., Audebert, P., Brambrink, E., Fernandez, J.C., Flippo, K.A., Gautier, D.C., Geissel, M., Heglich, B.M., Schreiber, J., Roth, M.: Phys. Plasmas **15**, 053101 (2008)

Snavely, R.A., Key, M.H., Hatchett, S.P., Cowan, T.E., Roth, M., Philips, T.W., Stoyer, M.A., Henry, E.A., Sangster, T.C., Singh, M.S., Wilks, S.C., MacKinnon, A., Offenberger, A., Pennington, D.M., Yasuike, K., Langdon, A.B., Lasinski, B.F., Johnson, J., Perry, M.D., Campbell, E.M.: Phys. Rev. Lett. **85**, 2945 (2002)

Tahir, N.A., Hoffmann, D.H.H., Kozereva, A., Tauschwitz, A., Shutov, A., Maruhn, J.A., Spiller, P., Neuner, U., Jacoby, J., Roth, M., Bock, R., Juranek, H., Redmer, R.: Phys. Rev. E **63**, 016402 (2001)

Tahir, N.A., Juranek, H., Shutov, A., Redmer, R., Piriz, A.R., Temporal, M., Varentsov, D., Udrea, S., Hoffmann, D.H.H., Deutsch, C., Lomonoson, I.V., Fortov, V.E.: Phys. Rev. B **67**, 184101 (2003)

Tahir, N.A., Deutsch, C., Fortov, V.E., Gryaznov, V., Hoffmann, D.H.H., Kulish, M., Lomonosov, I.V., Mintsev, V., Ni, P., Nikolaev, D., Piriz, A.R., Shilkin, N., Spiller, P., Shutov, A., Temporal, M., Ternovoi, V., Udrea, S., Varentsov, D.: Phys. Rev. Lett. **95**, 035001 (2005a)

Tahir, N.A., Kain, V., Schmidt, R., Shutov, A., Lomonosov, I.V., Gryaznov, V., Piriz, A.R., Temporal, M., Hoffmann, D.H.H., Fortov, V.E.: Phys. Rev. Lett. **94**, 135004 (2005b)
Tahir, N.A., Adonin, A., Deutsch, C., Fortov, V.E., Grandjouan, N., Geil, B., Gryaznov, V., Hoffmann, D.H.H., Kulish, M., Lomonosov, I.V., Mintsev, V., Ni, P., Nikolaev, D., Piriz, A.R., Shilkin, N., Spiller, P., Shutov, A., Temporal, M., Ternovoi, V., Udrea, S., Varentsov, D.: Nucl. Inst. Meth. **544**, 16 (2005c)
Tahir, N.A., Spiller, P., Shutov, A., Lomonosov, I.V., Gryaznov, V., Piriz, A.R., Wouchuk, G., Deutsch, C., Fortov, V.E., Hoffmann, D.H.H., Schmidt, R.: Nucl. Instr. Meth. **577**, 238 (2007)
Taylor, G.I.: Proc. R. Soc. Lond. **201**, 192 (1950)
Wigner, E., Huntigton, H.B.: J. Chem. Phys. **3**, 764 (1935)
Wouchuk, G.: Phys. Plasmas **8**, 2890 (1994)
Yang, Y., et al.: Phys. Fluids **8**, 1666 (1994)

Astrophys Space Sci (2009) 322: 189–193
DOI 10.1007/s10509-009-9985-x

ORIGINAL ARTICLE

Magnetically accelerated foils for shock wave experiments

S. Neff · J. Ford · S. Wright · D. Martinez · C. Plechaty · R. Presura

Received: 11 June 2008 / Accepted: 12 January 2009 / Published online: 28 January 2009

Abstract Many astrophysical phenomena involve the interaction of a shock wave with an inhomogeneous background medium. Using scaled experiments with inhomogeneous foam targets makes it possible to study relevant physics in the laboratory to better understand the mechanisms of shock compression and to benchmark astrophysical simulation codes. First experiments on Zebra at the Nevada Terawatt Facility (NTF) have demonstrated flyer acceleration to sufficiently high velocities (up to 5 km/s) and that laser shadowgraphy can image sound fronts in transparent targets. Based on this, we designed an optimized setup to improve the flyer parameters (higher speed and mass) to create shock waves in transparent media. Once x-ray backlighting with the Leopard laser at NTF is operational, we will switch to foam targets with parameters relevant for laboratory astrophysics.

Keywords Flyer acceleration · Shock physics · Laboratory astrophysics

1 Introduction

Many astrophysical processes involve the interaction of a shock wave with an inhomogeneous background medium, e.g. the interaction of supernova remnants with the interstellar medium and the triggering of gravitational collapse by shock waves, resulting in star formation in gas clouds. Most laboratory studies have focused on the interaction of a shock wave with a single clump, yet in many instances the interaction with multiple clumps is essential (Remington et al. 2006).

Scaled laboratory experiments using foam targets with inhomogeneities have been proposed for the Z machine at Sandia (Drake 2002), using its large discharge current to magnetically accelerate a metallic flyer plate. Upon impact on a low density foam target this flyer generates a shock wave in the target and the dynamics of the shock can be studied. While the Z machine is unmatched in its discharge current (up to 20 MA), the high costs for experiments on Z impose strict limits on the number of experiments. Smaller pulsed-power machines, such as the Zebra generator at the Nevada Terawatt Facility (NTF), offer lower costs and a higher repetition rate. Experiments on a small machine can thus map a wide range of experimental parameters and are also useful as a testbed for experiments on Z.

A successful experiment requires sufficiently high flyer velocities (several kilometers per second), suitable foam targets with controlled inhomogeneity, and suitable diagnostics, including x-ray backlighting and velocity interferometry. We have carried out first experiments at NTF demonstrating flyer acceleration. These experiments are the focus of this paper. Future campaigns will demonstrate shock creation in transparent solids with low sound speed and study the interaction of a shock wave with clumps in these targets. Once x-ray backlighting is operational at NTF, we will use it to study shock waves in low density foams. In addition, we plan to install a VISAR diagnostic (Bloomquist and Sheffield 1983) to obtain more accurate velocity measurements for the flyers.

S. Neff (✉) · J. Ford · S. Wright · D. Martinez · C. Plechaty · R. Presura
University of Nevada, Reno, NV 89506, USA
e-mail: neff@physics.unr.edu

2 Acceleration mechanism

Figure 1 shows a schematic of our setup for flyer acceleration. The load is short-circuited and consists of a central rod and a current strip made of copper that carries the return current. The strip has holes for the flyer plates, which are glued to the side of the band facing the central electrode. In this configuration the current in the central conductor and the current through the flyer are anti-parallel, so that a $\mathbf{j} \times \mathbf{B}$ force accelerates the flyer away from the rod. Our setup is asymmetric, which increases the magnetic pressure that can be created by a given discharge current but has the disadvantage that the mechanical stress on the electrode is high. Due to the enormous forces acting on it and Ohmic heating, the copper strip disintegrates during the experiment. However, due to its high mass, this takes place on a much longer timescale than the flyer acceleration and thus does not influence the experiment. Nevertheless, care must be taken to protect windows and sensitive equipment from the resulting debris.

If the effect of flyer ablation is neglected, the total acceleration of the flyer can be calculated by integrating the $\mathbf{j} \times \mathbf{B}$ force over the volume of the flyer. The force acting on a small element of the flyer (mass density ρ, velocity $\mathbf{v}$, current density in the flyer $\mathbf{j}$, and magnetic field $\mathbf{B}$ at the position of the flyer) is given by

$$\rho \frac{d\mathbf{v}}{dt} = \mathbf{j} \times \mathbf{B} \qquad (1)$$

Since both the current density and the magnetic field are proportional to the total current I, the total force acting on the flyer is proportional to I^2. The current density, magnetic

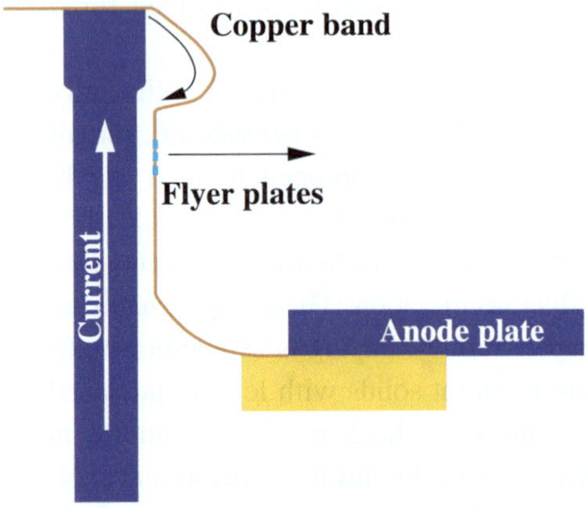

Fig. 1 Schematic of the short-circuited load used in our flyer experiments. The plot shows the x–z plane. The current flows through the central rod and the copper band to the anode plate. A $\mathbf{j} \times \mathbf{B}$ force accelerates the flyer outwards

field and total force have been calculated numerically for our setup. The flyer velocity is given by Lemke et al. (2003)

$$v_{\mathrm{fl}} = \frac{\mu_0}{2\rho_0 \, D \, S^2} \int_0^{t_{\max}} I^2(\hat{t}) d\hat{t} \qquad (2)$$

where S is a scale factor, that depends on the load geometry, defined by $B = \mu_0 \, I(t)/S$ at the location of the target, D the flyer thickness, and $I(t)$ the current flowing through the flyer. As can be seen from (2), in this simplified model the final flyer velocity is inversely proportional to the flyer mass per unit area. The biggest uncertainty in this calculation results from estimating the acceleration time $t_{\max}$. A self-consistent modelling of the flyer acceleration requires a magnetohydrodynamical simulation, which also takes ablation into account.

The current in the flyer flows only in a thin surface layer on the side facing the central conductor; its thickness is roughly given by the skin depth (Jackson 1998)

$$\delta_{\mathrm{skin}} = 1/\sqrt{\mu_0 \, \sigma \, \omega} \qquad (3)$$

where μ_0 is the vacuum permeability, σ the conductivity of the flyer, and ω the frequency of the current. Therefore the current density in the return conductor and the flyer is approximately (assuming that they have similar conductivities and that they are much thicker than the skin depth)

$$j = I/(\delta_{\mathrm{skin}} \, W) \qquad (4)$$

where W is the width of the return conductor. Like the current density, the acceleration of the flyer does not depend on its height (as long as variations of the magnetic field are small). The acceleration is inversely proportional to the width of the return conductor and the thickness of the flyer enters only through its total mass. Since the total current going through the flyer is independent of its thickness, the acceleration of the flyer is inversely proportional to its thickness.

Besides accelerating the flyer, the current also leads to Ohmic heating of the flyer and ablation of material. Depending on the energy deposited in the skin layer, the layer might melt or even turn into plasma. It is therefore important to choose a flyer that is thick enough so that the front of the flyer (the side facing the target) is still solid. While only the skin layer is heated directly, heat conduction leads to temperature increase and ablation in other parts of the flyer. For a part of the flyer to be solid upon impact, the heat diffusion must take place on a longer timescale than the time to impact.

3 Experiment

The pulsed-power generator Zebra at the NTF was originally built and operated at the Los Alamos National Laboratories for a wire pinch experiment (HDZP-II (Shlachter 1990)) before it came to the University of Nevada, Reno. It provides a maximum current of 1 MA with a current rise time of 90 ns. The current is measured with Bdot probes as shown in Fig. 2. A measured curve is shown in Fig. 3. The load inductance is about 6 nH, whereas the inductance of the entire load region is of the order of 60 nH, so that the discharge current is insensitive to minor changes of the load configuration.

Figure 2 shows the setup of our experiment in the Zebra load chamber. The load is mounted at the center of the chamber; the debris shields protect a port window from the flyers and debris from the copper strip. We used a 532 nm laser with a pulse duration of 0.2 ns to image the load a few microseconds after the current maximum. As the acceleration phase is much shorter (of the order of 100 ns), the velocity of the flyer can be estimated by measuring the distance traversed by the flyer. The error in this measurement

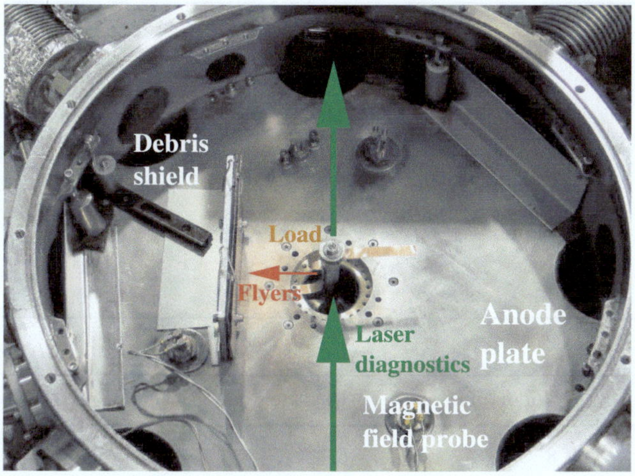

Fig. 2 The experimental setup for our flyer velocity measurements. Shown is the open Zebra vacuum chamber. The load is in the center and the flyers in this setup are accelerated towards the left-hand side. A green laser is used to image the flyers

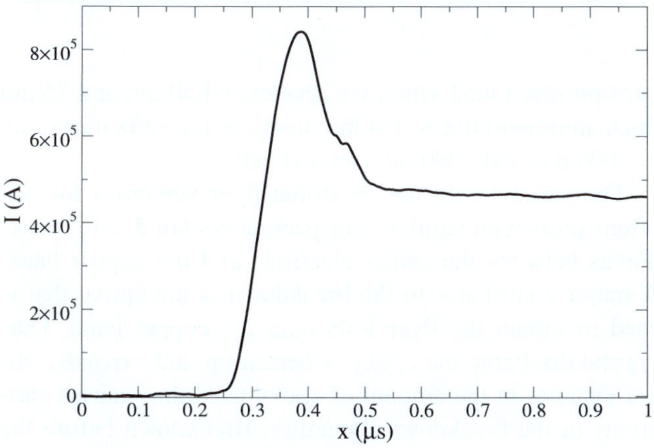

Fig. 3 Measured Zebra current for a flyer shot

is determined by the uncertainty in the time of release of the flyer. Typical imaging delays are of the order of 2 µs after the current peak; assuming an uncertainty in the time of release of 200 ns, corresponding to a measurement error of 10%. Figure 5 shows a typical shadowgram image of a setup with two flyers with different thickness accelerated in parallel. The diagnostics images the flyer front and the ablated material behind it. It cannot determine if the front of the flyer is still solid; future campaigns will use a VISAR diagnostic (Bloomquist and Sheffield 1983), which can determine the state of flyer front by measuring the reflectivity and will in addition provide a highly accurate, time resolved velocity measurement.

Two campaigns measuring flyer velocities have been carried out so far. The first campaign used a simple load configuration and was a proof-of-principle experiment. Based on the results from this campaign, an optimized load was designed for the second campaign to increase the flyer velocities.

The first load design consisted of a cylindrical stainless steel rod (diameter 15 mm) as the central conductor and a 10 mm wide copper strip acting as the return current conductor. The copper band used had 2 mm diameter holes drilled in it, with aluminum flyers glued to the inside of the copper band (facing the central rod) covering the holes. Up to three flyers were installed in this fashion, making it possible to carry out multiple measurements in each shot. The magnetic field profile and the resulting were calculated for a simplified model with the numerical software package FEMLAB (COMSOL 2004), using the COMSOL Multiphysics module using a harmonic analysis for a simplified, two-dimensional setup, yielding the current density distribution and magnetic field profile. By using a two-dimensional model we neglect end effects and assume that the flyer is a thin strip (instead of the circular flyer used in the experiment). The harmonic analysis uses a sinusoidal current profile $I_0 \cdot \sin(2\pi \nu t)$. For the calculation we assumed a current amplitude of 1 MA and a frequency of 10 MHz. In the model, the flyer was 20 µm thick and placed 1.5 mm from the surface of the central conductor. The width of the flyer is 2 mm and the surrounding copper bands have a width of 4 mm and a thickness of 1 mm. Figure 4 shows the calculated current density and contour lines for the magnetic field. The calculated magnetic field at the flyer surface is 105 T, resulting in an accelerating $\mathbf{j} \times \mathbf{B}$ force per unit length of 4×10^6 N/m at the peak current.

The second load design used a center conductor with a flat surface (10 mm wide) facing an 8 mm wide copper strip. The diameter of the flyers was increased to 6.5 mm. The modified design resulted in an increased magnetic field of 115 T and an increased current density, resulting in an accelerating force of 10^7 N/m at the peak of the current.

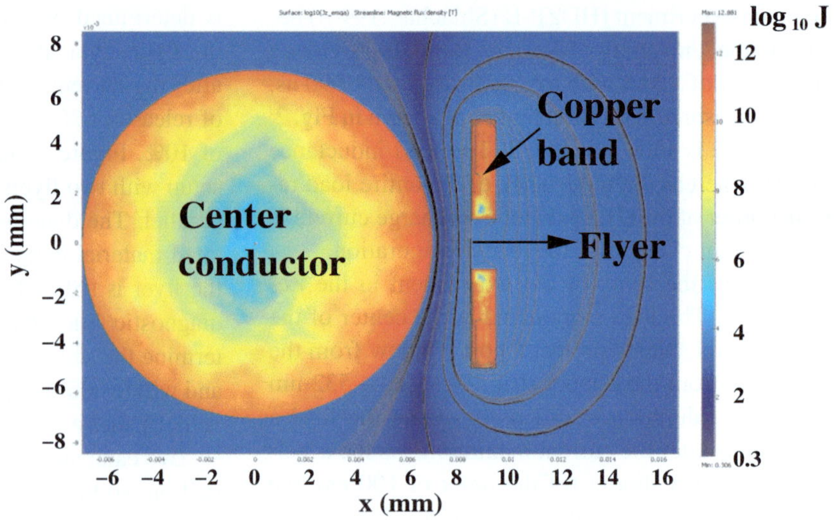

Fig. 4 Calculated current densities and magnetic field contours. The current density is plotted logarithmically, ranging from $10^{0.3}$ A/m^2 (*dark blue*) to 10^{12} A/m^2 (*dark red*). The non-zero current density between the electrodes is due to numerical error. The plot shows a cut through the load; the z-axis is perpendicular to this plane

Table 1 Measured flyer velocities in the original setup. Listed are the flyer thickness, the distance between the flyer and the center electrode, the width of the copper band, the diameter of the flyer, and the measured flyer speed

Thickness (μm)	Δx (mm)	Width (mm)	Diameter (mm)	Speed (m/s)
20	1.3	12	2.0	2300
20	1.3	12	2.0	2000
50	1.3	12	2.0	1300
50	1.4	10	2.0	1700
130	1.4	10	2.0	800
130	1.4	11	2.0	700
130	1.7	10	2.0	800
130	1.7	10	2.0	900

Table 2 Measured flyer velocities in the modified setup. The listed quantities are identical to Table 1

Thickness (μm)	Δx (mm)	Width (mm)	Diameter (mm)	Speed (m/s)
50	2.4	8	6.4	5600
50	2.4	8	6.4	3000
50	1.6	8	5.6	4800
75	2.4	8	6.4	2200
75	2.4	8	6.4	3000
75	1.6	8	5.6	2900

4 Results and discussion

In the first campaign, we accelerated aluminum flyers (2 mm diameter) with thicknesses of 20 μm, 50 μm, and 130 μm, achieving speeds of up to 2300 m/s, 1700 m/s, and 900 m/s, respectively (see Table 1). In addition to the velocity measurements we also carried out proof-of-principle experiments that impacted the flyers onto a transparent plastic target. While the flyer velocities were insufficient to create a shock wave, we were able to image a sound front in the target, thus demonstrating the suitability of our laser diagnostics for studies of shock waves in transparent targets with low sound speeds. In the second campaign (Table 2), using our optimized load setup, we accelerated 50 μm and 75 μm thick aluminum flyers (6.4 mm diameter) to velocities of up to 5600 m/s and 3000 m/s, respectively.

The variations in the measured flyer velocities for different shots with similar flyer parameters are due to breakdowns between the center electrode and the copper band. A major contributor to the breakdown is the epoxy that is used to mount the flyer foils onto the copper band. During the discharge the epoxy is heated up and expands. Minor changes in the amount of epoxy used thus causes variations in the breakdown dynamics. Breakdown before the flyer loses electrical contact with the copper band results in reduced terminal velocities. To minimize the risk of break-

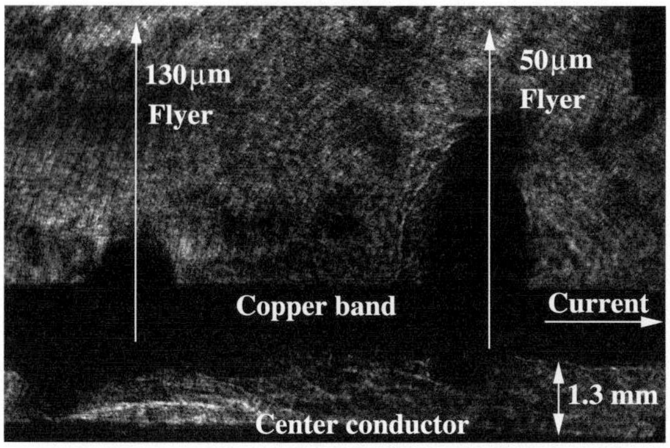

Fig. 5 Shadowgraphic image 2.5 μs after the current peak.The image is shown sideways. The z-axis is horizontal, while the x-axis is vertical. The edge of the central rod is visible at the bottom of the image. Above, separated by a distance of 1.3 mm, is the copper strip. In the setup shown in the image, two aluminum flyers were mounted on the copper band, with thicknesses of 50 μm and 130 μm, respectively

downs and improve the electrical contact, we will use conducting epoxy in future campaigns. This will result in more efficient and more reproducible flyer acceleration.

The experiments so far have demonstrated the possibility to accelerate flyers to velocities high enough for shock experiments in foam targets. These experiments will require a knowledge of the state of the flyer upon impact. For this purpose, future experiments will use velocity interferometry to determine the flyer velocity more accurately and time resolved; the reflectivity of the flyer will also show if its front is still solid. A multi-spot measurement could be used to study the flyer flatness.

Foam target experiments will also require x-ray backlighting for diagnostics. Backlighting for Zebra using the 50 TW short-pulse laser Leopard is currently being implemented. Once both the velocity interferometry and the backlighting are operational it will be possible to carry out shock wave studies that complement the experiments at bigger facilities like the Z machine.

Acknowledgement This work was supported by DOE/NNSA under UNR grant DE-FC52-06NA27616.

References

Bloomquist, D.D., Sheffield, S.A.: Optically recording interferometer for velocity measurements with subnanosecond resolution. J. Appl. Phys. **54**, 1717–1722 (1983)

COMSOL, FEMLAB 3. COMSOL (2004)

Drake, R.P.: Design of flyer-plate-driven compressible turbulent mix experiments using z. Phys. Plasmas **9**, 3545–3551 (2002)

Jackson, J.D.: Classical Electrodynamics. Wiley, New York (1998)

Lemke, R.W., Knudson, M.D., Hall, C.A., Haill, T.A., Desjarlais, P.M., Asay, J.R., Mehlhorn, T.A.: Characterization of magnetically accelerated flyer plates. Phys. Plasmas **10**, 1092–1099 (2003)

Remington, B.A., Drake, R.P., Ryutov, D.D.: Experimental astrophysics with high power lasers and z pinches. Rev. Mod. Phys. **78**, 755–807 (2006)

Shlachter, J.: Solid D2 fiber experiments on HDZP-II. Plasma Phys. Control. Fusion **32**(11), 1073–1081 (1990)

Astrophys Space Sci (2009) 322: 195–199
DOI 10.1007/s10509-009-9997-6

ORIGINAL ARTICLE

Penetration of plasma across a magnetic field

C. Plechaty · R. Presura · S. Wright · S. Neff · A. Haboub

Received: 15 June 2008 / Accepted: 30 January 2009 / Published online: 14 February 2009

Abstract Experiments were performed at the Nevada Terawatt Facility to investigate the plasma penetration across an externally applied magnetic field. In experiment, a short-pulse laser ablates a polyethylene laser target, producing a plasma which interacts with an external magnetic field. The mechanism which allows the plasma to penetrate the applied magnetic field in experiment will be discussed.

Keywords Plasma penetration · Strong magnetic field · Laser plasma · Laboratory astrophysics · Instabilities · Magnetic Rayleigh-Taylor instability · Electromagnetic flute instability

1 Introduction

The plasmas considered in this paper are sufficiently conductive to affect the topology of the magnetic field. The mechanism which allows such a plasma to penetrate through a magnetic field is an important topic of study with a wide variety of applications, for example the interaction of the solar wind with the Earth's magnetic field. Several explanations exist for the mechanism which allows the solar wind to penetrate the Earth's magnetosphere. Some of these explanations include: magnetic reconnection (Dungey 1961), diffusion (Axford and Hines 1961), impulsive penetration (most recently reviewed in Echim and Lemaire 2000), and Rayleigh-Taylor instabilities of the magnetopause (Gratton et al. 1996).

The role of instabilities in the plasma penetration across a magnetic field was investigated in the laboratory at the Nevada Terawatt Facility (NTF). The interaction of a plasma flow (produced by ablating a polyethylene target with a short-pulse laser) with an external magnetic field (created by a pulsed-power generator) was investigated. In Sect. 2, the experiment will be presented. In Sect. 3, the magnetohydrodynamic simulations performed to estimate some of the plasma parameters in experiment will be described. In Sect. 4 the role of instabilities in the plasma penetration will be discussed.

2 Experiment

An experiment was performed to investigate the interaction of a laser-produced plasma with an external magnetic field created by a pulsed power generator. A schematic of the experiment setup is shown in Fig. 1. A current pulse rising to 0.6 MA in 200 ns was driven by a 2 TW pulsed power generator through a cylindrical electrode to produce an azimuthal magnetic field. A polyethylene laser target was placed at 5–7 mm from the surface of the electrode, where the field

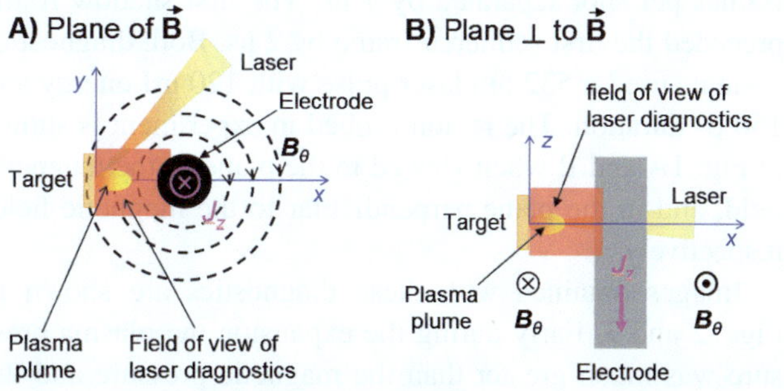

Fig. 1 Experimental setup used in Zebra, as viewed (**A**) in the plane parallel to the magnetic field, (**B**) in the plane perpendicular to the magnetic field

C. Plechaty (✉) · R. Presura · S. Wright · S. Neff · A. Haboub
Nevada Terawatt Facility, University of Nevada, Reno, USA
e-mail: plechaty@physics.unr.edu

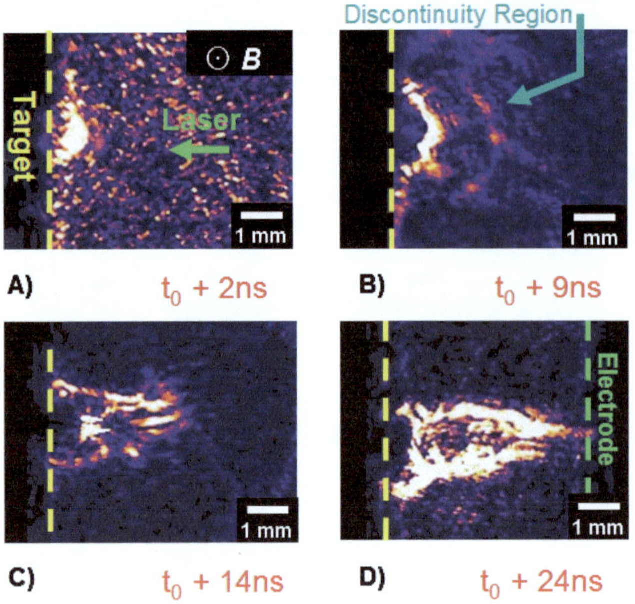

Fig. 2 Schlieren images taken in the plane perpendicular to the magnetic field. The ablation laser hits the target at the moment t_0. (**A**) Initially, the plasma expands hemi-spherically. (**B**) As the plasma is decelerated by the magnetic field, the density distribution shows the formation of a discontinuity region. (**C**) The plasma boundary becomes unstable. (**D**) Plasma penetrates the field. Frames **A** and **B** were obtained in the same shot, while **C** and **D** were obtained in different shots. Lighter areas denote larger plasma gradients

strength was up to 17 T. The target was ablated by a 5 ps laser pulse with 2–4 J and irradiance up to 10^{16} W/cm^2. Since the laser pulse duration was much shorter than the plasma evolution time, in the absence of a magnetic field the plasma expanded quasi-adiabatically.

A variety of diagnostics were used in this experiment. The magnetic field in vacuum was measured with inductive probes (B-dot). The laser-produced plasma was observed with diagnostics based on the plasma refractive index. Laser shadowgraphy was used to investigate the evolution of the critical plasma density region. Schlieren diagnostics were employed to image the plasma density gradients and to measure the plasma plume velocity and deceleration. Both shadow and schlieren diagnostics each allowed two frames per shot separated by 7 ns. The first shadow frame preceded the first schlieren frame by 2 ns. Both diagnostics systems used a 532 nm laser pulse with 120 mJ energy and 150 ps duration. The region probed in experiment is shown in Fig. 1A and B when viewed in the plane of the magnetic field, and in the plane perpendicular to the magnetic field, respectively.

Images obtained with these diagnostics are shown in Figs. 2 and 3. Early during the expansion the plasma pressure was much greater than the magnetic pressure and the ablation plume expanded hemi-spherically (Fig. 2A). As the plasma continued to expand, it was decelerated by the magnetic field, forming a discontinuity region with increased

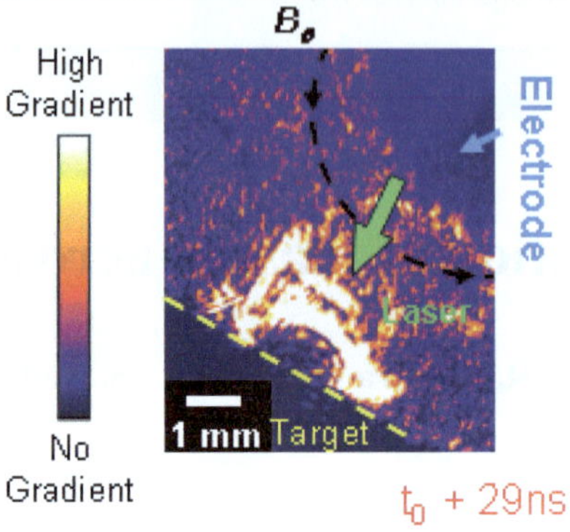

Fig. 3 Schlieren image taken in the plane of the magnetic field

density (Fig. 2B). Shortly after the formation of the discontinuity, flute-like structures were observed (Fig. 2C). The structures were determined to be two-dimensional plasma sheaths since they were not observed in schlieren images taken in the plane of the magnetic field (Fig. 3).

Identifying the mechanism which leads to the formation of the flute-like structures and eventually to the plasma penetration through the magnetic field (Fig. 2D) requires knowledge of the plasma parameters. For the discontinuity region, several plasma parameters were determined from the laser diagnostics and others were estimated by using a resistive, radiation magnetohydrodynamics (MHD) code, as indicated in Table 1.

3 MHD modeling

A one-dimensional, two-temperature MHD simulation was performed using the Magneto-Hydro-Radiative-Dynamics Research (MHRDR) code (Lindemuth 1979) to determine the conditions present in the discontinuity region for times relevant to the onset of the growth of the flute-like structures. An Eulerian grid with cylindrical symmetry was used. The inital simulation geometry is shown in Fig. 4. The computational domain consisted of two regions, the laser target region, and the expansion region. The polyethylene target material was treated with an ideal gas equation of state.

In the target region, the simulation was initiated with polyethylene at solid state density (1000 kg/m^3) and both the electrons and ions at room temperature (0.026 eV). Based on hydrodynamic simulations with the code MULTI (Ramis et al. 1988), in the MHRDR simulation the laser energy was considered to be deposited in a 13 µm thick region adjacent to the expansion region. The electron temperature was considered to decrease linearly in the target bulk from a maximum value of 22 eV. The expansion region was modeled as initially containing a polyethylene gas with a density of 10^{-9} kg/m^3 corresponding to a residual pressure $p = 6 \times 10^{-6}$ Torr. The simulations included a magnetic

Table 1 Plasma parameters for the discontinuity region 10 ns after target ablation

Parameter name	Symbol	Value	Derived from
Scale length (mm)	L	0.2	Schlieren imaging
Velocity (m/s)	u	1×10^5	Schlieren imaging
Deceleration (m/s^2)	g	$0.5–1 \times 10^{13}$	Schlieren imaging
Perturbation wavelength (mm)	λ	0.2–0.4	Schlieren imaging
Magnetic field strength (T)	B	10–20	B-Dot measurements
Electron number density (cm^{-3})	n_e	$0.5–1 \times 10^{19}$	MHRDR simulations
Electron temperature (eV)	T_e	30–45	MHRDR simulations
Ion temperature (eV)	T_i	25–45	MHRDR simulations

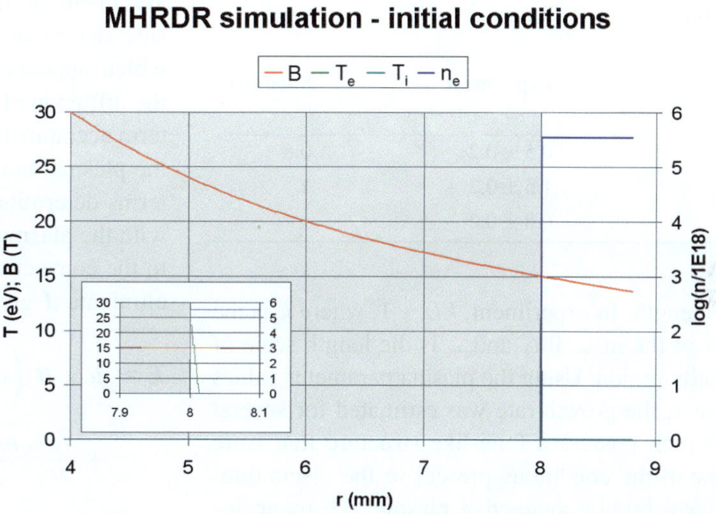

Fig. 4 Initial conditions used in the one-dimensional magnetohydrodynamic calculations with the code MHRDR. The *shaded area* indicates the expansion region and the non-shaded area the laser target region. The *inset* shows the initial electron temperature distribution in the heated region of the target

field with a $B \propto 1/r$ dependence, oriented perpendicular to the plane of calculation, and with a strength at the target surface in the range 10–20 T which included the experimental values.

As in experiment, in the simulation the heated region expanded and was decelerated by the magnetic field, forming a discontinuity region. An example of magnetic field, density and temperature distributions after 10 ns of expansion is shown in Fig. 5. The phase velocity of the critical density region in simulation matches that observed in experiment, as measured with shadow diagnostics. Additionally, the phase velocity of the discontinuity region in simulation matches the experimental value, measured by tracking the position of the high density gradient region with schlieren diagnostics. The MHRDR simulations were used to estimate the plasma density as well as the electron and ion temperature in the discontinuity region at the time the flute-like structures were observed in experiment (Fig. 2C). The parameters estimated with MHRDR are listed in Table 1 along with those measured in the experiment for the discontinuity region.

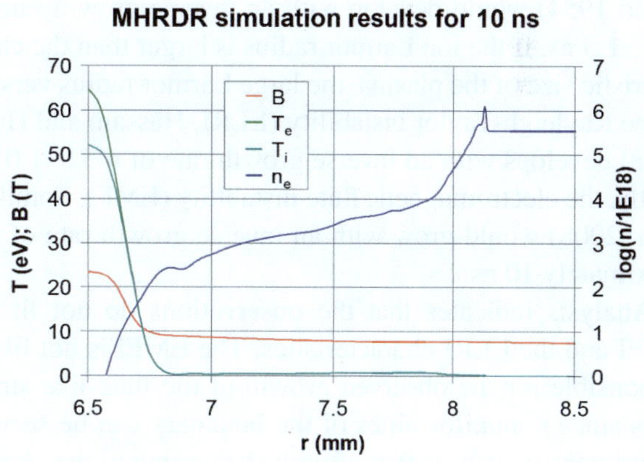

Fig. 5 Magnetic field, electron and ion temperature, and electron number density after 10 ns of expansion calculated with MHRDR for the initial conditions mentioned in the text

4 Results and discussion

The laser diagnostics indicated the formation of flute-like structures at the plasma—magnetic field interface, which ultimately allowed the plasma to penetrate the magnetic field (Figs. 2 and 3). The instability was initially seeded by imperfections in the laser beam profile or on the laser target surface. The instability can be classified using the growth

Table 2 Values of the coefficients in (4.2) and dimensionless plasma parameters estimated for the plasma parameter values given in Table 1. The magnetic Reynolds number is calculated using the Spitzer resistivity. β is the ratio between the plasma pressure and the magnetic pressure, $\rho_{i,d}$ is the ion gyroradius, and c/ω_{pi} is the ion collisionless skin depth. The solar wind parameters used: $n_i = n_e = 5 \times 10^6\ \mathrm{m}^{-3}$, $T_i = 10$ eV, $T_e = 20$ eV, $L = 1 \times 10^5$ m, $u = 4 \times 10^5$ m/s, and $B = 10$ nT

Coefficient	Experiment	Solar wind
Convective term	1	1
Diffusion term	1 ± 0.4	5×10^{-10}
Hall term	1 ± 0.5	0.25
Diagmagetic term	0.1 ± 0.05	0.05
Dimensionless plasma parameter	Experiment	Solar wind
β	0.5 ± 0.2	0.6
$\rho_{i,d}/L$	1.5 ± 0.2	1
$(c/\omega_{pi})/L$	0.8 ± 0.2	1

rate and wavelength. In experiment, $kL \sim 1$, where k is the wave number of the instability and L is the length scale of the discontinuity region. Using the plasma parameter values listed in Table 1, the growth rate was estimated for several instabilities which possess a flute-like structure that were likely to grow in the conditions present in the discontinuity region. For a highly conductive plasma, the magneto-Rayleigh-Taylor instability (MRTI, Kruskal and Schwarzschild 1954) would develop with an inverse growth rate of 3.6 ± 1.5 ns. If the ion Larmor radius is larger than the characteristic size of the plasma, the large Larmor radius version of the Rayleigh-Taylor instability (LLRI, Hassam and Huba 1988) develops with an inverse growth rate of 4.5 ± 1.0 ns. Lastly, the electromagnetic flute instability (EMFI, Sotnikov et al. 2008) would grow with an inverse growth rate of approximately 10 ns.

Analysis indicates that the observations do not fit the EMFI and the LLRI characteristics. The EMFI is not likely responsible for the observed growth of the flute-like structures since nonuniformities of the boundary can be seen in experiment as early as 9 ns after the beginning of the plasma expansion. Moreover, the wave number k does not evolve in time, as is characteristic of the EMFI (Sotnikov et al. 2008). The LLRI requires a charge separation at the plasma-field interface, produced by unmagnetized ions. In experiment, the ion Larmor radius is comparable with the scale length (Table 2) and, in addition, the high collisionality prevents charge separation. Therefore the LLRI is an unlikely explanation of the observations, as well.

To understand how the remaining candidate, the MRTI, leads to plasma penetration across the magnetic field, a more detailed analysis of the plasma parameters in the discontinuity region is necessary. The time evolution of the magnetic field in the discontinuity region is described by the Faraday equation, using a form of Ohm's law appropriate for the conditions of the plasma investigated. For this experiment, Ohm's law is expressed as:

$$\boldsymbol{E} = -\boldsymbol{u} \times \boldsymbol{B} + \eta \boldsymbol{J} + \frac{1}{n_e e}(\boldsymbol{J} \times \boldsymbol{B} - \nabla P_e) \qquad (4.1)$$

where e is the electron charge, η is the resistivity, n_e is the electron number density, P_e is the electron pressure (assumed to be isotropic), $\boldsymbol{u}$ is the plasma velocity, $\boldsymbol{J}$ is the current density vector, and $\boldsymbol{E}$ and $\boldsymbol{B}$ are the electric field and magnetic field vectors, respectively. On the right hand side (RHS) of (4.1), the first term is the convective term which appears in ideal MHD, the second term describes the diffusion of the magnetic field into the plasma, the third term accounts for the Hall effect, and the last term describes the plasma diamagnetism. The relative importance of these terms determines the characteristics of the field interaction with the plasma. Normalizing each of the terms on the RHS to the convective term $|\boldsymbol{u} \times \boldsymbol{B}|$, and using Ampère's law to eliminate $\boldsymbol{J}$ gives:

$$\boldsymbol{E} \approx |\boldsymbol{u} \times \boldsymbol{B}| \left(\hat{\boldsymbol{e}}_{convection} + \frac{1}{R_m}\hat{\boldsymbol{e}}_{diffusion} + Al\frac{c}{L\omega_{pi}}\hat{\boldsymbol{e}}_{Hall} + \frac{V_{e,th}}{V_A}\frac{\rho_{e,th}}{L}\hat{\boldsymbol{e}}_{diamagnetic} \right) \qquad (4.2)$$

where ∇ is approximated with the inverse scale length $1/L$. The $\partial \boldsymbol{E}/\partial t$ term in Ampère's law is small for the plasma parameters given in Table 1 and was neglected. In (4.2) R_m is the magnetic Reynolds number, Al is the Alfvén number, ω_{pi} is the ion plasma frequency, $V_{e,th}$ is the electron thermal velocity, V_A is the Alfvén velocity, $\rho_{e,th}$ is the thermal electron gyroradius, and $\hat{\boldsymbol{e}}_{convection}$, $\hat{\boldsymbol{e}}_{diffusion}$, $\hat{\boldsymbol{e}}_{Hall}$, and $\hat{\boldsymbol{e}}_{diamagnetic}$, are unit vectors indicating the direction of the respective contributions.

The values of the dimensionless coefficients from the RHS of (4.2), calculated for the plasma parameters in the discontinuity region (Table 1), are listed in Table 2. The magnitude of each of the coefficients is of the order of 1, so none of the effects included can be neglected.

As the MRTI grows, the density and temperature in the expanding plasma flutes decrease. The magnetic field diffuses in the plasma with a characteristic time of about 2 ns. The expansion of the flutes causes collisions to become less frequent and the Hall effect to become more important. Eventually, charge separation becomes possible and the flute-like structures become polarized. The corresponding electric field is oriented perpendicular to both the magnetic field and the expansion velocity at the tip of the flutes. This effect enables the plasma flutes to penetrate the magnetic field by $\boldsymbol{E} \times \boldsymbol{B}$ drift, where $\boldsymbol{E}$ is the polarization field, and $\boldsymbol{B}$ is the magnetic field in the plasma.

In Table 2, the coefficients in the RHS of (4.2) for the experiment are compared with those for the solar wind in the region of the Earth's bow shock. For the case of the solar wind, only the Hall effect and the convection are significant. However, the solar wind is magnetized even in the absence of diffusion. The main improvement required for this experiment to become a pertinent laboratory simulation of the solar wind is reducing the plasma collisionality while maintaining its magnetization.

5 Conclusion

This paper presents an experiment in which a laser-produced plasma expands in an external magnetic field. In the absence of the field, the plasma expands hemispherically. When present, the magnetic field decelerates the front of the plasma plume, causing the perturbations present at the formation of the plasma plume to be amplified due to the magneto-Rayleigh-Taylor instability. As the instability grows, the perturbations develop into flute-like structures. The plasma flutes become polarized and are able to penetrate the field due to $\boldsymbol{E} \times \boldsymbol{B}$ drift.

Acknowledgements The authors of this paper would like to acknowledge Vladimir Ivanov, Alexey Morozov, Mike Bakeman, Gary Pettee, and Alexey Astanovitskiy. This work was supported by DOE/NNSA grant DE-FC52-06NA27616.

References

Axford, W.I., Hines, C.O.: A unifying theory of high-latitude geophysical phenomena and geomagnetic storms. Can. J. Phys. **39**(10), 1433–1464 (1961)

Dungey, J.W.: Interplanetary magnetic field and the auroral zones. Phys. Rev. Lett. **6**(2), 47–48 (1961)

Echim, M.M., Lemaire, J.F.: Laboratory and numerical simulations of the impulsive penetration mechanism. Space Sci. Rev. **92**(3), 565–601 (2000)

Gratton, F.T., Farrugia, C.J., Cowley, S.W.H.: Is the magnetopause Rayleigh-Taylor unstable sometimes? J. Geophys. Res. **101**(A3) (1996)

Hassam, A.B., Huba, J.D.: Magnetohydrodynamic equations for systems with large Larmor radius. Phys. Fluids **31**, 318 (1988)

Kruskal, M., Schwarzschild, M.: Some instabilities of a completely ionized plasma. Proc. R. Soc. Lond., Ser. A, Math. Phys. Sci. **223**(1154), 348–360 (1954) (1934–1990)

Lindemuth, I.: The ANIMAL code. Lawrence Livermore National Laboratory Report, ucrl-52492 edition (1979)

Ramis, R., Schmalz, R., Meyer-Ter-Vehn, J.: MULTI—A computer code for one-dimensional multigroup radiation hydrodynamics. Comput. Phys. Commun. **49**, 475–505 (1988)

Sotnikov, V.I., et al.: Investigation of compressible electromagnetic flute mode instability in finite beta plasma in support of z-pinch and laboratory astrophysics experiments. Commun. Comput. Phys. **4**, 611 (2008)

Astrophys Space Sci (2009) 322: 201–204
DOI 10.1007/s10509-008-9943-z

ORIGINAL ARTICLE

Kelvin–Helmholtz instabilities actuated by an external magnetic field

Sandra Wright · Radu Presura · Andrey Esaulov · Stephan Neff · Christopher Plechaty · David Martinez · Abdelmoula Haboub

Received: 13 June 2008 / Accepted: 31 October 2008 / Published online: 27 November 2008

Abstract The solar wind crosses the magnetic field of the earth and mixes with the plasma of terrestrial origin. The fast particles from the solar wind are responsible for satellite damage, communication disruptions and power blackouts on earth. A better understanding of this penetration process is needed in order to be able to accurately predict it. One experiment attempting to accomplish this was performed at the Nevada Terawatt Facility. This experiment used a short-pulse laser to produce a plasma plume and a pulsed-power generator to independently produce a magnetic field. The magnetic field actuated a sheared flow along the boundary of the plasma plume which caused a Kelvin–Helmholtz instability similar to that found in the interaction between the solar wind and the magnetosphere that leads to the penetration of the fast particles. The instability produced in the experiment will be discussed along with its relevance to the solar wind/magnetosphere interaction.

Keywords Laboratory astrophysics · Kelvin–Helmholtz instability · Magnetized plasmas

1 Introduction

Understanding plasma transport into the earth's magnetosphere is of utmost importance. Data taken by the Cluster satellites located in the earth's magnetotail have conclusively shown that Kelvin–Helmholtz instabilities form along the flanks of the magnetotail (Hasegawa et al. 2004). Some of the most powerful Coronal Mass Ejections (CME) on record took place between mid-October 2003 and early November 2003. The CME that were earth directed had far reaching effects ranging from blackouts and communication disruptions to doses of solar radiation equivalent to a chest X-ray for astronauts and some air travelers. Some of the airlines redirected their high altitude flights to avoid the worst of the radiation (Rosen and Johnson 2004). Around 60% of NASA's earth and space science missions were in some way affected and aurorae were seen as far south as Spain and Florida. A better understanding of how the plasma transport occurs is required for better predictions of space weather.

One of the proposed mechanisms for solar wind transport into the magnetosphere is through magnetic reconnection within the Kelvin–Helmholtz vortices in the magnetotail. Measurements in the magnetosphere are done with satellites, so they depend on the current conditions of the magnetosphere at the location of the satellites. Controlled laboratory experiments are needed to constrain the models. One experiment addressing the generation of the Kelvin–Helmholtz instability was done at the Nevada Terawatt Facility. The following sections will outline the experimental setup, results and conclusions.

2 Experiment design

This experiment used a short-pulse laser to produce a plasma plume and a pulsed-power generator to independently produce a magnetic field. The pulsed power generator has a 0.6 MA current pulse with a rise time of 200 ns. The load for this generator was a short circuit load. The short pulse laser is a Nd:Glass laser with a pulse duration of 5 ps, wavelength of $\lambda = 1057$ nm, an energy of $E \approx 2$–4 J on target, and an

S. Wright (✉) · R. Presura · A. Esaulov · S. Neff · C. Plechaty · D. Martinez · A. Haboub
Nevada Terawatt Facility, University of Nevada, 5625 Fox Ave.. Reno, NV 89506, USA
e-mail: swright@physics.unr.edu

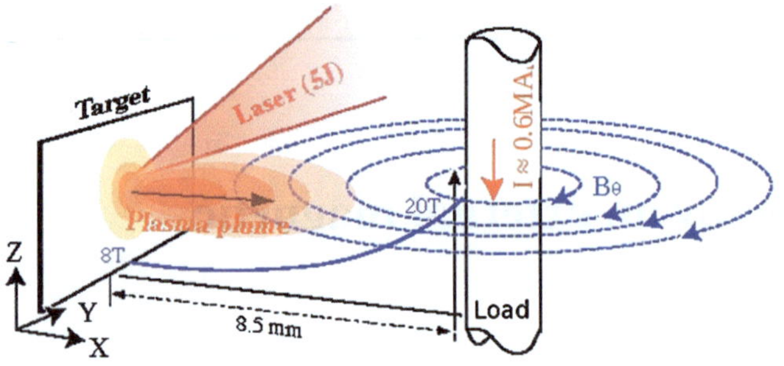

Fig. 1 This diagram indicates the concept for the experiment. The laser is incident on a target and creates a plasma as the load creates an external azimuthal magnetic field. Then the plasma evolution in the external magnetic field is investigated

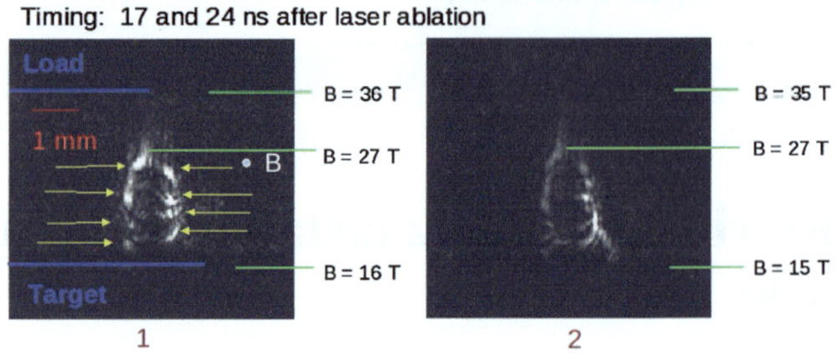

Fig. 2 Schlieren images of the plasma plume. Frame 1 is taken 17 ns after laser ablation, and frame 2 is taken 24 ns after laser ablation. KH instabilities are forming along the plasma flanks indicated by the *arrows* in frame 1. The laser intensity is 2 J on target

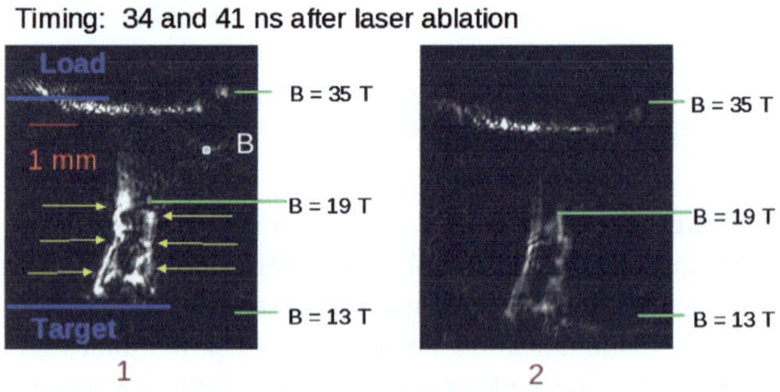

Fig. 3 Schlieren images of the plasma plume. Frame 1 is taken 34 ns after laser ablation, and frame 2 is taken 41 ns after laser ablation. *Arrows* indicate the instabilities. The laser intensity is 2 J on target

irradiance of $I > 10^{15}$ W/cm^2. The laser target is a 1 mm diameter polyethylene (CH_2) rod. Figure 1 shows the concept of the experiment. The high intensity short pulse laser was incident on a target resulting in a plasma plume which expanded across an externally generated magnetic field.

The laser ablated the CH_2 target at the current peak, and the resulting plasma expanded across the externally produced magnetic field. At the current peak the current varied by less than 5 percent for several tens of nanoseconds during the laser ablation and expansion.

3 Results

Multi-frame shadow and schlieren imaging where used to investigate the density evolution of a plasma plume expanding across a magnetic field. In the absence of a magnetic field, the plasma plume expands hemispherically. However, when a magnetic field is added, observations in the plane of the magnetic field show cases ranging from plasma guidance along the field lines to plasma penetration across the field lines depending on the plasma parameters.

Images in the plane perpendicular to the magnetic field show a periodic structure between 17 ns and 41 ns during the experiment. This structure is observed along the edge of the plasma plume in the direction perpendicular to the expansion front and is shown in Figs. 2 and 3 where arrows indicate the structures. The images in each figure were taken 7 ns apart. The wavelength, λ, is the distance between the vortices and is approximately 1 mm. This structure is most likely the result of an instability.

The plasma parameters are calculated for the interaction region where the instability is observed. The density is 10^{17}–10^{19}/cm^3 and determined by the laser parameters. The electron and ion temperature $T_e \approx T_i$ respectively is between 20 and 100 eV inferred from MULTI simulations (Ramis et al. 1988). The velocity is about 100 km/s determined by measuring the propagation of the expansion front of the plasma plume. The magnetic field is between 10 T and 60 T

Table 1 These are the values for the gyrofrequency, collision frequency, collision time, number of cyclotron gyrations before collision, and Larmor radius for both electrons and ions. These values were calculated using a magnetic field of 20 T, a density of $n = 10^{18}$ cm^{-3} and a temperature of 50 eV. Here A is the atomic mass number and Z is the ionization stage

	Electrons	Ions
ω_c (1/s)	3.5×10^{12}	$1.9 \times 10^9 Z/A$
$\nu_{e,i}$ (1/s)	1.3×10^{11}	$1.4 \times 10^9 Z^4/A$
$\tau_{e,i}$ (s)	7.6×10^{-12}	$7.3 \times 10^{-10} Z^{-4} A$
$\omega_c \tau_{e,i}$	27	$1.4Z^{-3}$
r_L (m)	2.8×10^{-8}	$5.2 \times 10^{-5} A/Z$

calculated from the current measured passing through the load. The system length is width of the high density gradient within the plasma plume and is $L = 100$ μm. Taking for example a magnetic field of 20 T, and number density of $n = 10^{18}$ cm^{-3}, and a temperature of 50 eV, we calculate the gyrofrequency, collision frequency, collision time, magnetization, and Larmor radius. The results are listed in Table 1. The number of cyclotron gyrations before collision determine whether the plasma is magnetized or not and the Larmor radius determines if the ions can escape the quasi neutral plasma region.

The cyclotron frequency of the electrons and ions is given by $\omega_{ce} = eB/m_e$ for electrons and $\omega_{ci} = ZeB/m_i$ for ions where Z is the ionization state of the plasma, e is the elementary charge, B is the magnetic field, and m_e and m_i are the masses of an electron and ion respectively. The collision frequency is given by $\nu_e = 2.91 \times 10^{-6} n_e Z \ln\Lambda T_e^{-3/2}$ for electrons and $\nu_i = 4.8 \times 10^{-8} Z^4 (A)^{-1/2} n_i \ln\Lambda T_i^{-3/2}$ for ions (Huba 2002), where A is the atomic number, n is the number density, T is the temperature, and the indices e and i mean electrons and ions respectively, and $\ln\Lambda$ is the Coulomb logarithm. The number density is in units of $1/\text{cm}^3$ and the temperature is in units of eV. The collision frequency is in units of $1/\text{s}$, and the collision time, τ is the inverse of the collision frequency.

The magnetic diffusion time is given by

$$\tau_D = \mu_0 L^2/\eta \tag{1}$$

where L is the characteristic length of the region of the plasma that is being considered, and η is the Spitzer resistivity. For the experimental plasma, $L \approx 100$ μm which is the approximate width of the high density gradient region and the diffusion time is approximately 4 ns.

Instabilities are observed along the boundary perpendicular to the deceleration front after 17 ns. By this time, using (1), the magnetic field has diffused into the plasma a distance of 140 μm.

This means that the diffusion time is sufficient for the magnetic field to trap and magnetize the electrons, $\omega_{ce}\tau_e = 35$, but not the ions, $\omega_{ci}\tau_i \approx 1$. The trapped electrons drag on the ions at the edge of the plasma plume slowing them down. As a result, a velocity shear forms between these slower moving ions and ions in the bulk plasma flow that have not been slowed. The instability that forms is the Kelvin–Helmholtz instability.

The linear growth rate for the Kelvin–Helmholtz instability is given by Chandrasekhar (1981)

$$\gamma = \frac{k\sqrt{\rho_1\rho_2}|u_1 - u_2|}{\rho_1 + \rho_2}. \tag{2}$$

For $\rho_1/\rho_2 = 0.1$, $|u_1 - u_2| = 100$ km/s, and $k = 2\pi/\lambda = 6.3\ \text{mm}^{-1}$, this gives a growth time of $\gamma^{-1} \approx 5$ ns for the linear stage. However, the Larmor radius is 50 μm for hydrogen and 150 μm for carbon and both are comparable with the width of the unstable layer (100 μm) so the instability observed may be the large Larmor radius variation of the Kelvin Helmholtz instability (Opp and Hassam 1990) with a growth time $\gamma^{-1} \approx 5$ ns for a wavelength of 1 mm. In experiment a velocity shear was not initially present. As the shear started to form, the Kelvin–Helmholtz instability (KHI) was in a linear stage and was not resolvable with diagnostics. It took almost 17 ns to observe the instability with existing diagnostics. At this point the KHI was in the nonlinear stage.

Table 2 This compares our experimental parameters to those found in the magnetosphere. The dimensionless parameters are $\omega_c\tau_{e,i}$ is the cyclotron frequency multiplied by the collision time, R_M is the magnetic Reynolds number, and the plasma β

	Magnetosphere plasma	Experimental plasma
n (cm^{-3})	8	10^{18}
T_i (eV)	150	50
T_e (eV)	25	50
B (nT)	15	20×10^9
v (km/s)	400	100
τ_D (s)	10^{14}	$4.3 \times 10^{-9}/Z$
$\omega_c\tau_e$	2×10^9	35
$\omega_c\tau_i$	6.3×10^7	1.4
R_M	10^{11}	$4.2/Z$
β	1	1

After 41 ns in the experiment, the KHI is no longer observed. This means that either the instability is no longer present, or that the instability has become fully turbulent. If the instability is fully developed turbulence, then it is not resolvable with current diagnostics.

Comparison with solar wind/magnetosphere parameters yield the dimensionless parameters for determining magnetization, magnetic diffusion and pressure balance. These along with experimental parameters are listed in Table 2. One main difference between our experiment and the space system is that the solar wind contains an embedded magnetic field.

For the laboratory system to be relevant to the space system it is necessary to have the same physics dominant in both systems so that the plasmas behave similarly. The magnetization of the plasma is determined by multiplying the cyclotron frequency with the collision time. Electrons and ions in the magnetosphere are magnetized. However, in experiment the electrons are magnetized but the ions are not. The magnetic Reynolds number is the ratio of the flow velocity to the magnetic diffusion velocity and determines how frozen the magnetic field lines are with respect to the plasma. This shows that the laboratory system is highly diffusive but the space system is not. Last is the plasma β that determines the balance between the thermal pressure and the magnetic field pressure. The pressure balance is the same for both systems.

4 3D MHD simulation

A three-dimensional ideal MagnetoHydroDynamic (MHD) simulation indicates the formation of the vortices along the flanks of the plasma plume as shown in Fig. 4.

For this simulation, the grid size is $d = 0.1$ mm. The laser energy ($E = 2$ J) deposited over a depth of 0.1 mm

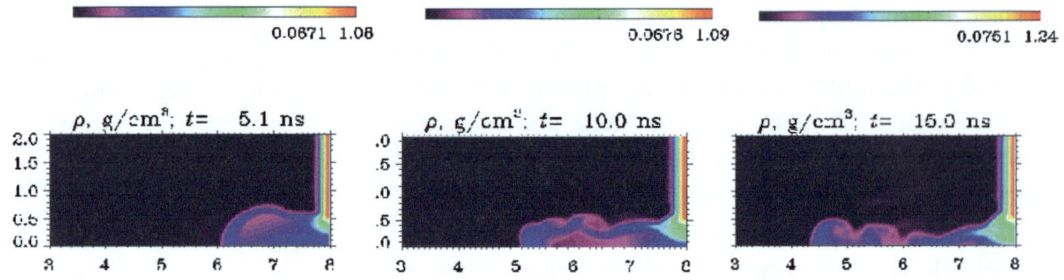

Fig. 4 An Ideal 3D MHD simulation clearly shows the formation of an instability. The grid size is $d = 0.1$ mm, $E = 2$ J, v_0, the initial velocity, is 0.0 m/s, $T_0 = 350$ eV, and the magnetic field profile is $B[T] = 100/R$ [mm]

at solid state density material leads to an initial temperature of $T_0 = 350$ eV. The plasma is initially at rest. The magnetic field profile is $B[T] = 100/R$ [mm]. These values are the same as in experiment. The images show the plasma expansion in a plane perpendicular to the plane of the magnetic field and clearly show the development of a periodic instability along the direction perpendicular to the target normal. This simulation predicts the development of the instability at about 10 ns while it is observed in experiment at 17 ns. Two factors contribute to the difference: the fact that the plasma in experiment is resistive and the sensitivity threshold of the laser diagnostics.

5 Conclusions

Work is still ongoing to determine if the plasma is in a kinetic or MHD regime in the density gradient region. Answering this question may shed light on the nature of the instability observed in experiment. Independent of the nature of the instability, there are a few differences between our experiment system and the space system such as the magnetic Reynolds number and the ion magnetization. These will be addressed in future experiments. Future experiments will create a hotter collisionless plasma in order to better approximate the solar wind. The hotter plasma will in turn raise the magnetic Reynolds number. An increase in temperature will increase the plasma β, so the magnetic field will be increased to compensate. The increase in magnetic field will magnetize the ions. With these modifications we will obtain a plasma/magnetic field interaction described by dimensionless parameters that is comparable to those of the solar wind/magnetosphere interaction.

Acknowledgements The authors would like to thank V. Ivanov, A. Morozov, M. Bakeman, A. Astanovitskiy, and G. Pettee for their contributions. This work was supported by DOE/NNSA DE-FC52-06NA27616.

References

Chandrasekhar, S.: Hydrodynamic and Hydromagnetic Stability. Dover, New York (1981), Chapter 11

Hasegawa, H., Fujimoto, M., Phan, T.-D., Reme, H., Balogh, A., Dunlop, M.W., Hashimoto, C., Tandokoro, R.: Nature **430** (2004)

Huba, J.: NRL plasma formulary. Naval Research Laboratory (2002)

Opp, E.N., Hassam, A.B.: Phys. Fluids B **3**, 885 (1990)

Ramis, R., Schmalz, R., Meyer-ter Vehn, J.: Comput. Phys. Commun. **49**, 475 (1988)

Rosen, R., Johnson, D.: Service assessment of intense space weather storms 19 October–7 November 2003. Technical report, U.S. Department of Commerce, National Oceanic and Atmospheric Association, National Weather Service (2004)

Astrophys Space Sci (2009) 322: 205–208
DOI 10.1007/s10509-008-9947-8

ORIGINAL ARTICLE

Generation of shear flow in conical wire arrays with a center wire

David Martinez · Radu Presura · Sandra Wright · Chris Plechaty · Stephan Neff · Lucas Wanex · David J. Ampleford

Received: 15 June 2008 / Accepted: 10 November 2008 / Published online: 29 November 2008

Abstract At the Nevada Terawatt Facility we investigated the generation of a sheared plasma flow using conical wire arrays with an additional wire located on the axis of the pinch. The additional center wire generates axial current carrying plasma that serves as a target for the plasma accelerated from the outer wires, generating a sheared plasma flow which leads to the growth of the Kelvin-Helmholtz instability. These experiments were conducted on Zebra, a 2 TW pulse power device capable of delivering a 1 MA current in 100 ns. This paper will focus on the implosion dynamics that lead to shear flow and the development of the Kelvin Helmholtz instability.

Keywords Shear flow · Kelvin-Helmholtz instability · Conical wire array · *Z*-pinch

1 Introduction

Many astrophysical systems have transverse variations in velocity and these sheared flows are susceptible to the Kelvin-Helmholtz instability (Chandrasekhar 1961). Instability growth leads to energy, momentum, and particle transport (or fluid mixing) across the shear interface. This is of particular interest when magnetic fields are present, as in many natural systems. For example, the Kelvin-Helmholtz instability (KHI) is a prominent candidate for explaining the transport from the solar wind to the magnetosphere and its role in magnetic storms (Hasegawa et al. 2004). In certain settings, such as the expansion of the solar coronal plumes through the interplume plasma, the velocity shear triggers drift KHI, which do not produce plasma inhomogeneity over large scales (Andries and Goossens 2001). In the case of protostellar jets, KHI may be responsible for the formation of a mixing layer and internal shocks (Cantó and Raga 1991; Raymond et al. 1994), which may lead to jet deformation and breakup (Keppens and Tóth 1999). In all these examples, the shear layer forms in magnetized plasma and both observations and theory suggest that the field has a significant effect upon the development of the KHI and its consequences. The role of magnetic fields upon the stability of boundaries with sheared flows was further investigated by sophisticated three-dimensional simulations (Keppens and Tóth 1999; Ryu et al. 2000). Few laboratory experiments investigate the effects of sheared flows in magnetized plasma, for example Shumlak et al. (2001). Here we present a proof of principle experiment that demonstrates the excitation of the Kelvin-Helmholtz instability in magnetized plasma at a velocity shear layer created by a conical wire array with an additional central conductor.

2 Generation of shear flow

Conical wire arrays have been previously shown to produce axially flowing plasma (Lebedev et al. 2002; Ampleford et al. 2005; Lebedev et al. 2004). The conical wire array is a modified wire array *Z*-pinch, where the wires are ohmically heated and ablated to form a coronal plasma around each individual wire. The current then flows through this coronal plasma along each wire, continuing to heat it. The global magnetic field generated by all wires causes a Lorentz

D. Martinez (✉) · R. Presura · S. Wright · C. Plechaty · S. Neff · L. Wanex
University of Nevada, Reno, USA
e-mail: dmartine@unr.nevada.edu

D.J. Ampleford
Sandia National Laboratories, Albuquerque, NM 87108, USA

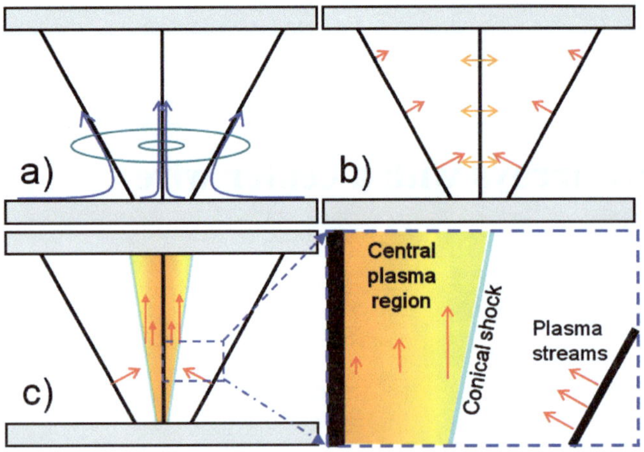

Fig. 1 (**a**) Current (*blue arrows*) flows through the wires, generating a local magnetic field around each wire and a global magnetic field (*green*) around the whole array. (**b**) Ohmic heating causes all wires to ablate. Plasma originating from the outer wires will be accelerated perpendicular to the wire surface toward the array axis (*red arrows*). The plasma from the center wire will expand (*orange*) and will be confined by its own magnetic field. (**c**) The interaction of the plasma streaming from the outer wires (*red*) with the center wire plasma (*orange yellow*) leads to the formation of a conical shock (*delineated* in *light blue*). Once it traverses the shock, the plasma originating from the outer wires will be redirected in axial direction, creating a sheared flow (*red*)

$\boldsymbol{j} \times \boldsymbol{B}$ force on the plasma coronas, which accelerates them towards the axis of the array. The acceleration is perpendicular to each wire, creating plasma streams that have radial and axial momentum. For a cylindrical wire array, the kinetic energy of the radial motion is thermalized at stagnation on the axis of the array, creating a hot precursor plasma column. This precursor column is modified for the conical array configuration, as described in Lebedev et al. (2002), Ampleford et al. (2007), to create a conical shock (Cantó et al. 1988). For the estimated values of the plasma parameters at the time of the conical shock development, the ion mean free path is of the order of 10 μm, thus the shocked material is redirected to the axial direction producing an axially flowing plasma. When introducing a wire on axis, it is ablated (Fig. 1a) due to heating by the current it carries. In this case, the plasma flow interacts with a central plasma stationary in axial direction. Assuming a no-slip boundary condition on the surface of the center plasma, a quasi-parabolic velocity profile results (Wanex 2005)

$$v_z(r) = V \frac{\ln(r_w) r^2 - r_w^2 \ln(r)}{R^2 \ln(r_w) - r_w^2 \ln(R)}, \tag{1}$$

where v_z is the axial velocity, r_w is the radius of the axial plasma, R is the flow radius, and V is the flow velocity at the edge.

For the laboratory experiments presented here, the conical wire arrays (Fig. 1) consisted of eight 15 μm Al wires, equally spaced and had a cathode (bottom) diameter of 6 mm and an anode (top) diameter of 18 mm, giving the array an opening angle of 16.7°, measured from the z axis to the outer wires (this array configuration is identical to the Zebra experiments described in Ampleford et al. 2007 and Presura et al. 2001). The center wire was a 10 μm diameter Al wire. Both the conical array and the center wire use the alloy Al 5056 (95% Al, 5% Mg) such that a small amount of magnesium is present for spectroscopic analysis. These arrays were imploded using the Zebra Z-pinch generator, which is capable of delivering 1 MA of current to a load in 90 ns (Shlachter 1990).

Considering a cylindrical wire array with 12 mm diameter as a proxy for the conical wire arrays used in experiment, the current through the center wire I_{in} can be estimated by using Velikovich et al. (2002)

$$\frac{I_{in}}{I} = \frac{\omega(L_{out} - M) + R_{out}}{\omega(L_{in} + L_{out} - 2M) + R_{in} + R_{out}}, \tag{2}$$

where I is the total array current, ω is the frequency of the current pulse, L_{in} is the inductance of the center wire, L_{out} is the inductance of the outer array, M is the mutual inductance between the center wire and the outer array, R_{in} is the resistance of the center wire, and R_{out} is the resistance of the outer array. This equation predicts that the center wire will carry about 6% of the total current while the wires remain cold.

The plasma dynamics were observed using multi-frame shadow imaging (Ivanov et al. 2006) with a 532 nm wavelength, 150 ps diagnostic laser. In addition, Faraday rotation images were also taken to qualitatively assess the presence of an azimuthal magnetic field inside the wire array. Time-gated X-ray pinhole images were used to obtain the phase velocity of the zippering pinch. Time-integrated spectroscopy was used to obtain estimates of the electron temperature and density.

3 Evidence of shear flow

Shadow images show evidence of a conical shock formed between the center wire and the outer wires Fig. 2a. At this shock, the streams from the outer wires change direction from a conical converging flow to an axially flowing plasma. The velocity of the material ablated from the outer wires can be estimated based on work by Ampleford et al. (2007). Measuring the time dependence of the stagnated pinch height in the time-gated X-ray pinhole images, indicates a "zippering" velocity of about 400 km/s. Using a purely geometrical analysis and considering that the delay between the plasma arrival at the array center and the X-ray emission is short compared with the characteristic time for plasma motion, the ablation velocity can be estimated as

$$v_{abl} = v_{zip} \sin(\alpha), \tag{3}$$

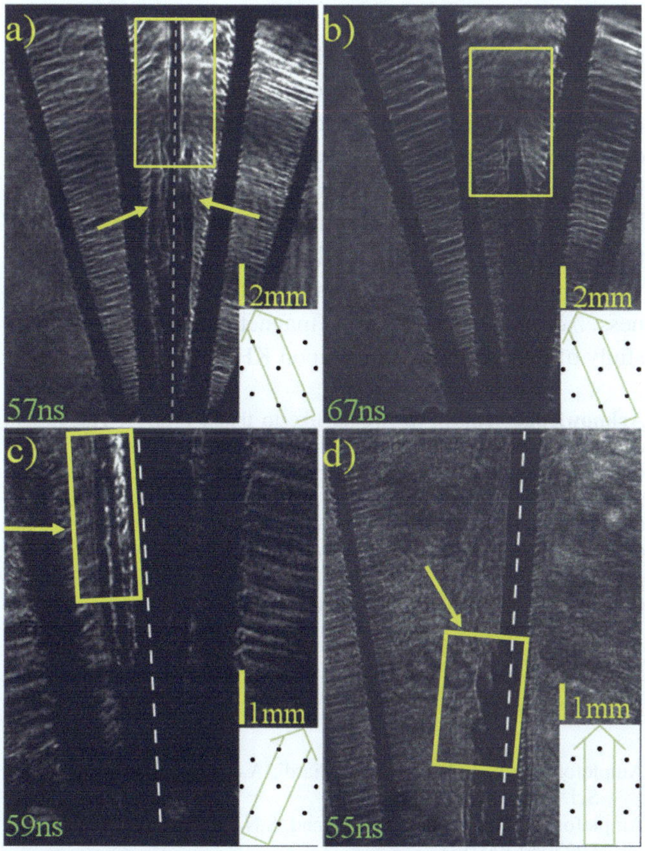

Fig. 2 Plasma images obtained with laser diagnostics. (**a**) Shadow image of a conical wire array with a center wire. In this image we see the development of the conical shock (*arrows*). The center wire is seen to expand uniformly, even above the conical shock. (**b**) Conical wire array without a center wire shows a lack of dense plasma above the conical shock in contrast to Fig. 2a. (**c**) Faraday image indicating that a magnetic field is present near the center of the array. (**d**) Shadow image showing the nonlinear stage of the Kelvin-Helmholtz instability. Time labels are zeroed to 10% current peak. The orientation of the probing laser beam through the wire array is shown in each image. The location of the center wire is denoted with a *dashed line*

where α is the opening angle of the array, v_{zip} is the observed zippering velocity and v_{abl} is the ablation velocity. For the measured value of the zippering velocity, (4) gives an ablation velocity of about 110 km/s. Assuming that the axial momentum is conserved in the pinching process, the resulting axial plasma velocity is about 33 km/s. However if any accelerating mechanism is active, the axial plasma velocity would exceed this estimate. In fact, in several shots, the axial velocity of features in the central plasma column could be measured, giving values around 100 km/s, which are in good agreement with theoretical results of Cantó et al. (1988) for strong adiabatic shocks.

This plasma flow interacts with an axially stationary center plasma produced by ablation of the center wire. For the case of a conical wire array with a center wire, a uniform column of dense plasma forms on axis along the full height of the array, even above the top of the conical shock Fig. 2a, in contrast with the case of conical wire arrays without a central wire Fig. 2b. The Faraday rotation images show brightening of the left side of the array and darkening of the right Fig. 2c compared with a reference image, indicating the presence of an azimuthal magnetic field and supporting the hypothesis that the center wire is carrying sufficient current to ablate the wire. The interaction of the axial plasma flow with this axially stationary plasma produces a radial dependence of the axial velocity.

4 Development of vortices

As described in the previous section, when the plasma streams from the outer wires cross the conical shock, they change direction from a conical converging flow to an axially flowing plasma. Across the boundary, the axial velocity may increase (in the case of an adiabatic flow) or remain unchanged (if the axial momentum is conserved). In the former case, the resulting shear layer would be too narrow for the diagnostics to detect. However, in both cases, a shear layer with continuously variable velocity forms when the axial flow interacts with the axially static center plasma and this layer is prone to the Kelvin-Helmholtz instability. The shadow images indicate development of vortices characteristic of the nonlinear stage of the KHI Fig. 2d close to the dense center plasma column, in the interior of the conical shock.

The Kelvin-Helmholtz instability grows from the same perturbations of the dense center plasma that would produce the $m = 0$ current-driven instability or the Rayleigh-Taylor instability in the absence of an axial flow, as observed in cylindrical wire arrays. As shown in Fig. 2d, the instability produces roll-up structures, in contrast to the bubble and spike structure expected for typical Z-pinch instabilities (Peterson et al. 1996).

Assuming that the shear layer is an infinitely thin step boundary, the linear growth rate can be calculated from (Chandrasekhar 1961)

$$\gamma = \frac{k\sqrt{\rho_1 \rho_2}|U_1 - U_2|}{\rho_1 + \rho_2}, \tag{4}$$

where k is the wave number of the mode considered, and ρ_i and U_i are the density and velocity of the i-th fluid, respectively. The wavelength is measured on the shadowgram to be ~1 mm. Assuming that the central wire plasma is one order of magnitude denser than the plasma streaming from the outer wires, such that $\rho_1 \approx 10\rho_2$, and using the flow velocity of 100 km/s as $|U_1 - U_2|$, a KHI growth time of about 5 ns results. If the actual density ratio is higher, the growth time increases approximately linearly with the square root of the ratio. The pinch evolution time includes several instability growth times, thus allowing the KHI to reach the non-linear stage. It is important to note that the KHI has been observed in a relatively small number of shots and never with azimuthal symmetry. The latter may be due to the low number of wires of the conical array, which leads the formation of a corrugated conical shock. In this case, the individual oblique shocks formed for each outer wire may have different parameters, so that not all of them meet the correct criteria for exciting the instability. Alternately, as the instability grows the fastest when the density ratio is one, the density-sensitive laser diagnostics used here may not always detect it.

5 Scaling

Values of dimensionless parameters can be used to assess the relevance of this experiment to astrophysical and space physics phenomena. Order of magnitude values of the Peclet, Reynolds and magnetic Reynolds numbers (Ryutov et al. 1999; Ryutov et al. 2001) were obtained from approximate plasma parameters. Time integrated Mg line spectra (Vainshtein et al. 1980; Vinogradov et al. 1975) were used to estimate the plasma temperature around 100 eV and the plasma density about 10^{19} cm^{-3}. A charge state $Z = 5$, a flow velocity of 100 km/s, and a plasma column radius of 0.5 mm were considered. Assuming that 10% of the array current halfway through the rise time flows through the center plasma, the magnetic field is of the order of 20 T. With these values, the dimensionless parameters can be calculated: $Pe \sim 10$, $Re \sim 10^3$, and $Re_M \sim 10$. Based on the accuracy of the plasma parameters used, these values are correct within a factor of 2. Although these numbers are by many orders of magnitude smaller than in those characteristic to the solar wind or protostellar jets, they are significantly larger than one, indicating that the same transport processes dominate.

6 Conclusions and future work

This proof of principle experiment demonstrated that a conical wire array with a center wire is capable of generating a sheared flow, as evidenced by laser imaging of the plasma. This shear flow can excite the Kelvin-Helmholtz instability, as indicated by the formation of vortices, which are characteristic of the nonlinear stage of the KHI. A simple hydrodynamic analysis of the instability growth rate indicates that the vortices seen can indeed be generated by the KHI. However, the appearance of the KHI is non-periodic and does not present cylindrical symmetry, which requires further investigation.

In future experiments, the configuration will be optimized to investigate the details of the KHI growth in magnetized plasma shear layers. The flow velocity, which determines the growth rate, can be varied by changing the conical wire array opening angle. In addition, this type of experiment presents the advantage that using wires with different atomic numbers changes the amount of energy lost by radiation. This radiative cooling process (Ampleford et al. 2005) can be used to modify plasma density and flow velocity. To investigate the effect of the magnetic field upon the instability growth, the intensity of the current flowing through the center wire can be varied by changing the number of wires in the wire array or changing the central wire thickness. Adding an external axial magnetic field will enable the study of its stabilizing effect upon KHI.

Acknowledgements We would like to thank Prof. V. Ivanov and A. Haboub for their help with the laser diagnostics. This work would not have been possible without the assistance of Zebra's technical staff. This work was supported by the DOE/NNSA grant DE-FC52-06NA27616. Sandia is a multiprogram laboratory operated by Sandia Corporation, a Lockheed Martin Company, for the United States Department of Energy's National Nuclear Security Administration under Contract No. DE-AC04-94AL85000.

References

Ampleford, D.J., Lebedev, S.V., Ciardi, A., et al.: Astrophys. Space Sci. **298**, 241 (2005)
Ampleford, D.J., Lebedev, S.V., Bland, S.N., et al.: Phys. Plasmas **14**, 102704 (2007)
Andries, J., Goossens, M.: Astron. Astrophys. **368**, 1083 (2001)
Cantó, J., Tenorio-Tagle, G., Rózyczka, M.: Astron. Astrophys. **192**, 287 (1988)
Cantó, J., Raga, A.C.: Astrophys. J. **372**, 646 (1991)
Chandrasekhar, S.: Hydrodynamic and Hydromagnetic Stability. Oxford University Press, Oxford (1961)
Hasegawa, H., Fujimoto, M., Phan, T.-D., et al.: Nature **430**, 755 (2004)
Ivanov, V.V., et al.: Phys. Plasmas **13**, 012704 (2006)
Keppens, R., Tóth, G.: Phys. Plasmas **6**, 1461 (1999)
Lebedev, S.V., Chittenden, J.P., Beg, F.N., et al.: Astrophys. J. **564**, 113 (2002)
Lebedev, S.V., Ampleford, D.J., Ciardi, A., et al.: Astrophys. J. **616**, 988 (2004)
Peterson, D.L., Bowers, R.L., Brownell, J.H., et al.: Phys. Plasmas **3**, 368 (1996)
Presura, R., Bauer, B.S., Kantsyrev, V.L., et al.: Bull. Am. Phys. Soc. **46**, 318 (2001)
Raymond, J.C., Morse, J.A., Hartigan, P., et al.: Astrophys. J. **434**, 232 (1994)
Ryu, D., Jones, T.W., Frank, A.: Astrophys. J. **545**, 475 (2000)
Ryutov, D., Drake, R.P., Kane, J., et al.: Astrophys. J. **518**, 821 (1999)
Ryutov, D.D., Remington, B.A., Robey, H.F., et al.: Phys. Plasmas **8**, 1804 (2001)
Shlachter, J.S.: Plasma Phys. Controlled Fusion **32**, 1073 (1990)
Shumlak, U., Golingo, R.P., Nelson, B.A., DenHartog, D.J.: Phys. Rev. Lett. **87**, 205005 (2001)
Vainshtein, L.A., et al.: In: Basov, N.G. (ed.) Proceedings of the Lebedev Physics Institute, vol. 119, p. 13. Nova Science, New York (1980)
Velikovich, A.L., Sokolov, I.V., Esaulov, A.A.: Phys. Plasmas **9**, 1366 (2002)
Vinogradov, A.V., et al.: Quantum Electron. Sov. J. **5**, 630 (1975)
Wanex, L.F.: Astrophys. Space Sci. **298**, 337 (2005)

Astrophys Space Sci (2009) 322: 209–213
DOI 10.1007/s10509-009-9998-5

ORIGINAL ARTICLE

Investigation of flute instability in application to laboratory astrophysics and Z-pinch experiments

V.I. Sotnikov · V.V. Ivanov · R. Presura · E. Yassin · J. Kindel · J.N. Leboeuf · O.G. Onishchenko · B.V. Oliver · B. Jones · T.A. Mehlhorn · C. Deeney

Received: 13 June 2008 / Accepted: 27 January 2009 / Published online: 21 February 2009

Abstract Interaction of plasma flows with magnetic fields can lead to excitation of flute-like oscillations. These oscillations can also be excited in Z-pinch plasmas. We have developed understanding of nonlinear dynamics of compressible flute mode turbulence with spatial scales comparable to the ion Larmor radius in a high beta plasma, when these modes become electromagnetic and in the presence of a non-uniform magnetic field. It is shown that the flute waves are responsible for generation of large scale structures of streamer and zonal flow types as well as Kolmogorov type spectra in the short scale region. The relevance of the obtained results to laboratory astrophysics and Z-pinch experiments is discussed.

Keywords Flute instability · High beta plasma · Nonlinear wave dynamics · Laboratory astrophysics experiments · Z-pinches

V.I. Sotnikov (✉) · V.V. Ivanov · R. Presura · E. Yassin · J. Kindel
University of Nevada at Reno, Reno, NV 89557, USA
e-mail: sotnikov@physics.unr.edu

J.N. Leboeuf
JNL Scientific, Casa Grande, AZ 85294, USA

O.G. Onishchenko
Institute of Physics of the Earth, 123995 Moscow, Russia

B.V. Oliver · B. Jones · T.A. Mehlhorn
Sandia National Laboratories, Albuquerque, NM 87123, USA

C. Deeney
Department of Energy, Washington, DC 20585, USA

1 Introduction

In the past decade the problem of formation of large scale coherent structures from random perturbations as well as coexistence of these structures with short scale turbulent spectra has attracted considerable attention (e.g. Newman et al. 1993; Horton 1999). However, in application to flute modes this problem was investigated only in the limit of low-beta plasma and in the presence of homogeneous magnetic field (Andrushchenko and Pavlenko 2002; Sandberg and Shukla 2004; Mikhailovskii et al. 2007).

During recent experiments on interaction of laser ablated plasma flows with magnetic field created by the Zebra pulse power generator strong wave activity was detected in the region of plasma flow deceleration by the magnetic field (Presura et al. 2005). Similar phenomena can take place during Novae explosions (Zakharov 2003) and artificial magnetospheric releases, similar to the AMPTE magnetotail release (Bernhardt et al. 1987; Ripin et al. 1993).

To study the linear excitation and nonlinear evolution of compressible electromagnetic flute modes in a high-β plasma, we use a recently customized (originally derived in Sotnikov et al. 2006, 2008) nonlinear set of equations for electrostatic potential, magnetic field, and density. We modified these equations using the approach developed in Kuvshinov and Mikhailovskii (1996), Onishchenko et al. (2001, 2008). The nonlinear set of equations for the description of compressible electromagnetic flute modes is now applicable for an arbitrary value of the parameter $k\rho_i$ (k is the wave number and ρ_i is the ion Larmor radius). In the previous studies (Sotnikov et al. 2006, 2008) the area of applicability of the nonlinear set of equations was limited to values of $k\rho_i < 1$. A description of the experimental setup for laboratory astrophysics experiments can be found in Presura et al. (2005) and Sotnikov et al. (2007). A detailed description of Z-pinch experiments related to the excitation and nonlinear evolution of flute modes is presented in Ivanov et al. (2007). Our results show that flute modes can be responsible for the intermittent convective-like transport across the magnetic field associated with large-scale coherent structures and formation of Kolmogorov type turbulent spectra in the short scale region. Both of these features help explain observed experimental results in laboratory astrophysics and in cylindrical Z-pinch wire implosion experiments at the Nevada Terawatt Facility (NTF).

2 Basic nonlinear equations

The two-fluid macroscopic equations which will be used to describe low frequency compressible electromagnetic flute modes ($\omega \ll \Omega_{ci}$, ω and $\Omega_{ci} = ZeB_0/M_i$ are the wave and ion cyclotron frequencies, Ze and M_i are the ion charge and mass) in an inhomogeneous high-β plasma with external magnetic field $B_{0z}(x)$ can be written in dimensionless form as follows:

$$\frac{\partial}{\partial t}\Delta_\perp \Phi - \frac{T_i}{T_e}\frac{(\kappa_N+\kappa_B)\rho_s}{1+\frac{\beta}{2}}\frac{\partial}{\partial y}\Delta_\perp\Phi + \frac{g}{\Omega_{ci}^2\rho_s}\left[1+\frac{\beta_i/2}{1+\beta/2}\Delta_\perp\right]\frac{\partial \delta n}{\partial y} = \frac{T_i}{T_e}\nabla_\perp\{\nabla_\perp\Phi, \delta n\} - \{\Phi, \Delta_\perp\Phi\} \tag{1}$$

$$\left(1+\frac{\beta}{2}\right)\frac{\partial \delta n}{\partial t} - \frac{\beta_e}{2}\frac{g}{\Omega_{ci}^2\rho_s}\frac{\partial \delta n}{\partial y} + (\kappa_B+\kappa_N)\rho_s\frac{\partial\Phi}{\partial y} = \left(1+\frac{\beta}{2}\right)\{\delta n, \Phi\} \tag{2}$$

Here we use the local Cartesian system of coordinates with the z-axis directed along the external magnetic field, $B_0 = B_0(x)\hat{z}$, $\hat{z}$ is the unit vector along the z-axis, the x-axis in the direction of the plasma inhomogeneity and the direction of the gravitational acceleration $\mathbf{g} = g\hat{\mathbf{x}}$, and y-axis completing the triad. Furthermore, the normalizations for space-time scales are ρ_s and Ω_{ci}^{-1}, $\rho_s = c_s/\Omega_{ci}$ the ion-acoustic Larmor radius and $c_s = (T_e/M_i)^{1/2}$ the ion-acoustic velocity computed with the electron temperature T_e, $\delta n = \delta n_i/n_{0i} = \delta n_e/n_{0e}$ dimensionless particle number density, $\delta n_{i(e)}$ and $n_{0i,e}$ the ion (electron) density perturbation and equilibrium ion (electron) density, T_i the ion temperature, $\Phi = Ze\varphi/T_e$ the dimensionless electrostatic potential, g the gravitational acceleration, the plasma beta for ion and electron components is defined as $\beta_{i,e} = 8\pi n_{0i,e}T_{i,e}/B_0^2$, $\beta = \beta_i + \beta_e$, $\{A, B\} = (\partial A/\partial x)\partial B/\partial y - (\partial A/\partial y)\partial B/\partial x$ the Poisson brackets, $\kappa_N = -d\ln n_0/dx > 0$ and $\kappa_B = d\ln B_{0z}/dx > 0$. In the derivation of (1) and (2) we have used relation $\kappa_B - (\beta/2)\kappa_n - g/v_A^2 = 0$, that can be obtained from the force balance condition $\nabla n_0(T_i + T_e) + (8\pi)^{-1}\nabla B_0^2 = M_i n_0 \mathbf{g}$, v_A is the Alfven velocity, and relation $\delta B_z/B_0 = -(\beta/2)\delta n$ which connects the perturbed magnetic field δB_z. The last relation can be obtained from Ampere's law $(\nabla \times B)_x = \frac{4\pi}{c} j_x$, where j_x is the x-component of the perturbed electric current. The second term in square brackets of (1) describes the effects of finite ion Larmor radius on the flute waves. The terms in (1) and (2) proportional to β correspond to the magnetic field perturbations δB_z in flute waves.

3 Numerical solution of nonlinear equations

In order to numerically solve the nonlinear system of equations (1)–(2) the two-dimensional code FLUTE has been developed. Detailed description of the code can be found in Sotnikov et al. (2008). We first present below the evolution in time of density and electric potential. This will be followed by analysis of the evolution of the spectrum of the excited flute modes. In the series of figures below the evolution in time of initially randomly seeded density perturbations is presented.

The following plasma parameters which correspond to Z-pinch experimental setup at NTF were used: $n_{0e} = 2.0 \times 10^{19}$ cm^{-3}, $n_{0i} = 2.0 \times 10^{18}$ cm^{-3}, $T_i \simeq T_e = 50$ eV, $B_0 = 0.3$ MG and $Z = 10$. This corresponds to the ion plasma beta $\beta_i = 0.045$ and to the electron plasma beta $\beta_e = 0.45$, so the total plasma beta $\beta = 0.5$. The initial level of density perturbations presented in Fig. 1(a) is $\delta n/n_0 \sim 10^{-6}$.

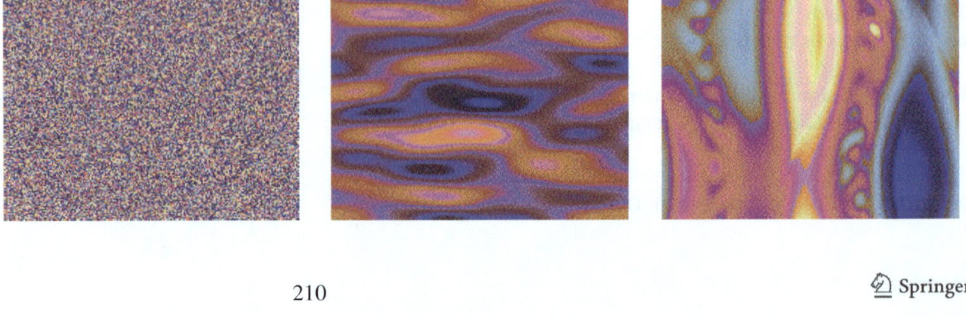

Fig. 1 Growth of random density perturbations (**a**) in the late linear—early nonlinear stage of flute mode instability are presented in (**b**). The density perturbations in the late nonlinear stage of flute mode instability are shown in (**c**)

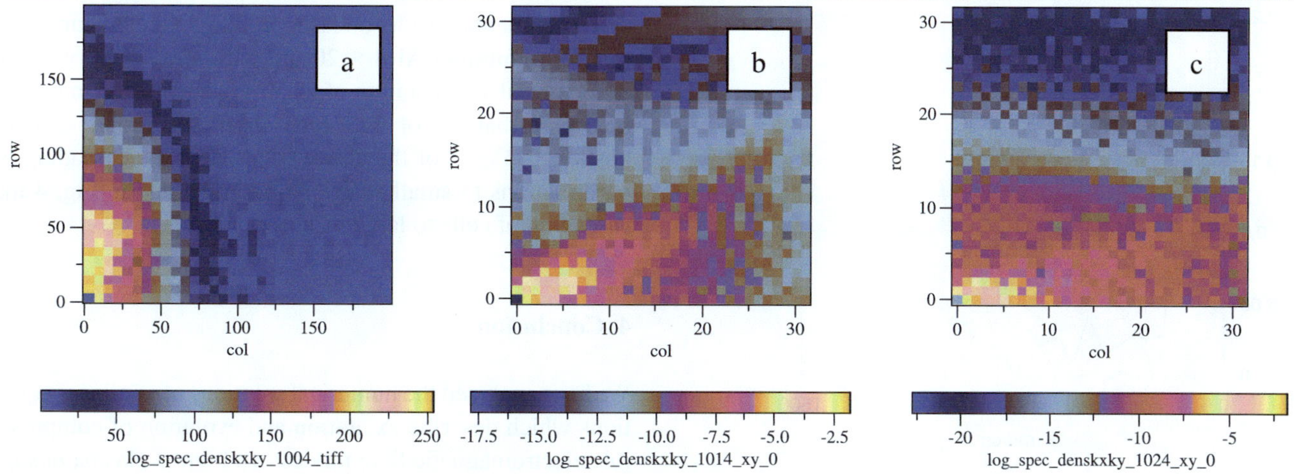

Fig. 2 Evolution in time of density spectra in (k_x, k_y) plane

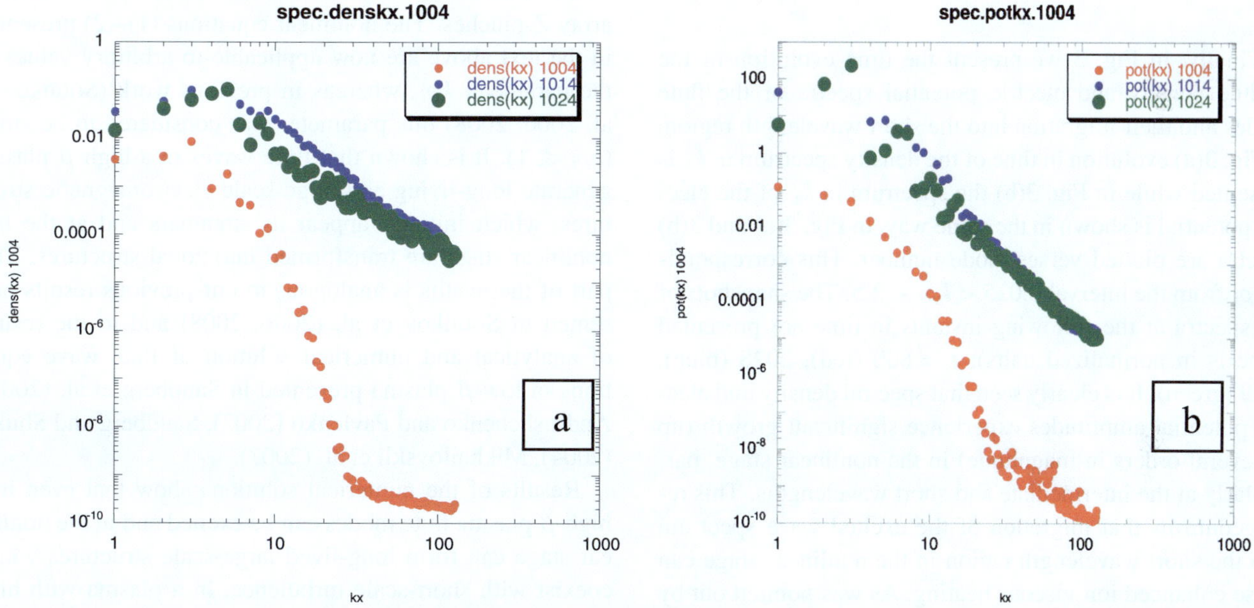

Fig. 3 Spectral distribution of density perturbations (**a**) and electric potential (**b**)

In Fig. 1(b) density perturbations at the late linear—early nonlinear stage of flute mode instability development are presented. It is clearly seen that perturbations are elongated along the x-axis and form so called streamers. This is directly connected with the fact that the growth rate is largest for waves propagating along the y-axis and for such waves $k_x \ll k_y$. This results in a much shorter scale for perturbations in the y-direction. In Fig. 1(c) the density perturbations in the late nonlinear stage of flute mode instability are shown. One can observe formation of large scale structures elongated in the y-direction (zonal flows) and on top of them development of short scale perturbations.

In Fig. 2 we present the evolution in time of the spectral distribution of the density perturbations.

In Fig. 2(a), at scaled time $t = 800$, the spectrum of density perturbations is elongated along the k_y direction. This spectrum corresponds to streamer-like structures, excited at the early nonlinear stage of the flute mode instability. With time this spectrum transforms into one with $k_x \sim k_y$ as can be seen in Fig. 2(b) at scaled time $t = 2028$. After that, the process of spectrum transformation with generation of spectral densities with larger k_y wave vector components continues. Finally, at the strongly nonlinear stage density spectra with wave vector components elongated along the x direction are formed as is apparent in Fig. 2(c) at scale time $t = 3099$. This corresponds to formation of zonal-like structures in real space.

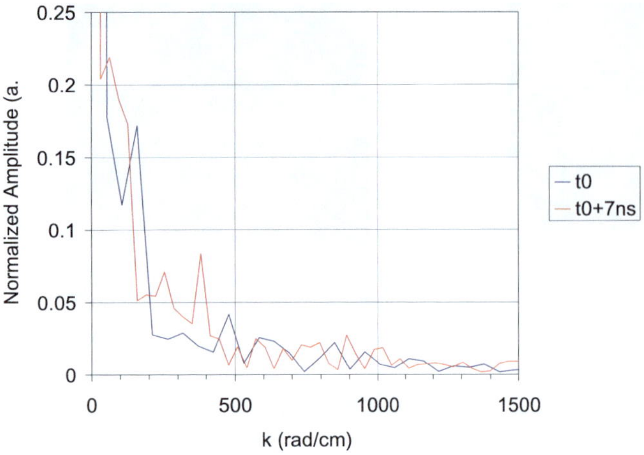

Fig. 4 Experimentally observed migration of the excited wave spectrum (in *blue* in the region between 100 and 200 dimensionless wave numbers) towards smaller and larger wave numbers (in *red*)

Finally, in Fig. 3 we present the time evolution of the excited density and electric potential spectra of the flute modes and their migration into the short wavelength region. In Fig. 3(a) evolution in time of the density spectrum in k_x is presented while in Fig. 3(b) the spectrum in k_x of the electric potential is shown in the same way. In Fig. 3(a) and 3(b) spectra are plotted verses mode number. This corresponds to $k\rho_i$ from the interval $0.025 < k\rho_i < 2.5$. The snapshots of the spectra at the following instants in time are presented (time is in normalized units): $t = 800$ (red), 2028 (blue), 3099 (green). It is clearly seen that spectral density and electric potential amplitudes experience significant growth (up to several orders in magnitude) in the nonlinear stage, particularly at the intermediate and short wavelengths. This result confirms that migration of the excited wave spectrum into the short wavelength region in the nonlinear stage can cause enhanced ion viscous heating. As was pointed out by Haines et al. (2006), ion viscous heating can play an important role in enhanced X-ray energy production in wire-array Z-pinches.

Wave spectral transfer was observed in laboratory astrophysics experiments (Presura et al. 2005). In these experiments, the expansion front of a laser ablation plasma plume became unstable when decelerated by an external magnetic field. Laser schlieren images of the plasma plume were recorded at several times during its evolution. The mode activity can be investigated by spectral analysis of contours traced along the unstable boundary. In Fig. 4, the normalized amplitude of the wave perturbations is plotted as a function of the wave number at two moments in time. It is clearly seen that, in time, the linearly excited wave spectrum (in blue between 100 and 200 rad/cm) migrates toward smaller and larger wave numbers (red curve). These experimental results support analytical and numerical findings presented above.

Experimental results on laser probing in the precursor region of aluminum Al 8×20 μm wire array (Ivanov et al. 2007) revealed generation of localized density and magnetic field perturbations of flute type observed by optical diagnostics. In Fig. 3 of the above cited paper the cascading of perturbations to small scales is presented and in Fig. 4 the cascading of cells to larger scales can be seen.

4 Conclusion

We have analyzed the numerical solutions of nonlinear equations which describe excitation and evolution of compressible electromagnetic flute modes in a high-β plasma in support of planned experiments in the NTF laboratory to investigate generation of flute modes in the region of plasma flow deceleration as well as experiments on cylindrical wire array Z-pinches. The nonlinear equations (1)–(2) presented in the text above are now applicable to arbitrary values of the parameter $k\rho_i$, whereas in previous work (Sotnikov et al. 2006, 2008) this parameter was considered to be small ($k\rho_i \ll 1$). It is shown that flute waves in a high-β plasma generate long-living and large-scale electromagnetic structures, which initially appear as streamers and at the late nonlinear stage are transformed into zonal structures. This part of the results is analogous to our previous results presented in Sotnikov et al. (2006, 2008) and to the results of analytical and numerical solution of flute wave equations in low-β plasma presented in Sandberg et al. (2005), Andrushchenko and Pavlenko (2002), Sandberg and Shukla (2004), Mikhailovskii et al. (2007).

Results of the numerical solutions show that even in a high-β plasma flute modes can be excited and in the nonlinear stage can form long-lived large-scale structures which coexist with short-scale turbulence. In a plasma with high beta value linear dispersion of the flute mode is modified, but the nonlinear terms in (1) and (2) remain the same. The main effect related to appearance of finite plasma beta in comparison with the low beta case examined in the above cited papers is connected with transformation of electrostatic flute modes to electromagnetic ones and generation of large scale structures not only in density but in magnetic field as well.

Spectral analysis of the numerical solutions also shows that streamer-like structures are formed initially and that structures associated with zonal flows are formed at the late nonlinear stage. This can be seen both from density and electric potential spectra. Analysis of the spectra also clearly shows that nonlinear decay processes of flute waves lead to excitation of the spectral energy flux to short wavelength scales in the late nonlinear stage. This in turn can lead to enhanced ion heating by flute turbulence. It is important to note that (1) and (2) are applicable for the description of short scale turbulence with arbitrary values of the parameter

$k\rho_i$ and reveal self-consistent generation of the Kolmogorov type turbulent spectra in the short-scale region. Obtained results support experimentally observed spectral evolution of excited flute modes.

Acknowledgements This work was supported by the United States Department of Energy under the following grants: Grant No. DE-FC52-01NV14050 at the University of Nevada at Reno, Grant No. DE-AC04-94AL85000 at Sandia National Laboratories, and the Grant ISTC N 3520.

References

Andrushchenko, Z.N., Pavlenko, V.P.: Turbulent generation of large-scale flows and nonlinear dynamics of flute modes. Phys. Plasmas **9**, 4512–4519 (2002)

Bernhardt, P.A., et al.: Observations and theory of the AMPTE magnetotail barium releases. J. Geophys. Res. A **92**, 5777–5794 (1987)

Haines, M., LePell, P.D., Coverdale, C.A., Jones, B., Deeney, C., Apruzese, J.P.: Ion viscous heating in a magnetohydrodynamically unstable z-pinch at over 2×10^9 kelvin. Phys. Rev. Lett. **96**, 075003 (2006)

Horton, W.: Drift waves and transport. Rev. Mod. Phys. **71**, 735–778 (1999)

Ivanov, V.V., Sotnikov, V.I., Sarkisov, G., Astanovitskiy, A.L., Laca, P.J., Cowan, T.E., Jones, B., Deeney, C., Oliver, B.V., Mehlhorn, T.A., Leboeuf, J.N.: Experimental study of the dynamics of large- and small-scale structures in the plasma column of wire array Z-pinches. IEEE Trans. Plasma **35**(4), 1170–1177 (2007)

Kuvshinov, B.N., Mikhailovskii, A.B.: MHD-model, which takes into account sub-Larmor scales for the plasma with gradient of temperature. Plasma Phys. Rep. **22**(6), 582–587 (1996)

Mikhailovskii, A.B., Kovalishen, E.A., Shirokov, M.S., Tsypin, V.S., Galvao, R.M.O.: Effect of magnetic field curvature on the generation of zonal flows by drift-Alfven waves. Plasma Phys. Rep. **33**(5), 407–419 (2007)

Newman, D.E., Terry, P.W., Diamond, P.H., Liang, Y.M.: The dynamics of spectral transfer in a model of drift wave turbulence with two nonlinearities. Phys. Fluids B **5**, 1140–1153 (1993)

Onishchenko, O.G., Pokhotelov, O., Sagdeev, R.Z., Pavlenko, V.P., Stenflo, L., Shukla, P.K.: Decay instability and Kolmogorov spectra of ion-drift waves in low-β dusty plasmas. Phys. Plasmas **8**, 4351 (2001)

Onishchenko, O.G., Krasnoselskikh, V.V., Pokhotelov, O.A.: Drift-Alfven vortices at the ion Larmor radius scale. Phys. Plasmas **15**, 022905 (2008)

Presura, R., Ivanov, V., Sentoku, Y., Sotnikov, V., Le Galloudec, N., Kemp, A., Mancini, R., Ruhl, H., Astanovitskiy, A., Laca, P., Cowan, T., Ditmire, T., Chiu, C., Horton, W., Valanju, P., Keely, S.: Laboratory simulation of magnetospheric plasma shocks. Astrophys. Space Sci. **298**, 299–304 (2005)

Ripin, B.H., Huba, J.D., McLean, E.A., Manka, C.K., Peyser, T., Burris, H.R., Grun, J.: Sub-alfvenic plasma expansion. Phys. Fluids B **5**, 3491 (1993)

Sandberg, I., Isliker, H., Pavlenko, V.P., Hizanidis, K., Vlahos, L.: Generation and saturation of large-scale flows in flute turbulence. Phys. Plasmas **12**, 032503 (2005)

Sandberg, I., Shukla, P.K.: Magnetic-curvature-driven interchange modes in dusty plasmas. Phys. Plasmas **11**, 542–547 (2004)

Sotnikov, V.I., Ivanov, V.V., Cowan, T.E., Leboeuf, J.N., Oliver, B.V., Coverdale, C.A., Jones, B.M., Deeney, C., Mehlhorn, T.A., Sarkisov, G.S., LePell, P.D.: Investigation of electromagnetic flute mode instability in a high beta Z-pinch plasma. IEEE Trans. Plasma **34**(5), 2239 (2006)

Sotnikov, V.I., Presura, R., Ivanov, V.V., Cowan, T.E., Leboeuf, J.N., Oliver, B.V.: Excitation of electromagnetic flute modes in the process of interaction of plasma flow with inhomogeneous magnetic field. Astrophys. Space Phys. **307**, 99–101 (2007)

Sotnikov, V.I., Ivanov, V.V., Presura, R., Leboeuf, J.N., Onishchenko, O.G., Oliver, B.V., Jones, B., Mehlhorn, T.A., Deeney, C.: Investigation of compressible electromagnetic flute mode instability in finite beta plasma in support of Z-pinch and laboratory astrophysics experiments. Commun. Comput. Phys. **4**, 611–623 (2008)

Zakharov, Y.P.: Collisionless laboratory astrophysics with lasers. IEEE Trans. Plasma Sci. **31**, 1243–1251 (2003)

GPSR Compliance

The European Union's (EU) General Product Safety Regulation (GPSR) is a set of rules that requires consumer products to be safe and our obligations to ensure this.

If you have any concerns about our products, you can contact us on ProductSafety@springernature.com

In case Publisher is established outside the EU, the EU authorized representative is:

Springer Nature Customer Service Center GmbH
Europaplatz 3
69115 Heidelberg, Germany

Batch number: 10786594

Printed by Printforce, the Netherlands